Problem Books in Mathematics

Series Editor
Peter Winkler, Department of Mathematics, Dartmouth College, Hanover, NH, USA

Books in this series are devoted exclusively to problems - challenging, difficult, but accessible problems. They are intended to help at all levels - in college, in graduate school, and in the profession. Arthur Engels "Problem-Solving Strategies" is good for elementary students and Richard Guys "Unsolved Problems in Number Theory" is the classical advanced prototype. The series also features a number of successful titles that prepare students for problem-solving competitions.

Corneliu Mănescu-Avram

Selection Tests in Number Theory for Mathematical Olympiads

 Springer

Corneliu Mănescu-Avram
Ploiesti, Romania

ISSN 0941-3502　　　　　　　ISSN 2197-8506　(electronic)
Problem Books in Mathematics
ISBN 978-3-031-59741-1　　　ISBN 978-3-031-59742-8　(eBook)
https://doi.org/10.1007/978-3-031-59742-8

Mathematics Subject Classification: 11-XX; 97F60; 11Bxx; 11Dxx

© The Editor(s) (if applicable) and The Author(s), under exclusive license to Springer Nature Switzerland AG 2024

This work is subject to copyright. All rights are solely and exclusively licensed by the Publisher, whether the whole or part of the material is concerned, specifically the rights of translation, reprinting, reuse of illustrations, recitation, broadcasting, reproduction on microfilms or in any other physical way, and transmission or information storage and retrieval, electronic adaptation, computer software, or by similar or dissimilar methodology now known or hereafter developed.

The use of general descriptive names, registered names, trademarks, service marks, etc. in this publication does not imply, even in the absence of a specific statement, that such names are exempt from the relevant protective laws and regulations and therefore free for general use.

The publisher, the authors and the editors are safe to assume that the advice and information in this book are believed to be true and accurate at the date of publication. Neither the publisher nor the authors or the editors give a warranty, expressed or implied, with respect to the material contained herein or for any errors or omissions that may have been made. The publisher remains neutral with regard to jurisdictional claims in published maps and institutional affiliations.

This Springer imprint is published by the registered company Springer Nature Switzerland AG
The registered company address is: Gewerbestrasse 11, 6330 Cham, Switzerland

If disposing of this product, please recycle the paper.

About the Book

This book contains some selection tests proposed to the national IMO teams in many countries. Most solutions are original or adapted by the author. The material is organized chronologically, in order to have an image of the evolution of the phenomenon. The index allows the direct access of a theme.

The book is a possible preparation material for the interested students or other mathematics lovers. Further readings are indicated in the references.

Contents

1 Problems 1
    1.1   1968 1
    1.2   1969 1
    1.3   1973 1
    1.4   1975 1
    1.5   1976 2
    1.6   1977 2
    1.7   1978 2
    1.8   1979 3
    1.9   1980 3
    1.10   1981 3
    1.11   1982 4
    1.12   1983 4
    1.13   1984 4
    1.14   1986 4
    1.15   1987 5
    1.16   1988 5
    1.17   1989 6
    1.18   1990 6
    1.19   1991 7
    1.20   1992 7
    1.21   1993 8
    1.22   1994 9
    1.23   1995 9
    1.24   1996 10
    1.25   1997 10
    1.26   1998 11
    1.27   1999 11
    1.28   2000 12
    1.29   2001 13

1.30	2002	13
1.31	2003	15
1.32	2004	16
1.33	2005	17
1.34	2006	19
1.35	2007	21
1.36	2008	22
1.37	2009	24
1.38	2010	26
1.39	2011	27
1.40	2012	28
1.41	2013	30
1.42	2014	32
1.43	2015	33
1.44	2016	35
1.45	2017	37
1.46	2018	40
1.47	2019	42
1.48	2020	44
1.49	2021	46
1.50	2022	48
1.51	2023	50
1.52	2024	53
2	**Solutions**	**55**
2.1	1968	55
2.2	1969	55
2.3	1973	56
2.4	1975	56
2.5	1976	57
2.6	1977	57
2.7	1978	59
2.8	1979	60
2.9	1980	62
2.10	1981	64
2.11	1982	64
2.12	1983	65
2.13	1984	66
2.14	1986	66
2.15	1987	66
2.16	1988	68
2.17	1989	69
2.18	1990	71
2.19	1991	73
2.20	1992	76

2.21	1993	79
2.22	1994	81
2.23	1995	84
2.24	1996	86
2.25	1997	87
2.26	1998	91
2.27	1999	92
2.28	2000	96
2.29	2001	99
2.30	2002	102
2.31	2003	107
2.32	2004	113
2.33	2005	119
2.34	2006	125
2.35	2007	133
2.36	2008	140
2.37	2009	147
2.38	2010	155
2.39	2011	161
2.40	2012	169
2.41	2013	175
2.42	2014	183
2.43	2015	190
2.44	2016	198
2.45	2017	207
2.46	2018	220
2.47	2019	230
2.48	2020	239
2.49	2021	250
2.50	2022	258
2.51	2023	265
2.52	2024	277

Appendix 279

Bibliography 289

Index 291

About the Author

Corneliu Mănescu-Avram is graduated from the Faculty of Mathematics, University of Bucharest, Romania. In 2010, he obtained the QTS certification. He is an active problem solver and proposer in many mathematical journals, such as *Mathematical Reflections*, *Crux Mathematicorum*, *Pi Mu Epsilon Journal*, *Pentagon Kappa Mu Epsilon Journal*, *Mathematical Excalibur*, *School Science and Mathematics*, *Mathematical Recreations* (in Romanian), and *Mathematical Review from Timişoara* (in Romanian).

He was a computer programmer and a high school teacher of mathematics, computer science, and astronomy. At the present, he is retired, but remains active.

Notations

μ	Möbius function
$\nu_p(n)$	Exponent of the largest power of prime p dividing n
π	Prime-counting function
σ	Sum of divisors
τ	Number of divisors
φ	Euler totient function
ω	Number of distinct prime divisors
Ω	Number of prime factors counted with multiplicities

Chapter 1
Problems

1.1 1968

1. Let $n > 3$ be a positive integer. Prove that n is a prime number if and only if there exists a positive integer α such that $n! = n(n-1)(\alpha n + 1)$. (*Yugoslavia*)

1.2 1969

2. Let $a < b$ be two positive integers. Prove that in each set of b consecutive positive integers there are two numbers whose product is divisible by ab. (*Yugoslavia*)

1.3 1973

3. All sides of a rectangle are odd positive integers. Prove that there does not exist a point inside the rectangle whose distances to each of the vertices are all integers. (*Yugoslavia*)

1.4 1975

4. In a cone $(n-1)$, planes parallel to the basis are given such that the resulting corps have equal volumes. Let M_1 be the set of these planes and M_2 the set of the $(n-1)$ planes parallel to the basis which divide the lateral surface of the cone in n parts of equal area. Prove that the sets M_1 and M_2 are disjoint if and only if n is squarefree. (*Romania*)

1.5 1976

5. Show that among every n integers (not necessarily distinct) there are numbers whose sum is divisible by n. (*Romania*)

1.6 1977

6. Let (a_n) be a strictly increasing sequence of positive integers. Prove that there exist infinitely many terms a_m which can be written like $a_m = xa_p + ya_q$, with x, y strictly positive integers and $p \neq q$. (*Germany*)
7. When 4444^{4444} is written in decimal notation, the sum of its digits is A. Let B be the sum of the digits of A. Find the sum of the digits of B. (A and B are written in decimal notation.) (*Germany*)
8. Prove that the equation

$$x_1^2 + x_2^2 + \cdots + x_n^2 - \frac{(x_1 + x_2 + \cdots + x_n)^2}{n} = 1$$

has solutions in integers if and only if $n = 4$. (*Romania*)
9. Prove that among every ten consecutive positive integers, there is a number which is coprime with the other nine numbers. (*Romania*)
10. Determine all six-tuples (p, q, r, x, y, z), where p, q, and r are primes and x, y, and z are positive integers such that $p^{2x} = q^y r^z + 1$. (*Yugoslavia*)

1.7 1978

11. Let $c \neq 1$ be a positive rational number. Show that it is possible to partition $\mathbb{N}^*$ into two nonempty disjoint subsets A and B so that $x/y \neq c$ holds whenever x and y lie both in A or both in B. (*Austrian – Polish Competition*)
12. Let n be an integer greater than 1. Define

$$x_1 = n, y_1 = 1, x_{i+1} = \left\lfloor \frac{x_i + y_i}{2} \right\rfloor, y_{i+1} = \left\lfloor \frac{n}{x_{i+1}} \right\rfloor,$$

for $i = 1, 2, \ldots$. Prove that $\min\{x_1, x_2, \ldots, x_n\} = \lfloor \sqrt{n} \rfloor$. (*Germany*)

1.8 1979

13. Solve in positive integers the equation

$$3^x + 4^y = 5^z.$$

   (*Romania*)

14. Prove that if $p > 2$ is a prime number, then

$$(-1)^{\lfloor \sqrt{p} \rfloor} \sin \frac{p\pi}{2} \sin \frac{p\pi}{3} \cdots \sin \frac{p\pi}{p-1} > 0.$$

   (*Romania*)

15. Find all integers n with $1 < n < 1979$ having the following property: if m is an integer coprime with n and $1 < m < n$, then m is a prime number. (*Yugoslavia*)

1.9 1980

16. Let n be an odd positive integer. Prove that the number

$$\frac{2 \cdot (3n)!}{n! \cdot (2n)!}$$

   is divisible by $3(3n-1)(3n-2)$. (*Romania*)

17. Let b_n be the last digit of the number $1^1 + 2^2 + 3^3 + \ldots + n^n$. Prove that $b_{n+100} = b_n$ for every positive integer n. (*Romania*)

18. Prove that among every three positive distinct integers, there are two, say a and b, such that $a^3 b - ab^3$ is a multiple of 10. (*Romania*)

1.10 1981

19. Let p be a prime number greater than 2 and the sequence $(a_k)_{k \geq 1}$ defined by $a_1 = p$ and $a_{k+1} = 2a_k + 1$, $k \geq 1$. Prove that the first p terms of the sequence cannot be all prime numbers. (*Romania*)

20. Let m be a positive integer which is not divisible by 3. Prove that there are infinitely many positive integers n such that $s(n)$ and $s(n+1)$ are divisible by m, where $s(k)$ is the sum of the digits of the positive integer k. (*Romania*)

1.11 1982

21. Find all pairs (n, m) of positive integers such that

$$\gcd\left((n+1)^m - n, (n+1)^{m+3} - n\right) > 1.$$

   (*Austrian – Polish Competition*)

22. Let $p > 2$ be a prime number. For $k = 1, 2, \ldots, p-1$ denote by a_k the remainder when k^p is divided by p^2. Prove that

$$a_1 + a_2 + \ldots + a_{p-1} = \frac{p^3 - p^2}{2}.$$

   (*Yugoslavia*)

1.12 1983

23. Prove that the proposition "There exist positive integers having n digits such that the sum of the digits is equal to the product of the digits in base 10" is true for infinitely many positive integers n and false for infinitely many positive integers n. (*Romania*)

1.13 1984

24. Let A be the set of four-digit natural numbers having exactly two distinct digits, none of which is zero. Interchanging the two digits of $n \in A$ yields a number $f(n) \in A$ (for instance, $f(3111) = 1333$). Find those $n \in A$ with $n > f(n)$ for which $\gcd(n, f(n))$ is the largest possible. (*Austrian – Polish Competition*)

1.14 1986

25. Let k, n be integers with $0 < k < \frac{n^2}{4}$ such that k has no prime divisors greater than n. Prove that k divides $n!$. (*Austrian – Polish Competition*)

1.15 1987

26. Find all positive integers n such that the equation $x^3 + y^3 + z^3 = nx^2y^2z^2$ has positive integer solutions. (*China*)
27. Find positive integers A such that

$$A = \left(m - \frac{1}{n}\right)\left(n - \frac{1}{p}\right)\left(p - \frac{1}{m}\right)$$

with $m \geq n \geq p \geq 1$ positive integers. (*Romania*)
28. For the set $A = \{1, 2, \ldots, n\}$, find the maximum number of elements of a subset $B \subset A$ such that for every distinct $x, y \in B$, $x + y$ is not divisible by $x - y$. (*Romania*)
29. Show that for every positive integer $n \geq 2$, the number

$$\frac{3^n - 2^n}{n}$$

is not an integer. (*Romania*)
30. Let a, b, c be integers such that $a + b + c$ divides $a^2 + b^2 + c^2$. Show that $a + b + c$ divides $a^n + b^n + c^n$ for infinitely many positive integers n. (*Romania*)

1.16 1988

31. Prove that for every positive integer n, the number

$$\prod_{k=1}^{n} k^{2k-n-1}$$

is an integer. (*Romania*)
32. Let a be a positive integer. The sequence $(x_n)_{n \geq 1}$ is defined by $x_1 = 1$, $x_2 = a$, and $x_{n+2} = ax_{n+1} + x_n$ for all $n \geq 1$. Prove that (y, x) is a solution of the equation

$$\left|y^2 - axy - x^2\right| = 1$$

if and only if there exists a rank k such that $(y, x) = (x_{k+1}, x_k)$. (*Romania*)

1.17 1989

33. Let $(a_n)_{n \geq 2}$ be the sequence of positive integers defined by
$$a_n = n^6 + 5n^4 - 12n^2 - 36.$$

   (a) Prove that any prime number divides at least one term of this sequence.
   (b) There exists a positive integer which does not divide any term of the sequence.
   (c) Determine the least number n such that $1989 | a_n$. (*Romania*)

34. Let a, b, c be integer numbers different from zero and relatively prime. Prove that for all relatively prime integer numbers u, v, w such that $au + bv + cw = 0$, there exist integer numbers m, n, p such that $a = nw - pv$, $b = pu - mw$, and $c = mv - nu$. (*Romania*)

35. Let $(a_n)_{n \geq 1}$ be the sequence defined by $a_1 = a_2 = 1$, $a_3 = 199$ and $a_{n+1} = \frac{1989 + a_n a_{n-1}}{a_{n-2}}$ for all $n \geq 3$. Prove that all the terms of the sequence are positive integers. (*Romania*)

36. There are $n \geq 2$ weights such that each weight is a positive integer less than n and their total weight is less than $2n$. Prove that there is a subset of these weights such that their total weight is equal to n. (*Turkey*)

1.18 1990

37. Prove that for every integer power of 2, there exists a multiple of it with all digits (in decimal expression) not zero. (*China*)

38. Prove that for any positive integer n, the least common multiple of the numbers $1, 2, \ldots, n$ and the least common multiple of the numbers
$$\binom{n}{1}, \binom{n}{2}, \ldots, \binom{n}{n}$$
are equal if and only if $n + 1$ is a prime number. (*Romania*)

39. Let p, q be prime numbers and suppose that $q > 5$. Prove that if $q | 2^p + 3^p$, then $q > p$. (*Romania*)

40. Find the least m such that $m - v_2(m!) = 1990$. Here $v_2(n)$ denotes the greatest power of 2 which divides n. (*Turkey*)

1.19 1991

41. Let a, b be positive integers. Prove that if the number
$$\frac{a+1}{b} + \frac{b}{a}$$
is an integer, then it is equal to 3. (*Romania*)

42. Let $(a_n)_{n \geq 1}$ be a sequence of positive integers such that $(a_n, a_m) = a_{(n,m)}$ for all n, m positive integers. Prove that there exists an unique sequence of positive integers $(b_n)_{n \geq 1}$ such that
$$a_n = \prod_{d|n} b_d.$$
(*Romania*)

43. Consider the sequence $(a_n)_{n \geq 1}$ of positive integers given by $a_1 = a_2 = 1$ and $a_{n+1} = a_n + a_{n-1} + k$, where k is a positive integer. Find the least number k for which a_{1991} is not relatively prime with 1991. (*Romania*)

44. For every positive integer n, we define $f(n)$ by the following rule: $f(1) = 1$ and for $n > 1$, $f(n) = 1 + \alpha_1 \cdot p_1 + \ldots + \alpha_k \cdot p_k$, where $n = p_1^{\alpha_1} \ldots p_k^{\alpha_k}$ is the canonical prime factorization of n. For every positive integer s, let $f_s(n) = f(\ldots f(f(n))\ldots))$, where on the right hand side there are exactly s symbols f. Show that for every given positive integer a, there is a positive integer s_0 such that for all $s > s_0$ the sum $f_s(a) + f_{s-1}(a)$ does not depend on s. (*Vietnam*)

1.20 1992

45. For any prime p, prove that there exists an integer x_0 such that $p | x_0^2 - x_0 + 3$ if and only if there exists an integer y_0 such that $p | y_0^2 - y_0 + 25$. (*China*)

46. Let a be a positive integer and $(x_n)_{n \geq 1}$ be the sequence defined by $x_1 = x_2 = 1$ and $x_{n+2} = (a^4 + 4a^2 + 2)x_{n+1} - x_n - 2a^2$ for $n \geq 1$. Show that x_n is a perfect square, and for $n \geq 3$ it is the element on the first line and the first column of the matrix
$$\begin{pmatrix} a^2+1 & a \\ a & 1 \end{pmatrix}^{n-2}.$$
(*Romania*)

47. Let $(a_n)_{n \geq 1}$ and $(b_n)_{n \geq 1}$ be sequences of positive integers defined by
$$a_{n+1} = na_n + 1 \text{ and } b_{n+1} = nb_n - 1 \text{ for } n \geq 1.$$

Show that the two sequences cannot have infinitely many common terms. (*Romania*)

48. Are there 14 consecutive positive integers such that each of these numbers is divisible by one of the prime numbers p, where $2 \leq p \leq 11$? (*Turkey*)
49. Let p be an odd prime number. Prove that there are unique positive integers x, y such that $x^2 = y(y + p)$, and give the formulae for x and y in terms of p. (*United Kingdom*)
50. Find all pairs (x, y) of positive integers satisfying the equation

$$x^2 + y^2 - 5xy + 5 = 0.$$

(*Vietnam*)

1.21   1993

51. For all primes p, define $F(p) = \sum_{k=1}^{\frac{p-1}{2}} k^{120}$ and $f(p) = \frac{1}{2} - \left\{\frac{F(p)}{p}\right\}$. Find the value of $f(p)$. Here $\{a\}$ denotes the fractional part of the real number a. (*China*)
52. Find all integer solutions of the equation $2x^4 + 1 = y^2$. (*China*)
53. Suppose that p, q are prime numbers such that

$$\sqrt{p^2 + 7pq + q^2} + \sqrt{p^2 + 14pq + q^2}$$

is an integer. Show that $p = q$. (*Italy*)
54. Show that if x, y, z are positive integers such that

$$x^2 + y^2 + z^2 = 1993,$$

and then $x + y + z$ is not a perfect square. (*Romania*)
55. Show that there exists an infinite arithmetic progression of positive integers such that the first term is 16 and the number of positive divisors of each term is divisible by 5. Of all such sequences, find the one with the smallest common difference. (*Turkey*)
56. Let $m = \frac{4^p - 1}{3}$, where p is a prime number exceeding 3. Prove that 2^{m-1} has remainder 1 when divided by m. (*United Kingdom*)
57. Let $k > 1$ be an integer. For each integer $n > 1$, we put

$$f(n) = kn\left(1 - \frac{1}{p_1}\right)\left(1 - \frac{1}{p_2}\right)\cdots\left(1 - \frac{1}{p_r}\right),$$

where $p_1, p_2, \ldots, p_r$ are all distinct prime divisors of n. Find all values k for which the sequence (x_m) defined by $x_0 = a$ and $x_{m+1} = f(x_m)$, $m = 0, 1, 2, 3, \ldots$ is bounded for all integers $a > 1$. (*Vietnam*)

1.22 1994

58. Find all nonnegative integers x, y, and z satisfying the equation: $7^x + 1 = 3^y + 5^z$. (*Hong Kong*)
59. Let m and n be positive integers, where m has d digits in base 10 and $d \leq n$. Find the sum of all the digits (in base 10) of the product $(10^n - 1)m$. (*Hong Kong*)
60. Find all prime numbers for which $\frac{2^{p-1}-1}{p}$ is a perfect square. (*Italy*)
61. Prove that the sequence $(a_n)_{n \geq 1}$, $a_n = 3^n - 2^n$ does not contain three numbers in geometric progression. (*Romania*)
62. Find all pairs (a, b) of integers such that ab divides $a^2 + b^2 + 3$. (*Turkey*)
63. Find the first integer $n > 1$ such that the average of $1^2, 2^2, 3^2, \ldots, n^2$ is itself a perfect square. (*United Kingdom*)
64. Consider the equation $x^2 + y^2 + z^2 + t^2 - Nxyzt - N = 0$, where N is a given positive integer.

    (a) Prove that for infinitely many values of N, this equation has positive integral solutions.
    (b) Let $N = 4k(8m + 7)$, where k, m are nonnegative integers. Prove that the equation has no positive integral solutions. (*Vietnam*)

1.23 1995

65. Find the smallest prime number p that cannot be represented in the form $|3^a - 2^b|$, where a and b are nonnegative integers. (*China*)
66. Determine all triples (x, y, z) of integers greater than 1 with the property that x divides $yz - 1$, y divides $zx - 1$, and z divides $xy - 1$. (*Italy*)
67. Prove that for every positive integer n, the following conditions are equivalent:

    (a) $n \mid a^n - a$ for any positive integer a
    (b) for any prime divisor p of n, $p^2 \nmid n$, and $p - 1 \mid n - 1$ (*Turkey*)
68. Find all triples (a, b, c) of positive integers such that

$$\left(1 + \frac{1}{a}\right)\left(1 + \frac{1}{b}\right)\left(1 + \frac{1}{c}\right) = 2.$$

(*United Kingdom*)

69. Find all integers a, b, n greater than 1 which satisfy $(a^3 + b^3)^n = 4(ab)^{1995}$. (*Vietnam*)
70. For any nonnegative integer n, let $f(n)$ be the greatest integer such that $2^{f(n)} \mid n + + 1$. A pair (n, p) of nonnegative integers is called *nice* if $2^{f(n)} > p$. Find all triples (n, p, q) of nonnegative integers such that the pairs (n, p), (p, q), and $(n + p + q, n)$ are all nice. (*Vietnam*)

1.24 1996

71. Prove that there exists a set X of 1996 positive integers with the following properties:

    (a) The elements of X are pairwise coprime.
    (b) All elements of X and all sums of two or more distinct elements of X are composite numbers. (*Italy*)

72. The sequence $(a_n)_{n \geq 2}$ is defined as follows: if the distinct prime divisors of n are $p_1, p_2, \ldots, p_k$, then $a_n = \frac{1}{p_1} + \frac{1}{p_2} + \ldots + \frac{1}{p_k}$. Show that for any positive integer $N \geq 2$,

$$\sum_{n=2}^{N} a_2 a_3 \ldots a_n < 1.$$

(*Romania*)

73. Determine all sets of nonnegative integers x, y, and z which satisfy the equation

$$2^x + 3^y = z^2.$$

(*United Kingdom*)

1.25 1997

74. Prove that there exists $m \in \mathbb{N}$ such that there exists an integral sequence (a_n) which satisfies the following:

    (a) $a_0 = 1$, $a_1 = 337$.
    (b) $a_{n+1}a_{n-1} - a_n^2 + \frac{3}{4}(a_{n+1} + a_{n-1} - 2a_n) = m, \forall n \geq 1$.
    (c) $\frac{1}{6}(a_n + 1)(2a_n + 1)$ is a perfect square $\forall n \geq 1$. (*China*)

75. Determine all triples (x, y, p) with x, y positive integers and p a prime number verifying the equation $p^x - y^p = 1$. (*Italy*)

76. Let A be the set of positive integers represented by the form $a^2 + 2b^2$, where a and b are integer numbers and $b \neq 0$. Show that if p is a prime number and $p^2 \in A$, then $p \in A$. (*Romania*)

77. Let p, q, r be pairwise distinct prime numbers, and let A be the set of positive integers defined by

$$A = \{p^a q^b r^c \mid 0 \leq a, b, c \leq 5\}.$$

Find the least number n such that any set with n elements B, $B \subset A$, contains distinct elements x and y such that x is a divisor of y. (*Romania*)

78. Let $n \geq 2$ be an integer, and let

$$P(X) = X^n + a_{n-1}X^{n-1} + \cdots + a_1 X + 1$$

be a polynomial with positive integer coefficients. Suppose that $a_k = a_{n-k}$ for all k, $k = 1, 2, \ldots, n-1$. Prove that there exist infinitely many pairs (x, y) of positive integers such that $x | P(y)$ and $y | P(x)$. (*Romania*)

79. Let $a > 1$ be an integer. Show that the set of positive integers

$$\{a^2 + a - 1, a^3 + a^2 - 1, \ldots, a^{n+1} + a^n - 1, \ldots\}$$

contains an infinite subset of pairwise coprime numbers. (*Romania*)

80. Show that for each prime $p \geq 7$, there exist a positive integer n and integers x_i, y_i ($i = 1, 2, \ldots, n$) not divisible by p, such that $x_i^2 + y_i^2 \equiv x_{i+1}^2 \pmod{p}$, where $x_{n+1} = x_1$. (*Turkey*)

1.26 1998

81. Find $k \in \mathbb{N}$ such that
    (a) For any $n \in \mathbb{N}$, there does not exists $j \in \mathbb{Z}$ which satisfies the conditions $0 \leq j \leq n - k + 1$ and $\binom{n}{j}, \binom{n}{j+1}, \ldots, \binom{n}{j+k-1}$ forms an arithmetic progression.
    (b) There exists $n \in \mathbb{N}$ such that there exists j which satisfies $0 \leq j \leq n - k + 2$ and $\binom{n}{j}, \binom{n}{j+1}, \ldots, \binom{n}{j+k-2}$ forms an arithmetic progression. Find all n which satisfy part b. (*China*)

82. Find all positive integers x, n such that $x^n + 2^n + 1$ is a divisor of $x^{n+1} + 2^{n+1} + 1$. (*Romania*)

83. Suppose x, y, and z are positive integers satisfying the equation $\frac{1}{x} - \frac{1}{y} = \frac{1}{z}$, and let h be the highest common factor of x, y, and z. Prove that $hxyz$ and $h(y - x)$ are perfect squares. (*United Kingdom*)

84. Let $p_1, p_2, \ldots, p_k$ be all primes smaller than m. Prove that $\sum_{i=1}^{k} \left(\frac{1}{p_i} + \frac{1}{p_i^2}\right) > \ln(\ln m)$. (*Vietnam*)

1.27 1999

85. Find all prime numbers p which satisfy the following condition: for any prime $q < p$, if $p = kq + r$, $0 \leq r < q$, there does not exist an integer $a > 1$ such that $a^2 | r$. (*China*)

86. Prove that for any prime number p, the equation $2^p + 3^p = a^n$ has no solution (a,n) in integers greater than 1. (*Italy*)
87. (a) Show that it is possible to choose 1 number out of any 39 consecutive positive integers, having the sum of the digits divisible by 11.
    (b) Find the first 38 consecutive positive integers none of which having the sum of the digits divisible by 11. (*Romania*)
88. Show that for any positive integer n, the number
$$S_n = \binom{2n+1}{0} \cdot 2^{2n} + \binom{2n+1}{2} \cdot 2^{2n-2} \cdot 3 + \ldots + \binom{2n+1}{2n} \cdot 3^n$$
is the sum of two consecutive perfect squares. (*Romania*)
89. Given an odd prime number p such that $2^h \not\equiv 1 \pmod{p}$ for all integers h, $1 \le h < p-1$ and an even integer $a \in (\frac{p}{2}, p)$. Consider the sequence (a_n) defined by $a_0 = a$ and $a_{n+1} = p - b_n$ for $n = 0, 1, 2, \ldots$, where b_n is the greatest odd divisor of a_n. Show that (a_n) is periodic and find its least positive period. (*Vietnam*)

1.28 2000

90. We call *Pythagorean triple* a triple (x,y,z) of positive integers such that $x < y < z$ and $x^2 + y^2 = z^2$. Prove that for all positive integers n, the number 2^{n+1} is in exactly n Pythagorean triples. (*Bosnia* and *Herzegovina*)
91. Find all nonnegative integers x, y, and z such that $(x+1)^{y+1} + 1 = (x+2)^{z+1}$. (*France*)
92. Determine all triples (x,y,z) of positive integers such that
$$\frac{13}{x^2} + \frac{1996}{y^2} = \frac{z}{1997}.$$
(*Italy*)
93. Let n and k be arbitrary positive integers. Show that there exist positive integers $a_1 > a_2 > a_3 > a_4 > a_5 > k$ such that
$$n = \pm \binom{a_1}{3} \pm \binom{a_2}{3} \pm \binom{a_3}{3} \pm \binom{a_4}{3} \pm \binom{a_5}{3}.$$
(*Romania*)
94. Show that there exist infinitely many systems (x,y,z,t) of positive integers which do not have a common divisor greater than 1 and such that
$$x^3 + y^3 + z^2 = t^4.$$
(*Romania*)

95. Let $a > 1$ be an odd positive integer. Find the least positive integer n such that 2^{2000} is a divisor of $a^n - 1$. (*Romania*)

1.29 2001

96. Positive integers x, y, and z hold $\frac{1}{x^2} + \frac{1}{y^2} = \frac{1}{z^2}$. Prove that $xyz \geq 3600$. (*Bosnia and Herzegovina*)
97. Find the maximal value of a positive integer n such that there exists a subset of $S = \{1, 2, \ldots, 2001\}$ with n elements for which the equation $y = 2x$ does not have solutions in the set $S \times S$. (*Bosnia and Herzegovina*)
98. Prove that there exist infinitely many positive integers n such that the equation $(x + y + z)^3 = n^2 xyz$ has a solution (x, y, z) in the set $\mathbb{N}^3$. (*Bosnia and Herzegovina*)
99. Find the exponent of 37 in the representation of the number $\underbrace{111\ldots 11}_{3 \cdot 37^{2000} \text{ digits}}$ as product of prime powers. (*Estonia*)
100. A strictly increasing sequence (a_n) has the property that $\gcd(a_m, a_n) = a_{\gcd(m,n)}$ for all $m, n \in \mathbb{N}$. Suppose that k is the least positive integer for which there exist positive integers $r < k < s$ such that $a_k^2 = a_r a_s$. Prove that $r \mid k$ and $k \mid s$. (*India*)
101. Find all pairs (p, q) of prime numbers such that p divides $5^q + 1$ and q divides $5^p + 1$. (*Italy*)
102. Find all pairs (m, n) of positive integers, with $m, n \geq 2$, such that $a^n - 1$ is divisible by m for each $a \in \{1, 2, \ldots, n\}$. (*Romania*)
103. Find all ordered pairs (x, y) of integers such that $5^x = 1 + 4y + y^4$. (*Turkey*)
104. Let (a_n) be the sequence defined by $a_0 = 1$, $a_n = a_{n-1} + a_{\lfloor n/3 \rfloor}$ for $n \geq 1$. Show that for all primes $p \leq 13$, there are infinitely many positive integers k such that a_k is divisible by p. (*Vietnam*)

1.30 2002

105. Find all nonnegative integers m and n such that $(2^n - 1)(3^n - 1) = m^2$. (*China*)
106. Consider the set S of integers k which are products of four distinct primes. Such an integer $k = p_1 p_2 p_3 p_4$ has 16 positive divisors $1 = d_1 < d_2 < \ldots < d_{15} < d_{16} = k$. Find all elements of S less than 2002 such that $d_9 - d_8 = 22$. (*France*)
107. Let $p \geq 3$ be a prime number. Show that there exist p positive integers $a_1, a_2, \ldots, a_p$ not exceeding $2p^2$ such that the $\frac{p(p-1)}{2}$ sums $a_i + a_j$ ($i < j$) are all distinct. (*France*)

108. Determine the number of all numbers which are represented as $x^2 + y^2$ with $x, y \in \{1, 2, 3, \ldots, 1000\}$ and which are divisible by 121. (*Germany*)
109. Determine all $(x, y) \in \mathbb{N}^2$ which satisfy $x^{2y} + (x + 1)^{2y} = (x + 2)^{2y}$. (*Germany*)
110. Show that there is a set of 2002 consecutive positive integers containing exactly 150 primes. (You may use the fact that there are 168 primes less than 1000.) (*India*)
111. Let p be an odd prime and let a be an integer not divisible by p. Show that there are $p^2 + 1$ triples (x, y, z) of integers with $0 \leq x, y, z < p$ and such that $(x + y + z)^2 \equiv axyz \pmod{p}$. (*India*)
112. Given a prime p, show that there exists a positive integer n such that the decimal representation of p^n has a block of 2002 consecutive zeros. (*India*)
113. $\pi(n)$ is the number of primes that are not bigger than n. For $n = 2, 3, 4, 6, 8, 33, \ldots$, we have $\pi(n)|n$. Does exist infinitely many integers n such that $\pi(n)|n$? (*Iran*)
114. Prove that for each prime number p and positive integer n, p^n divides

$$\binom{p^n}{p} - p^{n-1}.$$

(*Italy*)

115. Prove that for any positive integer m, there exist an infinite number of pairs of integers (x, y) such that

   (a) x and y are relatively prime.
   (b) x divides $y^2 + m$.
   (c) y divides $x^2 + m$. (*Italy*)

116. Let $(a_n)_{n \geq 0}$ be the sequence defined as follows: $a_0 = a_1 = 1$ and $a_{n+1} = 14a_n - a_{n-1}$ for any $n \geq 1$. Show that the number $2a_n - 1$ is a perfect square, for all positive integers n. (*Romania*)
117. Let n be an even positive integer, and let S be the set of positive integers a such that $1 < a < n$ and $a^{a-1} - 1$ is divisible by n. Prove that if $S = \{n - 1\}$, then there is a prime p such that $n = 2p$. (*Romania*)
118. The smallest four divisors of the positive integer n are d_1, d_2, d_3, and d_4. Find n, if

$$d_1^2 + d_2^2 + d_3^2 + d_4^2 = n.$$

(*Switzerland*)

119. Prove that if a and b are distinct integers and $a^2 + ab + b^2$ divides $ab(a + b)$, then

$$|a - b| > \sqrt[3]{ab}.$$

(*Turkey*)

120. Find in explicit form all ordered pairs (m, n) of positive integers such that $mn - 1$ divides $m^2 + n^2$. (*USA*)

1.31 2003

121. Find all positive integers m and n such that
$$n(n+1)(n+2)(n+3) = m(m+1)^2(m+2)^3(m+3)^4.$$
(*Bulgaria*)

122. Let $x < y$ be positive integers and $P = \frac{x^3 - y}{1 + xy}$. Find all integer values that P can take. (*China*)

123. Sequence (a_n) satisfies $a_1 = 3, a_2 = 7, a_n^2 + 5 = a_{n-1}a_{n+1}$, and $n \geq 2$. If $a_n + (-1)^n$ is prime, prove that there exists a nonnegative integer m such that $n = 3^m$. (*China*)

124. Positive integer n cannot be divided by 2 and 3; there are no nonnegative integers a and b such that $|2^a - 3^b| = n$. Find the minimum value of n. (*China*)

125. Let n be a positive integer. Prove that if the number $\underbrace{99\ldots9}_{n}$ is divisible by n, then the number $\underbrace{11\ldots1}_{n}$ is also divisible by n. (*Estonia*)

126. Let $p_1, p_2, \ldots, p_n$ be distinct primes greater than 3. Show that $2^{p_1 p_2 \cdots p_n} + 1$ has at least 4^n divisors. (*France*)

127. Let N be a positive integer and $x_1, \ldots, x_n$ be positive integers less than N and such the least common multiple of any two of these n numbers is greater than N. Prove that
$$\sum_{i=1}^{n} \frac{1}{x_i} < 2.$$
(*Germany*)

128. Find all triples (a, b, c) of positive integers such that
   1. $a \leq b \leq c$.
   2. $\gcd(a, b, c) = 1$.
   3. $a^3 + b^3 + c^3$ is divisible by each of the numbers $a^2 b$, $b^2 c$, and $c^2 a$. (*India*)

129. Find all triples (a, b, p) of positive integers with p a prime number such that
$$2^a + p^b = 19^a.$$
(*Italy*)

130. Prove that the equation $\frac{1}{a}+\frac{1}{b}+\frac{1}{c}+\frac{1}{abc} = \frac{12}{a+b+c}$ has infinitely many solutions (a,b,c) in positive integers. (*Moldova*)
131. Let $\mathcal{P}$ be the set of all the primes, and let M be a subset of $\mathcal{P}$, having at least three elements and such that for any proper subset A of M all of the prime factors of the number

$$-1 + \prod_{p \in A} p$$

are found in M. Prove that $M = \mathcal{P}$. (*Romania*)
132. For every positive integer n, we denote by $d(n)$ the sum of its digits in the decimal representation. Prove that for each positive integer k, there exists a positive integer m such that the equation $x + d(x) = m$ has exactly k solutions in the set of positive integers. (*Romania*)
133. Find the largest positive integer n which divides $a^{25} - a$ for every positive integer a. (*Switzerland*)
134. For each integer $n > 1$, let $p(n)$ denote the largest prime factor of n. Determine all triples (x,y,z) of distinct positive integers satisfying the following:

   (i) x, y, and z are in arithmetic progression.
   (ii) $p(xyz) \leq 3$. (*United Kingdom*)

135. Let n be a positive integer. Prove that the number $2^n + 1$ has no prime divisor of the form $8k - 1$, where k is a positive integer. (*Vietnam*)

1.32 2004

136. Determine whether does exists a triangle with area 2004 and his sides positive integers. (*Bosnia and Herzegovina*)
137. Find all positive integers n, for which the number of all positive divisors of the number lcm $(1,2,\ldots,n)$ is equal to 2^k for some non-negative integer k. (*Estonia*)
138. If n is a positive integer, let $A = \{n, n+1, \ldots, n+17\}$. Does there exist some values of n for which we can divide A into two disjoint subsets B and C such that the product of the elements of B is equal to the product of the elements of C? (*France*)
139. Prove that for every positive integer n, there exists an n−digit number divisible by 5^n all of whose digits are odd. (*India*)
140. Find all triples (x,y,n) of positive integers such that $(x+y)(1+xy) = 2^n$. (*India*)
141. Show that the only solutions of the equation $p^k + 1 = q^m$ in positive integers $k, q, m > 1$ and prime p are as follows:

   1. $(p,k,q,m) = (2,3,3,2)$.
   2. $k = 1, q = 2,$ and $p = 2^m - 1$. (*India*)

142. An integer n is said to be *good* if $|n|$ is not the square of an integer. Determine all integers m with the following property: m can be represented, in infinitely many ways, as a sum of three distinct integers whose product is the square of an odd integer. (*India*)

143. A positive integer n is said to be a *perfect power* if $n = a^b$ for some integers a and b with $b > 1$.

    (a) Find 2004 perfect powers in arithmetic progression.
    (b) Prove that perfect powers cannot form an infinite arithmetic progression. (*Italy*)

144. Find nonnegative integers which can be reached by the expression

    $$\frac{a^2 + ab + b^2}{ab - 1}$$

    when a and b are nonnegative integers and $ab \neq 1$. (*Romania*)

145. Let a, b, and c be integers, b odd, and consider the sequence $x_0 = 4$, $x_1 = 0$, $x_2 = 2c$, $x_3 = 3b$, $x_n = ax_{n-4} + bx_{n-3} + cx_{n-2}$, for $n \geq 4$. Prove that if p is a prime and m a positive integer, then x_{p^m} is divisible by p. (*Romania*)

146. Prove that if n and m are integers and m is odd, then

    $$\frac{1}{3^m n} \sum_{k=0}^{m} \binom{3m}{3k} (3n-1)^k$$

    is an integer. (*Romania*)

147. Let $m \geq 2$ be an integer. A positive integer n is called m–good if for every positive integer a, relatively prime to n, one has $n | a^m - 1$. Show that any m–good number is at most $4m(2^m - 1)$. (*Romania*)

148. Let $p \geq 5$ be a prime number. Prove that there exist at least two distinct primes q_1 and q_2 satisfying $1 < q_i < p - 1$ and $q_i^{p-1} \not\equiv 1 \pmod{p^2}$, for $i = 1, 2$. (*Singapore*)

149. Find the largest positive integer n for which $4^{995} + 4^{1500} + 4^n$ is a perfect square. (*Switzerland*)

1.33 2005

150. Determine the smallest positive integer $a \geq 2$ for which there exist a prime number p and a positive integer $b \geq 2$ such that

    $$\frac{a^p - a}{p} = b^2.$$

    (*Austrian – Polish Competition*)

151. Let n be a positive integer such that $n \geq 2$. Let $x_1, x_2, \ldots, x_n$ be n distinct positive integers and S_i the sum of all these numbers excepting x_i, for $i = 1, 2, \ldots, n$. Let

$$f(x_1, x_2, \ldots, x_n) = \frac{\gcd(x_1, S_1) + \gcd(x_2, S_2) + \ldots + \gcd(x_n, S_n)}{x_1 + x_2 + \ldots + x_n}.$$

Determine the maximal value of $f(x_1, x_2, \ldots, x_n)$, while $(x_1, x_2, \ldots, x_n)$ is an element of the set of all n-tuples of distinct positive integers. (*Bosnia* and *Herzegovina*)

152. Let a, b, and c be positive integers such that $\frac{a}{b} + \frac{b}{c} + \frac{c}{a} = 3$. Prove that abc is a perfect cube of an integer. (*Bosnia* and *Herzegovina*)

153. Find all pairs (x, y) of positive integers satisfying the equation $(x + y)^x = x^y$. (*Estonia*)

154. Let x and y be two positive integers such that $3x^2 + x = 4y^2 + y$. Prove that $x - y$ is a perfect square. (*France*)

155. The transformation $n \to 2n - 1$ or $n \to 3n - 1$, where n is a positive integer, is called the "change" of n. Numbers a and b are called "similar" if there exists such positive integer that can be got by a finite number of "changes" from both a and b. Find all positive integers "similar" to 2005 and less than 2005. (*Georgia*)

156. Let A be a subset of the set of positive integers having the following properties:

    1. If a belong to A, then all the divisors of a belong to A.
    2. If a and b, $1 < a < b$, belong to A, then $1 + ab$ is also in A.

    Prove that if A contains at least three positive integers, then A contains all positive integers. (*Georgia*)

157. Let a, b, c, d, and n be positive integers such that $7 \cdot 4^n = a^2 + b^2 + c^2 + d^2$. Prove that the numbers a, b, c, and d are all $\geq 2^{n-1}$. (*Germany*)

158. A positive integer is called *nice* if the sum of its digits in the number system with base 3 is divisible by 3. Calculate the sum of the first 2005 nice integers. (*Germany*)

159. Prove that one can find a $n_0 \in \mathbb{N}$ such that $\forall m \geq n_0$; there exist three positive integers a, b, and c such that

    1. $m^3 < a < b < c < (m+1)^3$.
    2. abc is the cube of an integer. (*India*)

160. Let $\tau(n)$ denote the number of positive divisors of the positive integer n. Prove that there exist infinitely many positive integers a such that the equation $\tau(an) = n$ does not have a positive integer solution n. (*India*)

161. Determine all positive integers $n > 2$ such that $\frac{1}{2}\varphi(n) \equiv 1 \pmod{6}$. Here $\varphi(n)$ denotes the number of positive integers less than n and coprime to n. (*India*)

162. The function $\psi: \mathbb{N} \to \mathbb{N}$ is defined by $\psi(n) = \sum_{k=1}^{n} \gcd(k,n)$.

   (a) Prove that $\psi(mn) = \psi(m)\psi(n)$ for every two coprime $m, n \in \mathbb{N}$.
   (b) Prove that for each $a \in \mathbb{N}$ the equation $\psi(x) = ax$ has a solution. (*Italy*)

163. Let n be a positive integer and
$$A = 3\sum_{m=1}^{n^2}\left(\tfrac{1}{2} - \{\sqrt{m}\}\right).$$
   Find the largest k such that n^k divides $\lfloor A \rfloor$. (*Moldova*)

164. The integer N is positive. There are exactly 2005 ordered pairs (x, y) of positive integers satisfying $\frac{1}{x} + \frac{1}{y} = \frac{1}{N}$. Prove that N is a perfect square. (*United Kingdom*)

165. Let $p > 3$ be a prime number. Calculate the following:

   (a) $S = \sum_{k=1}^{\frac{p-1}{2}}\left(\left\lfloor\frac{2k^2}{p}\right\rfloor - 2\left\lfloor\frac{k^2}{p}\right\rfloor\right)$ if $p \equiv 1 \pmod 4$.
   (b) $T = \sum_{k=1}^{\frac{p-1}{2}}\left\lfloor\frac{k^2}{p}\right\rfloor$ if $p \equiv 1 \pmod 8$. (*Vietnam*)

1.34 2006

166. Find all integer solutions for $xy + yz + zx - xyz = 2$. (*Argentina*)
167. Let p be a prime and let $S = \{p - n^2 \mid n \in \mathbb{N}, n^2 < p\}$. Prove that S contains two elements a and b such that $a|b$ and $1 < a < b$. (*Argentina*)
168. A positive integer d is called *nice* if for all positive integers x and y holds as follows: d divides $(x+y)^5 - x^5 - y^5$ if d divides $(x+y)^7 - x^7 - y^7$.

   (a) Is 29 nice?
   (b) Is 2006 nice?
   (c) Prove that infinitely many nice numbers exist. (*Austrian – Polish Competition*)

169. Find all pairs (a, n) of positive integers such that $\frac{(a+1)^n - a^n}{n}$ is an integer. (*China*)
170. Prove that for any positive integers m and n, there is always a positive integer k so that $2^k - m$ has at least n different prime divisors. (*China*)
171. Find all solutions positive integers of the equation $3^x = 2^x y + 1$. (*Croatia*)
172. Let a and b be positive integers such that $b^n + n$ is a multiple of $a^n + n$ for all positive integers n. Prove that $a = b$. (*France*)

173. Let a, b, c, d, e, and f be positive integers and let $S = a + b + c + d + e + f$. Suppose that the number S divides $abc + def$ and $ab + bc + ca - de - ef - fd$. Prove that S is composite. (*Germany*)

174. For any positive integer n, let $w(n)$ denotes the number of different prime divisors of the number n. (For instance, $w(12) = 2$.) Show that there exist infinitely many positive integers n such that $w(n) < w(n + 1) < w(n + 2)$. (*Germany*)

175. Find all positive integers n such that there exists a unique integer a with the following property: $n! \mid a^n + 1$. (*Germany*)

176. Suppose that p is a prime number. Find all positive integers n such that $p \mid \varphi(n)$ and for all a such that $(a, n) = 1$ we have $n \mid a^{\frac{\varphi(n)}{p}} - 1$. (*Iran*)

177. Let n be a positive integer, and let A_n be the set of all positive integers $a \leq n$ such that $n \mid a^n + 1$.

   (a) Find all n such that $A_n \neq \emptyset$.
   (b) Find all n such that $|A_n|$ is even and nonzero.
   (c) Is there n such that $|A_n| = 130$? (*Italy*)

178. Solve in integers x and y the equation $x^3 - y^3 = 2xy + 8$. (*Lithuania*)

179. Determine all even positive integers n such that

$$\frac{1}{d_1} + \frac{1}{d_2} + \ldots + \frac{1}{d_k} = \frac{1620}{1003},$$

where $d_1, d_2, \ldots, d_k$ are all different divisors of n. (*Moldova*)

180. Let (a_n) be the Lucas sequence: $a_0 = 2$, $a_1 = 1$, $a_{n+1} = a_n + a_{n-1}$ for $n \geq 1$. Show that a_{59} divides $a_{30}^{59} - 1$. (*Moldova*)

181. Let $(a_n)_{n \geq 1}$ be the sequence given by $a_1 = 1$, $a_2 = 4$, and for all integers $n > 1$,

$$a_n = \sqrt{a_{n-1}a_{n+1} + 1}.$$

   (a) Prove that all the terms of the sequence are positive integers.
   (b) Prove that the number $2a_n a_{n+1} + 1$ is a perfect square for all integers $n \geq 1$. (*Romania*)

182. Let n be a positive integer such that $\sigma(n) = 2n + 1$. Prove that n is an odd perfect square. Here $\sigma(n)$ denotes the sum of positive divisors of the positive integer n. (*Singapore*)

183. Let n be a positive integer, and let $1 = d_1 < d_2 < \ldots < d_k = n$ be the positive divisors of n. Find all n such that $2n = d_5^2 + d_6^2 - 1$. (*Switzerland*)

184. Find all positive integers k such that $3^k + 5^k$ is the power of a positive integer with the exponent ≥ 2. (*Switzerland*)

185. For all integers $n \geq 1$, we define $x_{n+1} = x_1^2 + x_2^2 + \ldots + x_n^2$, where x_1 is a positive integer. Find the least x_1 such that 2006 divides x_{2006}. (*Turkey*)

186. Find all pairs (n, k) of integers such that n is not negative, k is greater than 1 and satisfying that the number $A = 17^{2006n} + 4 \cdot 17^{2n} + 7 \cdot 19^{5n}$ can be represented as the product of k consecutive positive integers. (*Vietnam*)
187. The real sequence (a_n) is defined by $a_0 = 1$ and $a_{n+1} = \frac{1}{2}\left(a_n + \frac{1}{3a_n}\right)$ for $n \geq 0$. Denote $A_n = \frac{3}{3a_n^2 - 1}$. Prove that A_n is a perfect square and it has at least n distinct prime divisors. (*Vietnam*)

1.35 2007

188. Let x, y, and z be distinct positive integers having exactly two digits in such a way that $x = 10a + b$, $y = 10b + c$, $z = 10c + a$ (a, b, and c are digits). Find all possible values of gcd (x, y, z). (*Argentina*)
189. Let $d_1, d_2, \ldots, d_r$ be the positive divisors of n, $1 = d_1 < d_2 < \ldots < d_r = n$. If $d_7^2 + +d_{15}^2 = d_{16}^2$, find all possible values of d_{17}. (*Argentina*)
190. For a positive integer n, we define $s(n)$ as the sum of digits of n in base 10. Does there exist a positive real constant c such that for all positive integers n we have $s(n) \leq c\, s(n^2)$? (*Argentina*)
191. Find all pairs (x, y) of integers such that $x(x + 2) = y^2(y^2 + 1)$. (*Bosnia and Herzegovina*)
192. Let $p = 4k + 3$ be a prime number. Find the number of different residues mod p of $(x^2 + y^2)^2$, where $(x, p) = (y, p) = 1$. (*Bulgaria*)
193. Let n be a positive integer, and let A be a subset of $\{1, 2, \ldots, n\}$ satisfying the follows: for any two numbers $x, y \in A$, the least common multiple of x, y is not more than n. Show that $|A| \leq 1.9\sqrt{n} + 5$. (*China*)
194. Find all integral solutions of the equation $(m^2 - n^2)^2 = 16n + 1$. (*Croatia*)
195. Prove that the sequence $a_n = \lfloor n\sqrt{2} \rfloor + \lfloor n\sqrt{3} \rfloor$ contains infinitely many even and infinitely many odd numbers. (*Croatia*)
196. Let $n \geq 2$ be a positive integer. Prove that $\frac{b^n - 1}{b - 1}$ is a prime power for some positive integer b and then n is a prime. (*Estonia*)
197. For a positive integer a, a' is the integer obtained by the following method: the decimal writing of a' is the inverse of the decimal writing of a (the decimal writing of a' can begin with zeros, but not the one of a); for instance, if $a = 2370$, $a'=0732$, that is 732. Let a_1 be a positive integer and $(a_n)_{n \geq 1}$ the sequence defined by a_1 and the following formula: $a_{n+1} = a_n + a'_n$, for $n \geq 1$. Can a_7 be a prime? (*France*)
198. Find all integer solutions of the equation $\frac{x^7 - 1}{x - 1} = y^5 - 1$. (*India*)
199. Let A be the largest subset of $\{1, 2, \ldots, n\}$ such that for each $x \in A$, x divides at most one other element in A. Prove that

$$\frac{2n}{3} \leq |A| \leq \left\lceil \frac{3n}{4} \right\rceil.$$

(*Iran*)

200. Find all pairs (x, y) of nonnegative integers satisfying $(14y)^x + y^{x+y} = 2007$. (*Singapore*)
201. Find the two smallest positive integers which can be represented by $7m^2 - 11n^2$ with m and n integers. (*Switzerland*)
202. Find all primes p for which there is an integer n such that there are no integers x and y with $x^3 + y^3 \equiv n \pmod{p}$. (*Ukraine*)
203. Prove that there are infinitely many positive integers n for which all the prime divisors of $n^2 + n + 1$ are not more than $\sqrt{n}$. (*Ukraine*)
204. Show that there are infinitely many pairs (m, n) of positive integers such that
$$\frac{m+1}{n} + \frac{n+1}{m}$$
is a positive integer. (*United Kingdom*)
205. Determine whether or not there exist positive integers a and b such that a does not divide $b^n - n$ for all positive integers n. (*USA*)

1.36 2008

206. Let $\tau(n)$ be the number of positive divisors of the positive integer n. Find all n such that $n = p\,\tau(n)$, where p is a prime number. (*Argentina*)
207. Find all pairs (m, n) of positive integers that satisfy both the following conditions:

    (a) $m^2 - n$ divides $m + n^2$
    (b) $n^2 - m$ divides $n + m^2$ (*Bosnia and Herzegovina*)

208. The sequence (x_n) is defined by
$$x_1 = 2, x_2 = 12 \text{ and } x_{n+2} = 6x_{n+1} - x_n, n = 1, 2, \ldots.$$

    Let p be an odd prime number, and let q be a prime divisor of x_p. Prove that if $q \ne 2, 3$, then $q \ge 2p - 1$. (*China*)

209. Let $n > 1$ be an integer such that n divides $2^{\varphi(n)} + 3^{\varphi(n)} + \ldots + n^{\varphi(n)}$, and let $p_1, p_2, \ldots, p_k$ be all distinct prime divisors of n. Show that $\frac{1}{p_1} + \frac{1}{p_2} + \ldots + \frac{1}{p_k} + \frac{1}{p_1 p_2 \cdots p_k}$ is an integer. (*China*)
210. Sequence (a_n) is defined by $a_0 = 0, a_1 = 1$ and $a_n = a_{n-1} + a_{n-2} + 1$ for every $n \ge 2$. Prove that for every positive integer m, there exist two consecutive terms in the sequence that are both divisible by m. (*Estonia*)
211. Prove that there is an integer k for which $k^3 - 36k^2 + 51k - 97$ is a multiple of 3^{2008}. (*Germany*)
212. Find all pairs (a, b) of positive integers such that $7^a - 3^b$ divides $a^4 + b^2$. (*Germany*)

213. For every integer $k \geq 2$, prove that 2^{3k} divides the number
$$\binom{2^{k+1}}{2^k} - \binom{2^k}{2^{k-1}},$$
but 2^{3k+1} does not. (*Germany*)

214. Given the equation $x^2 + y^2 - axy + 2 = 0$, where a is a positive integral parameter.
    (a) Show that for $a \neq 4$, there are no pairs (x, y) of positive integers satisfying the equation.
    (b) Show that for $a = 4$, there are infinitely many pairs (x, y) of positive integers satisfying the equation and determine those pairs. (*Greece*)

215. Find the total number of solutions to the following system of congruences:
$$\begin{cases} a^2 + bc \equiv a \pmod{37} \\ b(a+d) \equiv b \pmod{37} \\ c(a+d) \equiv c \pmod{37} \\ bc + d^2 \equiv d \pmod{37} \\ ad - bc \equiv 1 \pmod{37}. \end{cases}$$

Show that the congruence $y^{37} \equiv x^3 + 1 \pmod{p}$ is solvable for every prime $p < 100$. (*Hong Kong*)

216. Prove that there are infinitely many primes p such that the total number of solutions of the congruence $3x^3 + 4y^4 + 5z^3 - y^4z \equiv 0 \pmod{p}$ is p^2. (*Hong Kong*)

217. Let p be a prime number. Solve in $\mathbb{N}_0 \times \mathbb{N}_0$ the equation $x^3 + y^3 - 3xy = p - 1$. (*Moldova*)

218. Let p be a prime number and k, n positive integers so that $\gcd(p, n) = 1$. Prove that $\binom{np^k}{p^k}$ and p are coprime. (*Moldova*)

219. Let $a, b, c,$ and d be positive integers such that $a > b > c > d$ and $(a + b - c + d) | ac + bd$. Prove that if m and n are positive integers and n is odd, then $a^n b^m + c^m d^n$ is composite. (*Mongolia*)

220. Given an integer a. Let p a prime number such that $p | a$ and $p \equiv \pm 3 \pmod{8}$. Define the sequence (a_n) such that $a_n = 2^n + a$. Prove that the sequence (a_n) contains only finitely many perfect squares. (*Mongolia*)

221. We say that a positive integer is *happy* if it can be expressed in the form $\frac{a^2 b}{a-b}$, where $a > b > 0$ are integers. We also say that a positive integer m is *evil* if doesn't exist a happy integer n such that $d(n) = m$. Prove that all numbers happy and evil are a power of 4. Here $d(n)$ denotes the number of positive divisors of the positive integer n. (*Peru*)

222. Find all odd primes p, if any, such that p divides $\sum_{n=1}^{103} n^{p-1}$. (*Singapore*)

223. Find all triplets (a, b, c) of positive integers such that

$$a|bc-1, \quad b|ca-1, \quad c|ab-1.$$

(*Switzerland*)

224. Prove that there exist infinitely many pairs (a, b) of distinct positive integers greater than 1 such that $b^b + a$ is divisible by $a^a + b$. (*Ukraine*)

225. Let m and n be positive integers. Prove that $6m|(2m+3)^n + 1$ if and only if $4m|3^n + 1$. (*Vietnam*)

1.37 2009

226. Find all the natural numbers m and n such that $1 + 5 \cdot 2^m = n^2$. (*Albania*)
227. Find all positive integers n such that $20^n - 13^n - 7^n$ is divisible by 309. (*Argentina*)
228. Find all pairs (a, b) of positive integers such that $\frac{a^2(b-a)}{b+a}$ is the square of a prime. (*Bosnia* and *Herzegovina*)
229. Let n be a composite number. Prove that there exists a positive integer m satisfying: $m|n, m \leq \sqrt{n}$ and $\tau(n) \leq \tau^3(m)$. (*China*)
230. Let $a > b > 1$ be positive integers with b an odd number, and let n be a positive integer. Prove that if $b^n|a^n - 1$, then $a^b > \frac{3^n}{n}$. (*China*)
231. Prove that there are infinitely many positive integers n such that $n^2 + 1|n!$ and infinitely many of those for which $n^2 + 1 \nmid n!$. (*Croatia*)
232. Determine all triplets (a, b, c) of positive integers for which $|2^a - b^c| = 1$. (*Croatia*)
233. Call a finite set of positive integers *independent* if its elements are pairwise coprime and *nice* if the arithmetic mean of the elements of every non-empty subset of it is an integer.

   (a) Prove that for every positive integer n, there is an n-element set of positive integers which is both independent and nice.
   (b) Is there an infinite set of positive integers whose every independent subset is nice and which has an n-element independent subset for every positive integer n? (*Estonia*)

234. For any positive integer n, let $c(n)$ be the largest divisor of n not greater than $\sqrt{n}$, and let $s(n)$ be the least integer x such that $n < x$ and the product nx is divisible by an integer y where $n < y < x$. Prove that, for every n,

$$s(n) = (c(n) + 1)\left(\frac{n}{c(n)} + 1\right).$$

(*Estonia*)

235. Let (a_n) be the sequence defined by $a_1 = 1$ and $a_{n+1} = a_n^4 - a_n^3 + 2a_n^2 + 1$ for $n \geq 1$. Show that there is an infinite number of primes p such none of the a_n is divisible by p. (*Germany*)

236. Let n be a positive integer and let p be a prime number. Prove that if a, b, and c are integers satisfying the equations $a^n + pb = b^n + pc = c^n + pa$, then $a = b = c$. (*Germany*)

237. Suppose that a is an even positive integer and $A = a^n + a^{n-1} + \ldots + a + 1$, $n \in \mathbb{N}^*$ is a perfect square. Prove that $8 | a$. (*Greece*)

238. Let a be a fixed positive integer. Prove that the set of prime divisors of $2^{2^n} + a$ for $n = 1, 2, \ldots$ is infinite. (*Iran*)

239. Find all polynomials f with integer coefficients such that for every prime p and positive integers u, v with the condition $p | uv - 1$, we always have $p | f(u)f(v) - 1$. (*Iran*)

240. Let n be a positive integer. Prove that

$$3^{\frac{5^{2^n}-1}{2^{n+2}}} \equiv (-5)^{\frac{3^{2^n}-1}{2^{n+2}}} \pmod{2^{n+4}}.$$

(*Iran*)

241. Find all pairs (x, y) of integers such that $y^3 = 8x^6 + 2x^3y - y^2$. (*Italy*)

242. Let p be a prime divisor of $n \geq 2$. Prove that there exists a set of positive integers $A = \{a_1, a_2, \ldots, a_n\}$ such that the product of any two numbers from A is divisible by the sum of any p numbers from A. (*Moldova*)

243. Prove that there are infinitely many pairs of prime numbers (p, q) such that $p | 2^{q-1} - 1$ and $q | 2^{p-1} - 1$. (*Romania*)

244. Determine the smallest positive integer N such that there exist six distinct positive integers a_1, a_2, a_3, a_4, a_5, and a_6 satisfying the following:

   (a) $N = a_1 + a_2 + a_3 + a_4 + a_5 + a_6$.
   (b) $N - a_i$ is a perfect square for $i = 1, 2, 3, 4, 5, 6$. (*Singapore*)

245. Let $S = \{a + np | n = 0, 1, 2, 3, \ldots\}$, where a is a positive integer and p is a prime. Suppose that there exist positive integers x and y such that x^{41} and y^{49} are in S. Determine if there exists a positive integer z such that z^{2009} is in S. (*Singapore*)

246. Find all pairs (m, n) of positive integers for which

$$m | n^2 + 8, \quad n | m^2 + 8.$$

(*Switzerland*)

1.38 2010

247. Prove that if n is the product of distinct Mersenne's primes, then $\sigma(n)$ is a power of 2. Is the converse true? Here $\sigma(n)$ denotes the sum of positive divisors of the positive integer n. (*Albania*)
248. Let p and q be prime numbers. The sequence (x_n) is defined by $x_1 = 1, x_2 = p$ and $x_{n+1} = px_n - qx_{n-1}$ for all $n \geq 2$. Given that there is some k such that $x_{3k} = -3$, find p and q. (*Argentina*)
249. (a) Let p and q be distinct prime numbers such that $p + q^2$ divides $p^2 + q$. Prove that $p + q^2$ divides $pq - 1$.
    (b) Find all prime numbers p such that $p + 121$ divides $p^2 + 11$. (*Bosnia and Herzegovina*)
250. Find all positive integers $m, n \geq 2$ such that
    (a) $m + 1$ is a prime number of type $4k - 1$.
    (b) There is a prime number p and a nonnegative integer a such that
    $$\frac{m^{2^n-1} - 1}{m - 1} = m^n + p^a.$$
    (*China*)
251. Let $k > 1$ be an integer, set $n = 2^{k+1}$. Prove that for any positive integers $a_1, a_2, \ldots, a_n$, the number $\prod_{1 \leq i < j \leq n} (a_i + a_j)$ has at least $k + 1$ prime divisors.
    (*China*)
252. For integers $n > 1$, define $f(n)$ to be the sum of all positive divisors of n that are less than n. Prove that for any positive integer k, there exists a positive integer $n > 1$ such that $n < f(n) < f^2(n) < \ldots < f^k(n)$, where $f^i(n) = f(f^{i-1}(n))$ for $i > 1$ and $f^1(n) = f(n)$. (*China*)
253. Given positive integer k, prove that there exists a positive integer N depending only on k such that for any integer $n \geq N$, $\binom{n}{k}$ has at least k prime divisors.
    (*China*)
254. Prove that there exists an unbounded sequence of positive integers $a_1 \leq a_2 \leq a_3 \leq \cdots$ such that there exists a positive integer M with the following property: for any integer $n \geq M$, if $n + 1$ is not prime, then any prime divisor of $n! + 1$ is greater than $n + a_n$. (*China*)
255. For arbitrary positive integers a and b, denote $a \ominus b = \frac{a-b}{\gcd(a,b)}$. Let n be a positive integer. Prove that the following conditions are equivalent:
    (a) $\gcd(n, n \ominus m) = 1$ for every positive integer $m < n$.
    (b) $n = p^k$, where p is a prime number and k is a non-negative integer.
    (*Estonia*)

256. Call a positive integer *good* if either $N = 1$ or N can be written as the product of an even number of prime numbers, not necessarily distinct. Let $P(x) = (x + a)(x + b)$, where a and b are positive integers.

   (a) Show that there exist distinct positive integers a and b such that $P(1)$, $P(2)$, ..., $P(2010)$ are all good numbers.
   (b) Suppose a and b are such that $P(n)$ is a good number for all positive integers n. Prove that $a = b$. (*India*)

257. Is there a triangle with angles in ratio 1:2:4 and the lengths of its sides are integers with at least one of them is a prime number? (*Indonesia*)

258. Let $0 \leq k < n$ be integers and $A = \{a | a \equiv k \pmod{n}\}$. Find the smallest value of n for which the expression

$$\frac{a^m + 3^m}{a^2 - 3a + 1}$$

does not take any integer values for $(a, m) \in A \times \mathbb{Z}^+$. (*Turkey*)

259. Determine whether or not there exists a positive integer k such that $p = 6k + 1$ is a prime and

$$\binom{3k}{k} \equiv 1 \pmod{p}.$$

(*USA*)

1.39 2011

260. Find all prime numbers p such that $2^p + p^2$ is also a prime number. (*Albania*)
261. Determine all positive integers n such that the number $n(n + 2)(n + 4)$ has at most 15 positive divisors. (*Argentina*)
262. Let $n > 1$ be an integer, and let k be the number of distinct prime divisors of n. Prove that there exists an integer k, $1 < a < \frac{n}{k} + 1$, such that $n | a^2 - a$. (*China*)
263. Prove that if n and k are positive integers such that $1 < k < n - 1$, and then the binomial coefficient $\binom{n}{k}$ is divisible by at least two different primes. (*Estonia*)
264. Find all prime numbers p and q such that $p^4 + p^3 + p^2 + p = q^2 + q$. (*Greece*)
265. Find all positive integers n satisfying the conditions:

   (a) $n^2 = (a + 1)^3 - a^3$.
   (b) $2n + 119$ is a perfect square. (*India*)

266. Prove that for no integer, n is $n^7 + 7$ a perfect square. (*India*)

267. Find all pairs (m, n) of nonnegative integers for which
$$m^2 + 2 \cdot 3^n = m(2^{n+1} - 1).$$
(*India*)

268. Suppose that $f: \mathbb{N} \to \mathbb{N}$ is a function for which the expression $af(a) + bf(b) + 2ab$ for all $a, b \in \mathbb{N}$ is always a perfect square. Prove that $f(a) = a$ for all $a \in \mathbb{N}$. (*Iran*)

269. Let $v(n)$ be the order of 2 in $n!$. Prove that for any positive integers a and m, there exists a positive integer n such that $v(n) \equiv a \pmod{m}$. (*Mongolia*)

270. Let a and b be integers and let $P(x) = ax^3 + bx$. For any positive integer n, we say that the pair (a, b) is n – good if $n | P(m) - P(k)$ implies $n | m - k$ for all integers m, k. We say that (a, b) is *very good* if (a, b) is n-good for infinitely many positive integers n.

   (a) Find a pair (a, b) which is 51-good, but not very good.
   (b) Show that all 2010-good pairs are very good. (*Peru*)

271. Show that there are infinitely many positive integer numbers n such that $n^2 + 1$ has two positive divisors whose difference is n. (*Romania*)

272. Find all pairs (p, q) of prime numbers such that $3 \nmid p + 1$ and $\frac{p^3 + 1}{q}$ is the square of a positive integer. (*Switzerland*)

273. Let $t(n)$ be the sum of the digits in the binary representation of a positive integer n, and let $k \geq 2$ be an integer.

   (a) Show that there exists a sequence (a_i) of integers such that $a_m \geq 3$ is an odd integer and $t(a_1 a_2 \ldots a_m) = k$ for all $m \geq 1$.
   (b) Show that there is an integer N such that $t(3 \cdot 5 \cdot \ldots \cdot (2m + 1)) > k$ for all integers $m \geq N$. (*Turkey*)

274. Find all positive integers x and y such that $x + y + 1$ divides $2xy$ and $x + y - 1$ divides $x^2 + y^2 - 1$. (*United Kingdom*)

275. Let $(a_n)_{n \geq 0}$ be the sequence of integers satisfying $a_0 = 1$, $a_1 = 3$, and $a_{n+2} = 1 + \left\lfloor \frac{a_{n+1}^2}{a_n} \right\rfloor \forall n \geq 0$. Prove that $a_n a_{n+2} - a_{n+1}^2 = 2^n$ for every positive integer n. (*Vietnam*)

276. Find all positive integers n such that $A = 2^{n+2}(2^n - 1) - 8 \cdot 3^n + 1$ is a perfect square. (*Vietnam*)

1.40 2012

277. Find all pairs of positive integers (a, b) not relatively prime such that
$$\gcd(a, b) + 9 \operatorname{lcm}(a, b) + 9(a + b) = 7ab.$$
(*Albania*)

278. Prove that for all odd prime numbers p, there exist a positive integer $m < p$ and integers x_1, x_2, and x_3 such that $mp = x_1^2 + x_2^2 + x_3^2$. (*Bosnia and Herzegovina*)

279. Define a function $f : \mathbb{N} \to \mathbb{N}, f(1) = p + 1, f(n + 1) = f(1) \cdot f(2) \cdot \ldots \cdot f(n) + p$ where p is a prime number. Find all p such that there exists a positive integer k for which $f(k)$ is a perfect square. (*Bosnia and Herzegovina*)

280. Let $x_n = \binom{2n}{n}$ for all $n \in \mathbb{Z}^+$. Prove that there exist infinitely many finite sets A, B of positive integers, satisfying $A \cap B = \emptyset$ and

$$\frac{\prod_{i \in A} x_i}{\prod_{j \in B} x_j} = 2012.$$

(*China*)

281. We call a positive integer n a *good* number if $\tau(m) < \tau(n)$ for all $m < n$. Prove that for any positive integer k, there are only finitely many good numbers not divisible by k. (*China*)

282. Find all integers $k \geq 3$ with the following property: there exist integers m, n such that $1 < m < k$, $1 < n < k$, $\gcd(m,k) = \gcd(n.k) = 1$, $m + n > k$, and $k | (m - 1)(n - 1)$. (*China*)

283. For a given positive integer n, one has to choose positive integers $a_0, a_1, \ldots$ so that the following conditions hold as follows:

   1. $a_i = a_{i+n}$ for every i.
   2. a_i is not divisible by n for every i.
   3. a_{i+a_i} is divisible by a_i for every i.

   For which positive integers $n > 1$ is this possible only if the numbers $a_0, a_1, \ldots$ are all equal? (*Estonia*)

284. Let p be a prime number. Find all positive integers $a, b, c \geq 1$ such that

$$a^p + +b^p = p^c.$$

(*France*)

285. Find all triples (p, m, n) satisfying the equation $p^m - n^3 = 8$, where p is a prime number and m and n are nonnegative integers. (*Greece*)

286. Let $0 < x < y < z < p$ be integers, where p is a prime. Prove that the following statements are equivalent:

   (a) $x^3 \equiv y^3 \pmod{p}$ and $x^3 \equiv z^3 \pmod{p}$.
   (b) $y^2 \equiv xz \pmod{p}$ and $z^2 \equiv xy \pmod{p}$. (*India*)

287. Show that there exist infinitely many pairs (a, b) of positive integers with the property that $a + b$ divides $ab + 1$, $a - b$ divides $ab - 1$, $b > 1$ and $a > b\sqrt{3} - 1$. (*India*)

288. Find all positive integers $n \geq 2$ such that for all integers i, j that $0 \leq i, j \leq n$, $i+j$ and $\binom{n}{i} + \binom{n}{j}$ have the same parity. (*Iran*)

289. Let $\sigma(x)$ denote the sum of divisors of the positive integer x. For every positive integer n define $f(n)$ as the number of positive integers $m \leq n$ for which $\sigma(m)$ is odd. Prove that there are infinitely many positive integers n such that $f(n) | n$. (*Serbia*)

290. Let $S_r(n) = 1^r + 2^r + \ldots + n^r$, where r is a rational number and n is a positive integer. Find all triplets $(a, b, c) \in \mathbb{Q}_+ \times \mathbb{Q}_+ \times \mathbb{N}$ for which there exist infinitely many positive integers n satisfying $S_a(n) = (S_b(n))^c$. (*Turkey*)

291. Determine, with proof, whether or not there exist integers $a, b, c > 2010$ satisfying the equation: $a^3 + 2b^3 + 4c^3 = 6abc + 1$. (*USA*)

292. Find all positive integers $a, n \geq 1$ such that for all primes p dividing $a^n - 1$, there exists a positive integer $m < n$ such that $p | a^m - 1$. (*USA*)

293. Let $p \geq 17$ be a prime. Prove that $t = 3$ is the largest positive integer which satisfies the following condition: for any integers $a, b, c,$ and d such that abc is not divisible by p and $a + b + c$ is divisible by p, there exist integers $x, y,$ and z belonging to the set $\{0, 1, 2, \ldots, \lfloor \frac{p}{t} \rfloor - 1\}$ such that $ax + by + cz + d$ is divisible by p. (*Vietnam*)

1.41 2013

294. The sequence (a_n) is defined by $a_0 = a_1 = 1$ and $a_{n+1} = 14a_n - a_{n-1} - 4$ for all positive integers n. Prove that all terms of this sequence are perfect squares. (*Bosnia* and *Herzegovina*)

295. Find all primes p, q such that p divides $30q - 1$ and q divides $30p - 1$. (*Bosnia* and *Herzegovina*)

296. For a positive integer $N > 1$ with unique factorization $N = p_1^{\alpha_1} p_2^{\alpha_2} \ldots p_k^{\alpha_k}$, we define
$$\Omega(N) = \alpha_1 + \alpha_2 + \ldots + \alpha_k.$$
Let $a_1, a_2, \ldots, a_n$ be positive integers and $P(x) = (x + a_1)(x + a_2) \ldots (x + a_n)$ such that for all positive integers k, $\Omega(P(k))$ is even. Show that n is an even number. (*China*)

297. Find all prime numbers p for which one can find a positive integer m and nonnegative integers $a_0, a_1, \ldots, a_m$ less than p such that
$$\begin{cases} a_0 + a_1 p + \ldots + a_{m-1} p^{m-1} + a_m p^m = 2013, \\ a_0 + a_1 + \ldots + a_{m-1} + a_m = 11. \end{cases}$$

(*Estonia*)

298. Call a tuple $(b_m, b_{m+1}, \ldots, b_n)$ of integers *perfect* if both following conditions are fulfilled:

1. There exists an integer $a > 1$ such that $b_k = a^k + 1$ for all $k = m, m+1, \ldots, n$.
2. For all $k = m, m+1, \ldots, n$, there exist a prime number q and a nonnegative integer t such that $b_k = q^t$.

Prove that if $n - m$ is large enough, then there is no perfect tuples, and find all perfect tuples with the maximal number of components. (*Estonia*)

299. Find all pairs (m, n) of nonnegative integers satisfying

$$\frac{n(n+2)}{4} = m^4 + m^2 - m + 1.$$

(*Greece*)

300. For a prime p, a positive integer n and an integer a, we let $S_n(a, p)$ denote the exponent of p in the prime factorization of $a^{p^n} - 1$. For example, $S_1(4, 3) = 2$ and $S_2(6, 2) = 0$. Find all pairs (n, p) such that $S_n(2013, p) = 100$. (*India*)

301. An integer a is called *friendly* if the equation $(m^2 + n)(n^2 + m) = a(m - n)^3$ has a solution over the positive integers.

(a) Prove that there are at least 500 friendly integers in the set $\{1, 2, \ldots, 2012\}$.
(b) Decide whether $a = 2$ is friendly. (*India*)

302. Do there exist positive integers a, b, and c such that $a^2 + b^2 + c^2$ is divisible by $2013(ab + bc + ca)$? (*Iran*)

303. Let m be the number of ordered solutions (a, b, c, d, e) in positive integers of the equation $\frac{1}{a} + \frac{1}{b} + \frac{1}{c} + \frac{1}{d} + \frac{1}{e} = 1$. Prove that m is odd. (*Moldova*)

304. Show that for each positive integer a, the equation $x^3 + x + a^2 = y^2$ has at least a solution (x, y) in positive integers. (*North Korea*)

305. Determine all integers $m \geq 2$ such that every n with $\frac{m}{3} \leq n \leq \frac{m}{2}$ divides the binomial coefficient $\binom{n}{m - 2n}$. (*Peru*)

306. Find all pairs (n, p) of positive integers and p a prime number such that $\sqrt{n + \frac{p}{n}}$ is a positive integer. (*Puerto Rico*)

307. Given an integer $n \geq 2$, let a_n, b_n, c_n be integer numbers such that

$$\left(\sqrt[3]{2} - 1\right)^n = a_n + b_n \sqrt[3]{2} + c_n \sqrt[3]{4}.$$

Prove that $c_n \equiv 1 \pmod{3}$ if and only if $n \equiv 2 \pmod{3}$. (*Romania*)

308. If x, y, and z are positive integers and $z(xz + 1)^2 = (5z + 2y)(2z + y)$, prove that z is an odd perfect square. (*Taiwan*)

309. Starting from 37, adding 5 before each previous term, forms the following sequence: 37, 537, 5537, 55,537, etc. How many prime numbers are there in this sequence? (*Taiwan*)
310. Find all pairs (m, n) of positive integers such that

$$2^n + (n - \varphi(n) - 1)! = n^m + 1.$$

Here $\varphi(n)$ denotes the number of positive integers less than n and coprime to n. (*Turkey*)
311. (a) Prove that there are infinitely many positive integers t such that both $2012t + 1$ and $2013t + 1$ are perfect squares.
(b) Suppose that m and n are positive integers such that both $mn + 1$ and $mn + n + 1$ are perfect squares. Prove that $8(2m + 1)$ divides n. (*Vietnam*)

1.42 2014

312. Find all nonnegative integer numbers such that $7^x - 2 \cdot 5^y = -1$. (*Bosnia and Herzegovina*)
313. Given a fixed positive integer $a \geq 9$, prove that there exist only finitely many positive integers n satisfying: (1) $\tau(n) = a$; 2) $n | \varphi(n) + \sigma(n)$. Here $\tau(n)$ is the number of positive divisors of n, $\varphi(n)$ is the number of positive integers less than n and coprime to n and $\sigma(n)$ is the sum of positive divisors of n. (*China*)
314. Let $k > 3$ be a fixed odd integer. Prove that there exist infinitely many positive integers n, such that there are two positive integers d_1 and d_2 each dividing $\frac{n^2+1}{2}$ and $d_1 + d_2 = n + k$. (*China*)
315. Find all positive integers n such that the equation $x^2 + y^2 + z^2 = nxyz$ has solutions in positive integers. (*Estonia*)
316. Prove that there exist infinitely many positive integers n such that the largest prime divisor of $n^4 + n^2 + 1$ is equal to the largest prime divisor of $(n + 1)^4 + + (n + 1)^2 + 1$. (*France*)
317. Let $(x_n)_{n \geq 1}$ be the sequence of real numbers defined by

$$x_1 = 1, 2x_{n+1} = 3x_n + +\sqrt{5x_n^2 - 4}, n \geq 1.$$

(a) Prove that the sequence consists only of positive integers.
(b) Check if there are terms of the sequence divisible by 2011. (*Greece*)
318. Let p be an odd prime and r an odd positive integer. Show that $pr + 1$ does not divide $p^p - 1$. (*India*)

319. In an old maths book, Mary found the problem asking to determine three integers a, b, and c such that

$$\frac{1}{14}\left(a^2 + b^2 + c^2\right) = ab + bc + ca.$$

(a) Someone scribbled a solution in this book: $b = -1$, $c = 2$, but a is not legible. Find all possibilities for a in this case.
(b) Prove that for each solution (a, b, c) to the given equation, both sides of this equation are the square of an integer.
(c) Prove that for each square number S, there exists a solution (a, b, c) to the given equation such that both sides of the equation are equal to S. (*Ireland*)

320. Define $p(n)$ to be the product of all nonzero digits of n. For instance $p(5) = 5$, $p(27) = 14$, $p(101) = 1$, and so on. Find the greatest prime divisor of the number

$$p(1) + p(2) + p(3) + \ldots + p(999).$$

(*Moldova*)

321. Let r be a positive integer, and let N_r be the smallest positive integer such that the numbers $\frac{N_r}{n+r}\binom{2n}{n}, n = 0, 1, 2, \ldots$ are all integers. Show that $N_r = \frac{r}{2}\binom{2r}{r}$. (*Peru*)

322. Determine all positive integers n such that all positive integers less than n and coprime to n are powers of primes. (*Romania*)

323. Prove that there are no two positive integers for which the harmonic, geometric, arithmetic, and quadratic means are all integers. (*Switzerland*)

324. Given the polynomial $p(x) = x^2 + x - 70$, do there exist integers $0 < m < n$, so that $p(m)$ is divisible by n and $p(m+1)$ is divisible by $n+1$? (*Tajikistan*)

325. Find all pairs (m, n) of positive odd integers such that $n \mid 3m + 1$ and $m \mid n^2 + 3$. (*Turkey*)

1.43   2015

326. Prove that there exist infinitely many composite positive integers n such that n divides $3^{n-1} - 2^{n-1}$. (*Bosnia and Herzegovina*)
327. Prove that for every prime number p and positive integer a, there exists a positive integer n such that p^n contains a consecutive equal digits. (*Estonia*)
328. A positive integer n is called *naughty* if it can be written in the form $n = a^b + b$ with integers $a, b \geq 2$. Is there a sequence of 102 consecutive positive integers such that exactly 100 of those numbers are naughty? (*Germany*)

329. Solve in positive integers the following equation:

$$xy(x+y-10) - 3x^2 - 2y^2 + 21x + 16y = 60.$$

(Greece)

330. Find all positive integers a and b such that $\frac{a^2+b}{b^2-a}$ and $\frac{b^2+a}{a^2-b}$ are also integers. *(India)*

331. For a composite number n, let d_n denote its largest proper divisor. Show that there are infinitely many n for which $d_n + d_{n+1}$ is a perfect square. *(India)*

332. Let $n \geq 2$ be an integer and let A_n be the set $A_n = \{2^n - 2^k | k \in \mathbb{Z}, 0 \leq k < n\}$.
    Determine the largest positive integer that cannot be written as the sum of one or more (not necessarily distinct) elements of A_n. *(India)*

333. Find all triples (p,x,y) consisting of a prime number p and two positive integers x and y such that $x^{p-1} + y$ and $x + y^{p-1}$ are both powers of p. *(India)*

334. Let $b_1 < b_2 < b_3 < \ldots$ be the sequence of all positive integers which are sum of squares of two positive integers. Prove that there exist infinitely many positive integers m for which $b_{m+1} - b_m = 2015$. *(Iran)*

335. Determine all triples (a,b,c) of positive integers satisfying both of the following properties:

    (a) a, b, and c are three consecutive odd integers in increasing order.
    (b) The number $a^2 + b^2 + c^2$ consists of four equal digits. *(Ireland)*

336. (a) Find with proof all integers x and y such that $\frac{x^4+x^2y^2+y^4}{3}$ is a prime number.
    (b) Prove that if x and y are integers, then $\frac{x^4+x^2y^2+y^4}{5}$ is not a prime number.
    (Ireland)

337. Find with proof all positive integers k such that, for $n = 2^k$, every prime number which divides $n! + 1$ also divides $n + 1$. *(Ireland)*

338. For a positive integer n, we define D_n as the largest integer that is a divisor of $a^n + (a+1)^n + (a+2)^n$ for all positive integers n.

    1. Show that for all positive integers n, the number D_n is of the form $3^k, k \geq 0$.
    2. Show that for all integers $k \geq 0$, there exists a positive integer n such that $D_n = 3^k$. *(Netherlands)*

339. For which positive integers m does the equation $(ab)^{2015} = (a^2 + b^2)^m$ have positive integer solutions? *(New Zealand)*

340. Determine all positive integers n which have a divisor d with the property that $dn + 1$ is a divisor of $d^2 + n^2$. *(New Zealand)*

341. Find all pairs (a,b) of coprime positive integers such that $a^2 + a = b^3 + b$. *(Switzerland)*

342. Prove that there exists a positive integer m such that the equation $\varphi(n) = m$ has at least 2015 solutions in n. Here $\varphi(n)$ denotes the number of positive integers less than n and coprime to n. *(USA)*

1.44 2016

343. For an infinite sequence $a_1 < a_2 < a_3 < \ldots$ of positive integers, we say that it is *nice* if for every positive integer n holds $a_{2n} = 2a_n$. Prove the following statements:

    (a) If there is given a *nice* sequence and a prime number $p > a_1$, then there exists some term of the sequence which is divisible by p.
    (b) For every prime number $p > 2$, there exists a *nice* sequence such that no terms of the sequence are divisible by p. (*Bosnia and Herzegovina*)

344. Determine the largest positive integer n which cannot be written as the sum of three numbers bigger than 1 which are pairwise coprime. (*Bosnia and Herzegovina*)

345. Let P a finite set of primes, A an infinite set of positive integers, where every element of A has a prime factor not in P. Prove that there exists an infinite subset B of A such that the sum of elements in any finite subset of B has a prime factor not in P. (*China*)

346. Find all pairs (p, q) of prime numbers such that $p(p^2 - p - 1) = q(2q + 3)$. (*Croatia*)

347. Let $p > 10^9$ be a prime number such that $4p + 1$ is also prime. Prove that the decimal expansion of $\frac{1}{4p+1}$ contains all the digits $0, 1, \ldots, 9$. (*Croatia*)

348. Let p be a prime number. Find all triples (a, b, c) of integers (not necessarily positive) such that $a^b b^c c^a = p$. (*Estonia*)

349. Let n be a positive integer such that there exists a positive integer that is less than $\sqrt{n}$ and does not divide n. Let $(a_1, a_2, \ldots, a_n)$ be an arbitrary permutation of $1, \ldots, n$. Let $a_{i_1} < \ldots < a_{i_k}$ be its maximal increasing subsequence, and let $a_{j_1} > \ldots > a_{j_l}$ be its maximal decresing subsequence. Prove that tuples $(a_{i_1}, \ldots, a_{i_k})$ and $(a_{j_1}, \ldots, a_{j_l})$ altogether contain at least one number that does not divide n. (*Estonia*)

350. Find all positive integers n such that $(n^2 + 11n - 4) \cdot n! + 33 \cdot 13^n + 4$ is a perfect square. (*Estonia*)

351. Find all positive integers n such that $n, n^2 + 10, n^2 - 2, n^3 + 6$, and $n^5 + 36$ are all prime numbers. (*Hong Kong*)

352. Find all triples (m, p, q) such that $2^m p^2 + 1 = q^7$, where p and q are primes and m is a positive integer. (*Hong Kong*)

353. Find all prime numbers p and q such that $p^2 | q^3 + 1$ and $q^2 | p^6 - 1$. (*Hong Kong*)

354. Let p be a prime number greater than 5. Suppose there is an integer k satisfying that $k^2 + 5$ is divisible by p. Prove that there are positive integers m and n such that $p^2 = m^2 + 5n^2$. (*Hong Kong*)

355. Let m and n be positive integers such that $m > n$. Define $x_k = \frac{m+k}{n+k}$ for $k = 1, 2, \ldots, n+1$. Prove that if all the numbers $x_1, x_2, \ldots, x_{n+1}$ are integers, then $x_1 x_2 \ldots x_{n+1} - 1$ is divisible by an odd prime. (*India*)

356. Find the greatest common divisor of all numbers of the form
$$\left(2^{a^2} \cdot 19^{b^2} \cdot 53^{c^2} + 8\right)^{16} - 1,$$
where a, b, and c are positive integers. (*Israel*)

357. A quadruple (p, a, b, c) of positive integers is a *karaka quadruple* if it is as follows:
- p is an odd prime number.
- a, b and c are distinct.
- $ab + 1$, $bc + 1$, and $ca + 1$ are divisible by p.

(a) Prove that for every karaka quadruple (p, a, b, c), we have
$$p + 2 \le \frac{a+b+c}{3}.$$

(b) Determine all numbers p for which a karaka quadruple exists with
$$p + 2 = \frac{a+b+c}{3}.$$

(*New Zealand*)

358. Find all positive integers n for which the equation $(x^2 + y^2)^n = (xy)^{2016}$ has positive integer solutions. (*New Zealand*)

359. Given a positive integer k and an integer $a \equiv 3 \pmod 8$, show that $a^m + a + 2$ is divisible by 2^k for some positive integer m. (*Romania*)

360. Let $w(x)$ be the largest odd divisor of x. Let a and b be positive integers such that $(a, b) = 1$ and $a + w(b + 1)$ and $b + w(a + 1)$ are powers of two. Prove that $a + 1$ and $b + 1$ are powers of two. (*Serbia*)

361. Prove that for every nonnegative integer n, the number $7^{7^n} + 1$ is the product of at least $2n + 3$ (not necessarily distinct) primes. (*Switzerland*)

362. Find all positive integers n such that
$$\sum_{d \mid n,\, 1 \le d < n} d^2 = 5(n+1).$$
(*Switzerland*)

363. Find all ordered pairs (a, b) of positive integers that satisfy $a > b$ and the equation $(a - b)^{ab} = a^b b^a$. (*Taiwan*)

364. Find all positive integers a such that for any positive integer $n \ge 5$, we have
$$2^n - n^2 \mid a^n - n^a.$$

(*Ukraine*)

365. Suppose that p is a prime number and that there are different positive integers u and v such that p^2 is the mean of u^2 and v^2. Prove that $2p - u - v$ is a square or twice a square. (*United Kingdom*)
366. Suppose that n and k are positive integers such that $1 = \underbrace{\varphi(\varphi(\ldots \varphi(n) \ldots))}_{k \text{ times}}$.

    Prove that $n \leq 3^k$. (*USA*)
367. Find all $a, n \in \mathbb{Z}^+ (a > 2)$ such that each prime divisor of $a^n - 1$ is also a prime divisor of $a^{3^{2016}} - 1$. (*Vietnam*)

1.45 2017

368. Find all prime numbers a and b such that $20a^3 - b^3 = 1$. (*Belarus*)
369. Let $1 = d_1 < d_2 < \ldots < d_k = n$ be all positive divisors of the positive integer n. Find all possible values of k if $n = d_2 d_3 + d_2 d_5 + d_3 d_5$. (*Belarus*)
370. Prove that for any positive integers a and b, there exist infinitely many prime numbers p such that $ap + b$ is a composite number. (Using Dirichlet's theorem is not allowed.) (*Belarus*)
371. Let $\tau(n)$ be the number of positive divisors of n. Let $\tau_1(n)$ be the number of positive divisors of n which have remainder 1 when divided by 3. Find all possible integral values of the fraction $\frac{\tau(10n)}{\tau_1(10n)}$. (*Belarus*)
372. For a given positive integer n and a prime number p, find the minimum value of a positive integer m that satisfies the following property: for any polynomial

    $$f(x) = (x + a_1)(x + a_2) \ldots (x + a_n)$$

    ($a_1, a_2, \ldots, a_n$ are positive integers) and for any nonnegative integer k, there exists a nonnegative integer k' such that $v_p(f(k)) < v_p(f(k')) \leq v_p(f(k)) + m$. (*China*)
373. Let n be a positive integer. Let D_n be the set of all divisors of n, and let $f(n)$ denote the smallest positive integer m such that the elements of D_n are pairwise distinct mod m. Show that there exists a positive integer N such that for all $n \geq N$, one has $f(n) \leq n^{0.01}$. (*China*)
374. An integer $n > 1$ is given. Find the smallest positive integer m satisfying the conditions: for any set $\{a, b\} \subset \{1, 2, \ldots, 2n - 1\}$, there are nonnegative integers x and y (not both zero) such that $2n | ax + by$ and $x + y \leq m$. (*China*)
375. Do there exist two positive powers of 5 such that the number obtained by writing one after the other is also a power of 5? (*Estonia*)
376. Let n be a positive integer. In how many ways can an $n \times n$ table be filled with integers from 0 to 5 such that

    (a) The sum of each row is divisible by 2 and the sum of each column is divisible by 3.

(b) The sum of each row is divisible by 2, the sum of each column is divisible by 3 and the sum of each of the two diagonals is divisible by 6. (*Estonia*)

377. Prove that the number $A = \frac{(4n)!}{(2n)! \cdot n!}$ is an integer and divisible by 2^{n+1}, where n is a positive integer. (*Greece*)

378. Find the first digit after the decimal point of the number $\frac{1}{1009} + \frac{1}{1010} + \cdots + \frac{1}{2016}$. (*Hong Kong*)

379. Let n be a positive integer with the following property: $2^n - 1$ divides a number of the form $m^2 + 81$, where m is a positive integer. Find all possible n. (*Hong Kong*)

380. Find all positive integers $p, q, r, s > 1$ such that $p! + q! + r! = 2^s$. (*India*)

381. Let $a, b, c,$ and d be pairwise positive integers such that

$$\frac{a}{a+b} + \frac{b}{b+c} + \frac{c}{c+d} + \frac{d}{d+a}$$

is an integer. Prove that $a + b + c + d$ is not a prime number. (*India*)

382. Define a sequence of integers $a_0 = m$, $a_1 = n$ and $a_{k+1} = 4a_k - 5a_{k-1}$ for all $k \geq 1$. Suppose that $p > 5$ is a prime with $p \equiv 1 \pmod{4}$. Prove that it is possible to choose m and n such that $p \nmid a_k$ for any $k \geq 0$. (*India*)

383. Let a be a positive integer which is not a perfect square, and consider the equation

$$k = \frac{x^2 - a}{x^2 - y^2}.$$

Let A be the set of positive integers k for which the equation admits a solution in $\mathbb{Z}^2$ with $x > \sqrt{a}$, and let B be the set of positive integers for which the equation admits a solution in $\mathbb{Z}^2$ with $0 \leq x < \sqrt{a}$. Show that $A = B$. (*India*)

384. We have arranged all the prime numbers in ascending order: $p_1 = 2 < p_2 < p_3 < \ldots$. Also assume that $n_1 < n_2 < \ldots$ is a sequence of positive integers that for all $i = 1, 2, 3, \ldots$; the equation $x^{n_i} \equiv 2 \pmod{p_i}$ has a solution for x. Is there always a number x that satisfies all the equations? (*Iran*)

385. A positive integer is said to be *near-square* if it is a product of two positive integers differing by 1. Prove that every near-square positive integer can be expressed as the ratio of two other near-square positive integers. (*Ireland*)

386. Determine all positive integers n of the form $n = [a, b] + [b, c] + [c, a]$, where $a, b,$ and c are positive integers and $[u, v]$ is the least common multiple of the integers u and v. (*Moldova*)

387. Let p be an odd prime. Prove that the number $\left\lfloor \left(\sqrt{5} + 2\right)^p - 2^{p+1} \right\rfloor$ is divisible by $20p$. (*Moldova*)

388. Find all ordered pairs (x, y) of nonnegative integers such that
$$x^4 - x^2y^2 + y^4 + 2x^3y - 2xy^3 = 1.$$
(*Moldova*)

389. Determine all pairs (p, q) of prime numbers such that $p^2 + 5pq + 4q^2$ is the square of an integer. (*Netherlands*)

390. Let a, b, and c be distinct positive integers, and suppose that $p = ab + bc + ca$ is a prime number.

   (a) Show that a^2, b^2, and c^2 give distinct remainders after division by p.
   (b) Show that a^3, b^3, and c^3 give distinct remainders after division by p.
   (*Netherlands*)

391. Compute the product of all positive integers n for which $3(n! + 1)$ is divisible by $2n - 5$. (*Netherlands*)

392. Find all prime numbers p such that $16p + 1$ is a perfect cube. (*New Zealand*)

393. Let a, b, c, d, and e be distinct positive integers such that $a^4 + b^4 = c^4 + d^4 = e^5$. Show that $ac + bd$ is composite. (*New Zealand*)

394. The product $1 \times 2 \times 3 \times \cdots \times n$ is written on the board. For what integers $n \geq 2$ we can add exclamation marks to some factors to convert them into factorials, in such a way that the final product can be a perfect square? (*Peru*)

395. Let k be a positive integer, and let n be the smallest positive integer having exactly k divisors. If n is a perfect cube, can the number k have a prime divisor of the form $3j + 2$? (*Serbia*)

396. Given $a, c \in \mathbb{N}$ and $b \in \mathbb{Z}$. Prove that there exists $x \in \mathbb{N}$ such that $a^x + x \equiv b \pmod{c}$. (*Switzerland*)

397. Find all triples (m, n, p) with m and n positive integers and p a prime number, satisfying
$$(m^3 + n)(n^3 + m) = p^3.$$
(*Turkey*)

398. Find all nonnegative integer solutions to $2^a + 3^b + 5^c = n!$. (*USA*)

399. Prove that there are infinitely many triples (a, b, p) of integers, with p prime and $1 < a \leq b < p$, for which p^5 divides $(a + b)^p - a^p - b^p$. (*USA*)

400. For each positive integer n, set $x_n = \binom{2n}{n}$.

   (a) Prove that if $\frac{2017^k}{2} < n < 2017^k$ for some positive integer k, then 2017 divides x_n.
   (b) Find all positive integers $h > 1$ for which there exist positive integers N, T such that $(x_n)_{n > N}$ is periodic mod h with period T. (*Vietnam*)

1.46 2018

401. A number n is *interesting* if 2018 divides $\tau(n)$. Determine all positive integers k such that there exists an infinite arithmetic progression with common difference k whose terms are all interesting. Here $\tau(n)$ denotes the number of positive divisors of the positive integer n. (*China*)

402. Let k and M be positive integers such that $k - 1$ is not square free. Prove that there exists a positive real α, such that $\lfloor \alpha k^n \rfloor$ and M are coprime for any positive integer n. (*China*)

403. Find all pairs (x, y) of positive integers such that $(xy + 1)(xy + x + 2)$ is a perfect square. (*China*)

404. Let p be a prime and k be a positive integer. Set S contains all positive integers a satisfying $1 \leq a \leq p - 1$, and there exists a positive integer x such that $x^k \equiv a \pmod{p}$. Suppose that $3 \leq |S| \leq p - 2$. Prove that the elements of S, when arranged in increasing order, does not form an arithmetic progression. (*China*)

405. Determine all integers $n \geq 2$ for which the number 11111 in base n is a perfect square. (*Cyprus*)

406. We call a positive integer n whose all digits are distinct *bright*, if either n is a one-digit number or there exists a divisor of n which can be obtained by omitting one digit of n and which is bright itself. Find the largest bright positive integer. (*Estonia*)

407. Let a *simple polynomial function* be a polynomial function $P(x)$ whose coefficients belong to the set $\{-1, 0, 1\}$. Let $n > 1$ be a positive integer. Find the smallest possible number of nonzero coefficients in a simple polynomial function of nth order whose values at all integral arguments are divisible by n. (*Estonia*)

408. Find all positive integers n such that $n^2 + 32n + 8$ is a perfect square. (*Hong Kong*)

409. Find infinitely many positive integers m such that the number $\frac{2^{m-1}-1}{8191m}$ is an integer. (*Hong Kong*)

410. Find all primes p and all positive integers a and m such that $a \leq 5p^2$ and $(p - 1)! + a = p^m$. (*Hong Kong*)

411. For a positive integer $k > 1$, define S_k to be the set of all triplets (n, a, b) of positive integers, with n odd and $\gcd(a, b) = 1$, such that $a + b = k$ and n divides $a^n + b^n$. Find all values of k for which S_k is finite. (*India*)

412. We say that distinct positive integers $a_1, a_2, \ldots, a_n$ are *good* if their sum is equal to the sum of all pairwise gcd's among them. Prove that there are infinitely many n such that n good numbers exist. (*Iran*)

413. The positive integers a and b satisfy the system $\begin{cases} a_{10}+b_{10}=a \\ a_{11}+b_{11}=b \end{cases}$, where $a_1 < a_2 < \ldots$ and $b_1 < b_2 < \ldots$ are the positive divisors of a and b. Find a and b. (*Moldova*)

414. Let $p > 3$ be a prime number and $k = \lfloor \frac{2p}{3} \rfloor$. Prove that

$$\binom{p}{1} + \binom{p}{2} + \ldots + \binom{p}{k}$$

is divisible by p^2. (*Moldova*)

415. Suppose that a, b, c, and d are four different integers. Explain why

$$(a-b)(a-c)(a-d)(b-c)(b-d)(c-d)$$

must be a multiple of 12. (*New Zealand*)

416. Find all pairs (a,b) of integers such that $a^2 + ab - b = 2018$. (*New Zealand*)

417. Let x, y, p, n, and k be positive integers such that $x^n + y^n = p^k$. Prove that if $n > 1$ is odd and p is an odd prime, then n is a power of p. (*New Zealand*)

418. Prove that there exist infinitely many positive integers n such that at least one of the numbers $2^{2^n} + 1$ and $2018^{2^n} + 1$ is not a prime. (*Serbia*)

419. Find all positive integers $n \geq 2$ such that for every integers $0 \leq i, j \leq n$ we have

$$i + j \equiv \binom{n}{i} + \binom{n}{j} \pmod{2}.$$

(*Switzerland*)

420. Prove that for all integers a and b, there exists a positive integer n such that the number $n^2 + an + b$ has at least 2018 different prime divisors. (*Turkey*)

421. The prime number $p > 2$ and the integer n are given. Prove that the number pn^2 has no more than a divisor d for which $n^2 + d$ is the square of a positive integer. (*Ukraine*)

422. Determine whether or not there is a positive integer m such that

$$(m+1)^3 + (m+2)^3 + \ldots + (2m)^3$$

is a square. (*United Kingdom*)

423. For which positive integers $b > 2$ do there exist infinitely many positive integers n such that n^2 divides $b^n + 1$? (*USA*)

424. Let $n \geq 2$ be a positive integer, and let $\sigma(n)$ denote the sum of the positive divisors of n. Prove that the nth smallest positive integer relatively prime to n is at least $\sigma(n)$, and determine for which n equality holds. (*USA*)

1.47 2019

425. Given the equation $a^b \cdot b^c = c^a$ in positive integers a, b, and c.
    (a) Prove that any prime divisor of a divides b as well.
    (b) Solve the equation under the assumption $b \geq a$.
    (c) Prove that the equation has infinitely many solutions. (*Belarus*)

426. Determine all pairs (m, n) of positive integers for which there exists a positive integer s such that sm and sn have an equal number of divisors. (*Belarus*)

427. Four positive integers x, y, z, and t satisfy the relations $xy - zt = x + y = z + t$. Is it possible that both xy and zt are perfect squares? (*Belarus*)

428. Prove that for $n > 1$, n does not divide $2^{n-1} + 1$. (*Belarus*)

429. The difference of two consecutive cubes is the square of a positive integer number. Prove that this number is the sum of two consecutive squares. (*Ecuador*)

430. Solve in positive integers the equation $m^2 + n^2 = 2018(m - n)$. (*Ecuador*)

431. Find all pairs (m, n) of positive integers such that $0 < m < n < 2018$ and
$$2018^2 + m^2 = 2017^2 + n^2.$$
(*Ecuador*)

432. Let p be a prime number greater than 10. Prove that there exist positive integers m and n such that $m + n < p$ and $5^m 7^n - 1$ is divisible by p. (*Hong Kong*)

433. Let $\{a_1, a_2, \ldots, a_m\}$ be a set of even positive integers and $\{b_1, b_2, \ldots, b_n\}$ be a set of odd positive integers such that
$$a_1 + a_2 + \ldots + a_m + b_1 + b_2 + \ldots + b_n = 2019.$$
Prove that $5m + 12n \leq 581$. (*India*)

434. For any positive integer n, define the subset S_n of the set of positive integers as follows: $S_n = \{x^2 + ny^2 | x, y \in \mathbb{Z}\}$. Find all positive integers n such that there exists an element of S_n which doesn't belong to any of the sets $S_1, S_2, \ldots, S_{n-1}$. (*Iran*)

435. For an integer $r \geq 2$, define $s(r)$ to be the smallest prime number that divides r. Show that for any integer $n \geq 2$,
$$\sum_{r=2}^{n} s(r) \geq 3n - 5.$$

(*Ireland*)

436. Prove that there exist infinitely many positive integers n such that $\frac{4^n+2^n+1}{n^2+n+1}$ is a positive integer. (*Kosovo*)

437. Let $p \geq 5$ be a prime number. Prove that there exist positive integers m and n with $m+n \leq \frac{p+1}{2}$ for which p divides $2^n \cdot 3^m - 1$. (*Moldova*)

438. Let $k \geq 0$ be an integer. The sequence (a_n) is defined as follows: $a_0 = k$ and for $n \geq 1$, a_n is the smallest integer greater than a_{n-1} so that $a_n + a_{n-1}$ is a perfect square. Prove that there are exactly $\lfloor \sqrt{2k} \rfloor$ positive integers that cannot be written as the difference of two terms of the sequence. (*Peru*)

439. Find all pairs (m,n) of integers such that $m^6 = n^{n+1} + n - 1$. (*Romania*)

440. Define the sequence $a_0, a_1, a_2, \ldots$ by $a_n = 2^n + 2^{\lfloor n/2 \rfloor}$. Prove that there are infinitely many terms of the sequence which can be expressed as a sum of (two or more) distinct terms of the sequence, as well as infinitely many of those which cannot be expressed in such a way. (*SAFEST*)

441. (a) Given 2019 different positive integers which have no odd prime divisor less than 37, prove that there exist two of these numbers such that their sum has no odd prime divisor less than 37.
    (b) Does the result hold if we change 37 to 38? (*Serbia*)

442. Solve in nonnegative integers the equation $2^x = 5^y + 3$. (*Serbia*)

443. Find the greatest prime number p for which there exist positive integers a and b such that
$$p = \frac{b}{2}\sqrt{\frac{a-b}{a+b}}.$$
(*Switzerland*)

444. Let (a_n) be the sequence defined by the following:
$$a_1 = 1, a_2 = 2 \text{ and } a_{n+2} = a_{n+1}^2 + +(n+2)a_{n+1} - a_n^2 - na_n \text{ for } n \geq 1.$$

   (a) Prove that the set of primes that divide at least one term of the sequence is infinite.
   (b) Find three different prime numbers that not divide any term of this sequence. (*Turkey*)

445. Let $p > 2$ be a prime number, and let $m > 1$ and n be positive integers such that $\frac{m^{pn}-1}{m^n-1}$ is a prime number. Show that $pn | (p-1)^n + 1$. (*Turkey*)

446. Find all triplets (x,y,z) of positive integers such that $2^x + 1 = 7^y + 2^z$. (*Vietnam*)

1.48 2020

447. For a nonempty finite set A of positive integers, let lcm(A) denote the least common multiple of elements in A, and let $d(A)$ denote the number of prime factors of lcm(A) (counting multiciplity). Given a finite set S of positive integers and

$$f_S(x) = \sum_{\emptyset \neq A \subset S} \frac{(-1)^{|A|} x^{d(A)}}{\text{lcm}(A)}.$$

Prove that, if $0 \leq x \leq 2$, then $-1 \leq f_S(x) \leq 0$. (*China*)

448. Show that the following equation has only finitely many solutions (t, A, x, y, z) in positive integers:

$$\sqrt{t(1-A^{-2})(1-x^{-2})(1-y^{-2})(1-z^{-2})} = (1+x^{-1})(1+y^{-1})(1+z^{-1}).$$

(*China*)

449. Let $a_0, a_1, a_2, \ldots$ be a sequence of positive integers and b_0, b_1, b_2 be the sequence defined by $b_n = \gcd(a_n, a_{n+1})$ for all $n \geq 0$. It is possible that the sequence (b_n) coincides with the sequence of positive integers in some order? (*France*)

450. Find all functions $f: \mathbb{N} \to \mathbb{N}$ such that for every positive integer n the following is valid: if $d_1, d_2, \ldots, d_s$ are all the positive divisors of n, then

$$f(d_1)f(d_2)\ldots f(d_s) = n.$$

(*Hong Kong*)

451. Let $s(n)$ be the sum of the decimal digits of the positive integer n. Given $k \in \mathbb{Z}$ prove that there exist infinitely many pairs of distinct positive integers such that

$$n + s(2n) = m + s(2m),$$
$$kn + s(n^2) = km + s(m^2).$$

(*Iran*)

452. Let p be an odd prime number. Find all $\frac{p-1}{2}$ tuples $\left(x_1, x_2, \ldots, x_{\frac{p-1}{2}}\right) \in \mathbb{Z}_p^{\frac{p-1}{2}}$ such that

$$\sum_{i=1}^{\frac{p-1}{2}} x_i \equiv \sum_{i=1}^{\frac{p-1}{2}} x_i^2 \equiv \ldots \equiv \sum_{i=1}^{\frac{p-1}{2}} x_i^{\frac{p-1}{2}} \pmod{p}.$$

(*Iran*)

453. Prove that for all positive integers m and n, the following inequality holds

$$\pi(m) - \pi(n) \leq \frac{(m-1)\varphi(n)}{n}.$$

When does the equality hold? Here $\pi(n)$ is the number of primes less than n, and $\varphi(n)$ is the number of positive integers less than n and coprime to n. (*Kosovo*)

454. All terms of the geometric progression $(b_n)_{n \geq 1}$ are terms of some arithmetic progression. It is known that b_1 is an integer. Prove that all terms of the geometric progression are integers. (*Moldova*)

455. Let a and $b \geq 2$ be positive integers with $\gcd(a, b) = 1$. Let r be the smallest positive value that $\frac{a}{b} - \frac{c}{d}$ can take, where c and d are positive integers satisfying $c \leq a$ and $d \leq b$. Prove that $\frac{1}{r}$ is an integer. (*Netherlands*)

456. Determine all pairs (a, b) of positive integers for which

$$a + b = \varphi(a) + \varphi(b) + \gcd(a, b).$$

Here $\varphi(n)$ denotes the number of positive integers less than n and coprime to n. (*Netherlands*)

457. For a positive integer n, let $\tau(n)$ be the number of positive divisors of n. Determine the positive integers k for which there exist positive integers a and b satisfying

$$k = \tau(a) = \tau(b) = \tau(2a + 3b).$$

(*Netherlands*)

458. Solve in positive integers $x^{100} - y^{100} = 100!$. (*Serbia, JBMO*)

459. The infinite sequence of (not necessarily distinct) integers has the following properties: $0 \leq a_i \leq i$ for all integers $i \geq 0$ and

$$\binom{k}{a_0} + \binom{k}{a_1} + \ldots + \binom{k}{a_k} = 2^k$$

for all integers $k \geq 0$. Prove that all integers $N \geq 0$ occur in the sequence (i.e., for all $N \geq 0$, there exists $i \geq 0$ with $a_i = N$). (*Taiwan*)

460. Find all triples (a, b, c) of positive integers such that $a^3 + b^3 + c^3 = (abc)^2$. (*Taiwan*)

461. Let a be a positive integer. We say that a positive integer b is a - good if $\binom{an}{b} - 1$ is divisible by $an + 1$ for all positive integers n with $an \geq b$. Suppose that b is a positive integer such that b is a-good, but $b + 2$ is not a-good. Prove that $b + 1$ is prime. (*Taiwan*)

462. Find all pairs (a, b) of positive integers satisfying the equation

$$\frac{a^3 + b^3}{ab + 4} = 2020.$$

(*Turkey*)

463. For a function $f: N^* \to N^*$, we denote $f \circ f \circ \ldots \circ f$ with f repeated l times by f_l. Find all functions $f: N^* \to N^*$ such that

$$(n-1)^{2020} < \prod_{l=1}^{2020} f_l(n) < n^{2020} + n^{2019}$$

for all $n \in N^*$. (*Turkey*)

464. Let $p(m)$ be the number of distinct prime divisors of a positive integer $m > 1$ and $f(m)$ the $\left\lfloor \frac{p(m)+1}{2} \right\rfloor$ th smallest prime divisor of m. Find all positive integers n satisfying the equation

$$f(n^2 + 2) + f(n^2 + 5) = 2n - 4.$$

(*Turkey, EGMO*)

1.49 2021

465. Given positive integer n. Prove that for any integers $a_1, a_2, \ldots, a_n$, at least $\left\lceil \frac{n(n-6)}{19} \right\rceil$ numbers from the set $\left\{ 1, 2, \ldots, \frac{n(n-1)}{2} \right\}$ cannot be represented as $a_i - a_j$, $1 \leq i, j \leq n$. (*China*)

466. Given positive integers a, b, and c, which are pairwise coprime. Let $f(n)$ denote the number of the nonnegative integer solutions of the equation

$$ax + by + cz = n.$$

Prove that there exist real constants α, β, and γ, such that for any nonnegative integer n,

$$|f(n) - (\alpha n^2 + \beta n + \gamma)| < \frac{a+b+c}{12}.$$

(*China*)

467. Prove that

$$\sum_{m=1}^{n} 5^{\omega(m)} \le \sum_{m=1}^{n} \left\lfloor \frac{n}{m} \right\rfloor (\tau(m))^2 \le \sum_{m=1}^{n} 5^{\Omega(m)}.$$

Here $\omega(n)$ is the number of prime divisors of n, $\tau(n)$ is the number of positive divisors of n, and $\Omega(n)$ is the number of prime divisors of n counted with multiplicities. (*China*)

468. Solve in positive integers $x^2 = 2^y + 2021^z$. (*Serbia, JBMO*)

469. Prove that if p and q are prime numbers such that

$$p + p^2 + \ldots + p^q = q + q^2 + \ldots + q^p,$$

then $p = q$. (*Moldova*)

470. Determine all positive integers n such that $\frac{a^2+n^2}{b^2-n^2}$ is a positive integer, for some nonnegative integers a, b. (*Moldova*)

471. Prove that $n! \cdot (n+1)! \cdot (n+2)!$ divides $(3n)!$ for every integer $n \ge 3$. (*Moldova*)

472. We call a positive integer *silly* if the sum of its positive divisors is a square. Prove that there are infinitely many silly numbers. (*Switzerland*)

473. For each positive integer n, define $V_n = \lfloor 2^n \sqrt{2020} \rfloor + \lfloor 2^n \sqrt{2021} \rfloor$. Prove that in the sequence $(V_n)_{n \ge 1}$, there are infinitely many odd integers, as well as infinitely many even integers. (*Taiwan*)

474. Given a positive integer k, show that there exists a prime p such that one can choose distinct integers $a_1, a_2, \ldots, a_{k+3} \in \{1, 2, \ldots, p-1\}$ such that $p | a_i a_{i+1} a_{i+2} a_{i+3} - i$ for all $i = 1, 2, \ldots, k$. (*Taiwan*)

475. Find all triples (x, y, z) of positive integers such that $x^2 + 4^y = 5^z$. (*Taiwan*)

476. Let S be a set of positive integers such that for every $a, b \in S$, there exists $c \in S$ such that $c^2 | a(a+b)$. Show that there exists an $a \in S$ that divides every element of S. (*Taiwan*)

477. Let $(a_n)_{n \ge 1}$ be the sequence of positive integers defined by

$$a_1 = 2021, \sqrt{a_{n+1} - a_n} = \lfloor \sqrt{a_n} \rfloor, n \ge 1.$$

Show that there are infinitely many odd numbers and infinitely many even numbers in this sequence. (*Taiwan*)

478. Prove that for all positive integers m and n, we have $\lfloor m\sqrt{2} \rfloor \cdot \lfloor n\sqrt{7} \rfloor < \lfloor mn\sqrt{14} \rfloor$. (*Thailand*)

479. Let n be a positive integer. Prove that

$$\frac{20 \cdot 5^n - 2}{3^n + 47}$$

is not an integer. (*Turkey*)

480. Define the sequence $(a_n)_{n \geq 1}$ as $a_1 = 1$, $a_{2n} = a_n$ and $a_{2n+1} = a_n + 1$ for all $n \geq 1$.

   (a) Find all positive integers n such that $a_{kn} = a_n$ for all integers $1 \leq k \leq n$.
   (b) Prove that there exist infinitely many positive integers m such that $a_{km} \geq a_m$ for all positive integers k. (*Vietnam*)

1.50 2022

481. Given a positive integer n, let D be the set of positive divisors of n, and let $f: D \to \mathbb{Z}$ be a function. Prove that the following are equivalent:

   (a) For any positive divisor m of n,
   $$n \left| \sum_{d \mid m} f(d) \left(\dfrac{\frac{n}{d}}{\frac{m}{d}} \right) \right.$$

   (b) For any positive divisor k of n,
   $$k \left| \sum_{d \mid k} f(d) \right.$$

   (*China*)

482. Given a positive integer n, let D be the set of all positive divisors of n. The subsets A and B of D satisfy that for every $a \in A$ and $b \in B$, it holds that $a \nmid b$ and $b \nmid a$. Show that
$$\sqrt{|A|} + \sqrt{|B|} \leq \sqrt{|D|}.$$

   (*China*)

483. Let n and m be positive integers such that $n(4n+1) = m(5m+1)$.

   (a) Show that $n - m$ is a perfect square.
   (b) Find a pair (n, m) of positive integers that satisfies the above relation. (*Cyprus*)

484. Find all pairs (m, n) of integers which satisfy
$$(2n^2 + 5m - 5n - mn)^2 = m^3 n.$$

   (*Cyprus*)

485. Determine for how many positive integers $n \in \{1, 2, \ldots, 2022\}$ it holds that 402 divides at least one of

$$n^2 - 1, n^3 - 1, n^4 - 1.$$

(*Cyprus*)

486. Let n be a positive integer. Anna writes $4n + 2$ distinct integers from the interval $[0, 5^n]$. Prove that among these integers there are three numbers a, b, and c such that

$$a < b < c \text{ and } c + 2a > 3b.$$

(*France*)

487. Find all positive integers n such that there is a permutation $(d_1, d_2, \ldots, d_k)$ of positive integers of n for which $d_1 + d_2 + \ldots + d_i$ is a perfect square, for all $i \leq k$. (*France*)

488. Let $(a_n)_{n \geq 1}$ be a sequence of positive integers such that a_{n+2m} divides $a_n + a_{n+m}$ for all $m, n \geq 1$. Prove that the sequence is periodic, that is, there are positive integers N, d such that $a_n = a_{n+d}$ for all $n \geq N$. (*France*)

489. Let S be an infinite set of positive integers, such that there exist four pairwise distinct $a, b, c, d \in S$ with $\gcd(a, b) \neq \gcd(c, d)$. Prove that there exist three pairwise distinct $x, y, z \in S$ such that $\gcd(x, y) = \gcd(y, z) \neq \gcd(z, x)$. (*Germany*)

490. For a positive integer n, let $\tau(n)$ and $\sigma(n)$ be the number of positive divisors and the sum of positive divisors of n, respectively. Let a and b be positive integers such that $\sigma(a^n) | \sigma(b^n)$ for all $n \in \mathbb{N}$. Prove that each prime factor of $\tau(a)$ divides $\tau(b)$. (*Iran*)

491. We call an infinite set $S \subseteq \mathbb{N}$ *good* if for all pairwise different integers a, b, and $c \in S$; all positive divisors of $\frac{a^c - b^c}{a - b}$ are in S. Prove that for each positive integer $n > 1$, there exists a good set S such that $n \notin S$. (*Iran*)

492. Find all positive integers a, b, and c such that $ab + 1$, $bc + 1$, and $ca + 1$ are all equal to factorials of some positive integers. (*Kosovo*)

493. Show that for every integer $n \geq 2$, the number

$$a = n^{5n-1} + n^{5n-2} + n^{5n-3} + n + 1$$

is a composite number. (*Moldova*)

494. Let $f: \mathbb{N} \to \mathbb{N}, f(n) = n^2 - 69n + 2250$ be a function. Find the prime numbers p for which the sum of the digits of the number $f(p^2 + 32)$ is as small as possible. (*Moldova*)

495. Let $(x_n)_{n \geq 1}$ be the sequence that verifies $x_1 = 1, x_2 = 7$, and $x_{n+1} = x_n + 3x_{n-1}$, for all $n \geq 2$. Prove that for every prime number p, the number $x_p - 1$ is divisible by $3p$. (*Moldova*)

496. Can every positive rational number q be written as

$$\frac{a^{2021} + b^{2023}}{c^{2022} + d^{2024}}$$

where a, b, c, d are positive integers? (*Romania*)

497. Which positive integers n make the equation

$$\sum_{i=1}^{n} \sum_{j=1}^{n} \left\lfloor \frac{ij}{n+1} \right\rfloor = \frac{n^2(n-1)}{4}$$

true? (*Thailand*)

498. Find all pairs (a, b) of integers satisfying the equation $a^7(a - 1) = 19b(19b + 2)$. (*Turkey, EGMO*)

499. Solve for prime pairs $2^p = 2^{q-2} + q!$ (*Turkey*)

500. For a polynomial $P(x)$ with integer coefficients and a prime number p, if there is no $n \in \mathbb{Z}$ such that $p \mid P(n)$, we say that P *excludes* p. Is there a polynomial of degree 5 with integer coefficients, without rational roots, that excludes exactly one prime? (*Turkey*)

1.51 2023

501. Find all positive integers n with the following property: There are only a finite number of positive multiples of n that have exactly n positive divisors. (*Brazil*)

502. A positive integer n is called *good* if there exist integers $1 = a_1, a_2, \ldots, a_{2023} = b$ such that $|a_{i+1} - a_i| = 2^i$ for all $i = 1, 2, \ldots, 2022$. Find the number of the good integers. (*Bulgaria*)

503. (1) Let a and b be coprime positive integers. Prove that there exist constants λ and β such that for all integers m,

$$\left| \sum_{k=1}^{m-1} \left\{ \frac{ak}{m} \right\} \left\{ \frac{bk}{m} \right\} - \lambda m \right| \leq \beta.$$

(2) Prove that there exists N such that for all prime numbers $p > N$ and any positive integers a, b, and c satisfying $p \nmid (a+b)(b+c)(c+a)$, there are at least $\left\lfloor \frac{p}{12} \right\rfloor$ solutions $k \in \{1, 2, \ldots, p-1\}$ such that

$$\left\{ \frac{ak}{p} \right\} + \left\{ \frac{bk}{p} \right\} + \left\{ \frac{ck}{p} \right\} \leq 1.$$

Here $\{x\}$ denotes the fractional part of the real number x. (*China*)

504. Given $m, n \in \mathbb{N}_+$, define
$$S(m,n) = \{(a,b) \in \mathbb{N}_+^2 \mid 1 \le a \le m, 1 \le b \le n, \gcd(a,b) = 1\}.$$

Prove that for any $d, r \in \mathbb{N}_+$, there exist $m, n \in \mathbb{N}_+$, $m, n \ge d$, such that
$$|S(m,n)| \equiv r \pmod{d}.$$

(*China*)

505. Does there exist a positive irrational number x such that there are only a finite number of positive integers n for which
$$\{kx\} \ge \frac{1}{n+1}, 1 \le k \le n?$$

Here $\{a\}$ denotes the fractional part of the real number a. (*China*)

506. Given a prime p and a real number $\lambda \in (0,1)$. Let s and t be positive integers such that $s \le t < \frac{\lambda p}{12}$. Let S and T be sets of s and t consecutive integers, respectively, which satisfy
$$|(x,y) \in S \times T \mid kx \equiv y \pmod{p}| \ge 1 + \lambda s.$$

Prove that there exist integers a and b such that
$$1 \le a \le \frac{1}{\lambda}, |b| \le \frac{t}{\lambda s}, ka \equiv b \pmod{p}.$$

(*China*)

507. Given a prime number p and integers x and y, find the remainder of the sum
$$x^0 y^{p-1} + x^1 y^{p-2} + \ldots + x^{p-1} y^1 + x^{p-1} y^0$$

upon division by p. (*Estonia*)

508. For each integer $k \ge 0$, denote by a_k the first digit of 2^k in base 10. Let $n \ge 1$ be an integer. Prove that among the digits $1, \ldots, 9$, there is one equal to at most $\frac{n}{17}$ of the digits $a_0, a_1, a_2, \ldots, a_{n-1}$. (*France*)

509. A positive integer n is called *nippon* if it has three divisors $d_1, d_2,$ and d_3 such that
$$1 \le d_1 < d_2 < d_3 \text{ and } d_1 + d_2 + d_3 = 2022.$$

Find the smallest nippon number. (*France*)

510. Find all pairs (k, n) of positive integers such that

$$1! + 2! + \ldots + k! = 1 + 2 + \ldots + n.$$

(*France*)

511. A two digit number s is *special* if s is the two common leading digits of the decimal expansion of 4^n and 5^n, where n is a positive integer. Given that there are two special numbers, find these two special numbers. (*Hong Kong*)

512. Find the length of the period of the fraction $\frac{39}{1428}$, using binary representations. (*Hong Kong*)

513. Let n be any positive integer, and let $S(n)$ denote the number of permutations τ of $\{1, \ldots, n\}$ such that $k^4 + (\tau(k))^4$ is prime for all $k = 1, \ldots, n$. Show that $S(n)$ is always a square. (*India*)

514. Let $\tau(n)$ be the number of positive divisors of the positive integer n. Prove that there is a positive integer n such that

$$\frac{\tau(n)}{\tau(n \pm i)} > 1401, i = 1, 2, \ldots, 1402.$$

(*Iran*)

515. For positive integers n, let $f_2(n)$ denote the number of divisors of n which are perfect squares, and $f_3(n)$ denotes the number of positive divisors which are perfect cubes. Prove that for each positive integer k, there exists a positive integer n for which $f_2(n) = kf_3(n)$. (*Israel*)

516. Find the largest constant $c > 0$ such that for every positive integer $n \geq 2$, there always exists a positive divisor d of n such that $d \leq \sqrt{n}$ and $\tau(d) \geq c\sqrt{\tau(n)}$, where $\tau(n)$ is the number of divisors of n. (*Malaysia*)

517. Let a, b, and c be distinct positive integers, and let r, s, and t be positive integers such that $ab + 1 = r^2$, $ac + 1 = s^2$, and $bc + 1 = t^2$. Prove that it is not possible that all three fractions $\frac{rt}{s}, \frac{rs}{t}$, and $\frac{st}{r}$ are integers. (*Moldova*)

518. Find all pairs (n, k) of positive integers for which

$$n + 2 \mid 1^{2k+1} + 2^{2k+1} + \cdots + n^{2k+1}.$$

(*Moldova*)

519. Let S be a non-empty set of positive integers such that for any $n \in S$, all positive divisors of $2^n + 1$ are also in S. Prove that S contains an integer of the form

$$(p_1 p_2 \cdots p_{2023})^{2023},$$

where $p_1, p_2, \ldots, p_{2023}$ are distinct prime numbers, all greater than 2023. (*Switzerland*)

520. Find all functions $f: \mathbb{N} \to \mathbb{N}$ satisfying that for all $m, n \in \mathbb{N}$, the nonnegative integer $|f(m+n) - f(m)|$ is a divisor of $f(n)$. (*Taiwan*)
521. Find all positive integers a, b, and c such that ab is a square and
$$a + b + c - 3\sqrt[3]{abc} = 1.$$
(*Taiwan*)
522. Find all pairs (p, q) of prime numbers that satisfy the equation
$$p(p^4 + p^2 + 10q) = q(q^2 + 3).$$
(*Turkey, EGMO*)

1.52 2024

523. Let n be a positive integer and p be a prime number of the form $8k + 5$. A polynomial Q of degree at most 2023 and nonnegative coefficients less than or equal to n will be called *cool* if
$$p \mid Q(2)Q(3)\ldots Q(p-2) - 1.$$
Prove that the number of cool polynomials is even. (*Israel*)
524. Solve in positive integers
$$x^{y^2+1} + y^{x^2+1} = 2^z.$$
(*Israel*)

Chapter 2
Solutions

2.1 1968

1. If $n! = n(n-1)(\alpha n + 1)$ with α integer, then after simplifying by $n(n-1)$, we conclude that $(n-2)! \equiv 1 \pmod{n}$. Conversely, if this congruence holds, then the given condition is satisfied, with

$$\alpha = \frac{(n-2)! - 1}{n}.$$

We will prove that the congruence is equivalent with the condition that n is a prime number. Indeed, if n is a prime, then by the Wilson's theorem, $(n-1)! + 1 \equiv 0 \pmod{n}$. Then

$$0 \equiv (n-2)! \cdot (n-1) + 1 \equiv -(n-2)! + 1 \pmod{n}.$$

On the other hand, if n is composite, then n divides $(n-2)!$; for $n = ab$, $1 \leq a < b \leq n-2$, the number $(n-2)!$ contains a, b as factors; for $n = q^2$, q prime, the product contains q, $2q$ as factors. In this case, the above congruence is impossible, since $1 \not\equiv 0 \pmod{n}$.

2.2 1969

2. Let n be the smallest of the given numbers. Then the set is

$$A = \{n, n+1, \ldots, n+b-1\}.$$

The set A contains exactly one element x divisible by b, namely, $x = n + b - r$, where

$$n \equiv r \pmod{b}, 0 \leq r \leq b-1.$$

Analogously, A contains at least one element y divisible by a; whence it follows that ab divides xy.

2.3 1973

3. Let $ABCD$ be the rectangle, with side lengths $AD = a$ and $AB = b$. Let P be the interior point, and let X and Y be the feet of the perpendiculars from P on CD and DA, respectively. Denote $DX = x$. Then $b(2x - b) = x^2 - (b-x)^2 = PD^2 - PC^2$ is an integer; therefore x is rational, $x = \frac{c}{d}$. Similarly, $DY = \frac{e}{f}$ and $\gcd(c,d) = \gcd(e,f) = 1$. Since $\left(\frac{c}{d}\right)^2 + \left(\frac{e}{f}\right)^2$ is an integer, we must have $d = f$; therefore $\frac{\sqrt{c^2+e^2}}{d}$ is an integer. The sum of two odd squares cannot be a square, so that one of c, e, say c, is even. Then d is odd, since c, d are coprime. If e is even, consider the length

$$PB = \frac{\sqrt{(bd-c)^2 + (ad-e)^2}}{d},$$

which is not an integer, since both terms inside the square root are odd. If e is odd, consider the length

$$PC = \frac{\sqrt{(bd-c)^2 + e^2}}{d} \notin \mathbb{N}.$$

In conclusion, there are no points with the required property.

2.4 1975

4. The $n - 1$ planes from M_1 determine $n - 1$ cones with the same vertex as the given cone. Let V be the volume of the given cone, and let $V_1, \ldots, V_{n-1}$ be the volumes of formed cones in decreasing order. Then

$$\frac{V_t}{V} = \frac{n-t}{n}, 1 \leq t \leq n-1.$$

A section containing the symmetry axis determine $n - 1$ triangles similar with the triangle in the given cone. Denote by $k_1, \ldots, k_{n-1}$ the quotients of similarity, then

$$\frac{V_t}{V} = k_t^3 = \frac{n-t}{n}, 1 \le t \le n-1.$$

For the set M_2, denote by S, S_t the lateral areas of the given cone, respectively, of the cones determined by the $n-1$ planes, and denote by k_t' the quotients of similarity of triangles determined by a section. Then

$$\frac{S_t}{S} = \frac{n-t}{n} = (k_t')^2, 1 \le t \le n-1.$$

It follows that $M_1 \cap M_2 \ne \emptyset$ if and only if there are $i, j \in \{1, 2, \ldots, n-1\}$ such that $k_i = k_j'$; whence $n(n-i)^2 = (n-j)^3$.

If there is a prime p which divides n and not divides $(n-i)^2$, then p^3 divides $(n-j)^3$, so that p^3 divides n. If every prime p which divides n, divides also $(n-i)^2$, then p^2 divides $(n-i)^2$.

2.5 1976

5. Consider the sets of integers

$$M = \{a_1, a_2, \ldots, a_n\}, M_1 = \{a_1\}, M_2 = \{a_1, a_2\}, \ldots, M_n = M,$$

and let s_m be the sum of elements of M_m, for $1 \le m \le n$. If there is m for which s_m is divisible by n, then M_m is the required set. Otherwise, the remainders modulo n of these sums belog to the set $\{1, 2, \ldots, n-1\}$. The Pigeonhole principle imples that there are two sums s_k, s_l, $1 \le k < l \le n$ with the same remainder; whence it follows that n divides $s_l - s_k$ and the required set is $\{a_{k+1}, \ldots, a_l\}$.

2.6 1977

6. Each positive integer has a remainder mod a_1 in the set $\{0, 1, \ldots a_1 - 1\}$. Since the sequence is strictly increasing, there is at least one r in this set such that $a_k \equiv r \pmod{a_1}$ for infinitely many k. Thus, we have infinitely many pairs (m, n) such that $m > n$ and $a_m = a_n + ya_1$.
7. From $4444^{4444} < 10,000^{4444}$, we deduce that 4444^{4444} has at most $4444 \cdot 4 + 1 < 20,000$ digits, so that $A < 9 \cdot 20,000 = 180,000$; whence $B < 9 \cdot 5 = 45$. If C is the sum of digits of B, then $C < 4 + 9 = 13$. The difference between a positive integer and the sum of its digits is divisible by 9; therefore, $4444^{4444} \equiv C \pmod 9$. From

$$4444 \equiv 7 \pmod 9, 4444^{4444} \equiv (-2)^{3 \cdot 1481} \cdot 7 \equiv (-8)^{1481} \cdot 7 \equiv 7 \pmod 9,$$

we deduce that $C = 7$.

8. For $n = 4$ the given equation has the solution $x_1 = x_2 = 1$, $x_3 = x_4 = 0$.
In the general case, the equation can be written as

$$\sum_{1 \leq i < j \leq n} (x_i - x_j)^2 = n.$$

Case 1. $n = 1$. The equation

$$x_1^2 - \frac{x_1^2}{1} = 1$$

is without solutions.

Case 2. $n = 2$. The equation $(x_1 - x_2)^2 = 2$ has no integer solutions.

Case 3. $n = 3$. The equation $(x_1 - x_2)^2 + (x_2 - x_3)^2 + (x_1 - x_3)^2 = 3$ has an integer solution if and only if $x_i - x_j = \pm 1$, $i < j$, but this is not possible if the unknowns are distinct. If, say, $x_2 = x_3$, then we have the equation $2(x_1 - x_2)^2 = 3$, without integer solutions. In this case there are no solution.

Case 4. $n \geq 5$. If the LHS of the equation has at least $n + 1$ nonzero terms, then it is $\geq n + 1$ and the equality is impossible. It follows that at least $\binom{n}{2} - n = \frac{n(n-3)}{2}$ terms vanish. The inequality $\frac{n(n-3)}{2} \geq \binom{n-2}{2}$ shows that in the LHS appear at most $n - 1$ unknowns. If in the terms which vanish appear all unknowns $x_1, \ldots, x_n$, then the equation has no solution. Finally we deduce that there is a unique unknown distinct from the others: $x_1 = \cdots = x_{n-1} = x$, $x_{n+1} \neq x$. Then we have

$$x_n - x = \pm \sqrt{\frac{n}{n-1}},$$

without integer solutions.

9. Denote $M = \{n, n+1, \ldots, n+9\}$. For $n = 1$, the required number is 7.

Suppose that $n \geq 2$. The set M contains five even numbers and five odd numbers. An odd number which is divisible by 3 is of the form $n_1 = 6k + 3$, $k \in \mathbb{N}^*$. If n_1, n_2, and n_3 are three consecutive odd numbers divisible by 3, then $n_2 = 6k + 9$, $n_3 = 6k + 15$. The set $N = \{n_1, n_1 + 1, \ldots, n_3\}$ of consecutive numbers contains $15 - 3 + 1 = 13$ numbers, which shows that M can contain at most two odd numbers divisible by 3. Analogously, M contais at most one number divisible by 5 and at most one number divisible by 7. It follows that M contains one number a which is not divisible by 2, 3, 5, and 7. This number is the required number.

Indeed, let $b \in M$ be a number such that $\gcd(a, b) = d > 1$. Then $d \mid |a - b|$ implies $d < 10$. In conclusion, $3|d, 5|d$ or $7|d$, contradiction.

10. Answer: $(p, q, r, x, y, z) = (2, 3, r, 1, 1, 0), (2, q, 3, 1, 0, 1), (2, 7, 3, 3, 1, 2),$
$(2, 3, 7, 3, 2, 1), (2, 3, 5, 2, 1, 1), (2, 5, 3, 2, 1, 1), (3, 2, r, 1, 3, 0), (3, q, 2, 1, 0, 3),$
$(3, 2, 5, 2, 4, 1), (3, 5, 2, 2, 1, 4), (3, 2, r, 1, 3, 0), (3, q, 2, 1, 0, 3), (5, 2, 3, 1, 3, 1),$
$(5, 3, 2, 1, 1, 3), (7, 2, 3, 1, 4, 1), (7, 3, 2, 1, 1, 4), (17, 2, 3, 1, 5, 2), (17, 3, 2, 1, 2, 5).$
Write the equation as $(p^x - 1)(p^x + 1) = q^y r^z$.

Case 1. $p = 2$. Both q and r are odd, $\gcd(2^x - 1, 2^x + 1) = 1$, $2^x + 1 > 2^x - 1$, therefore $2^x - 1 = 1$, with solutions $(2, 3, 3, 1, 1, 0), (2, 3, 3, 1, 0, 1); 2^x - 1 = q^y$ or r^z (the last case is similar).

If $x = 1$, we get the above solutions.

If $x > 1$, there are no solutions for $2^x - q^y = 1$ (Catalan theorem). Then $y = 1$ and $4^x - p \cdot r^z = 1$ with solution $(2, 7, 3, 3, 1, 2)$. If $z = 1$, then $2^x - 1$ and $2^x + 1$ are both powers of primes; whence the solution $(2, 3, 5, 2, 1, 1)$.

Case 2. $p = 3$. Then $q = 2$ or $r = 2$. If $q = 2$, then

$$(3^x - 1)(3^x + 1) = 2^y r^z.$$

For $z = 0$, r is arbitrary and $x = 1$, $y = 3$. For $z \geq 1$, $\gcd(3^x - 1, 3^x + 1) = 2$ implies $3^x - 1 = 2^{y-1}, 3^x + 1 = 2r^z$, or $3^x - 1 = 2r^z, 3^x + 1 = 2^{y-1}$. In the first case, we find $x = 2, y = 4, r^z = 5$; in the second case, we find $x = 1, y = 3, z = 0$.

Case 3. $p > 3$. Then $p \equiv \pm 1 \pmod{6}$ implies $\{q, r\} = \{2, 3\}$. We have $p^x - 1 = 2^{y-1}, p^x + 1 = 2r^z$, or $p^x - 1 = 2r^z, p^x + 1 = 2^{y-1}$. In the first case, we find $y = 3, z = 1, p^x = 5$, or $y = 5, z = 2, p^x = 17$. In the second case, we find $y = 4, z = 1, p^x = 7$.

2.7 1978

11. We can suppose that $c > 1$. Consider the following infinite algorithm: Initially, $A = B = \emptyset$.

    Step 1: Add 1 to A.
    ...
    Step n: If $\frac{n}{c} \notin A$, then add n to A; else add n to B.

    Thus every $n \in \mathbb{N}^*$ is added either to A or to B, and the resulting sets satisfy the requirements.

12. From the recurrence relations, it follows that

$$x_{i+1} = \left\lfloor \frac{x_i + \left\lfloor \frac{n}{x_i} \right\rfloor}{2} \right\rfloor = \left\lfloor \frac{x_i + \frac{n}{x_i}}{2} \right\rfloor \geq \lfloor \sqrt{n} \rfloor.$$

If $x_i > \lfloor \sqrt{n} \rfloor$ for some i, then $x_{i+1} < x_i$, which follows from the fact the last inequality is equivalent to

$$x_i > \frac{x_i + \frac{n}{x_i}}{2},$$

that is, $x_i^2 > n$. Therefore $x_i = \lfloor \sqrt{n} \rfloor$ holds for at least one $i \leq n - \lfloor \sqrt{n} \rfloor + 1$.

Note that if $n + 1$ is a square; then $x_i = \lfloor \sqrt{n} \rfloor$ implies $x_{i+1} = \lfloor \sqrt{n} \rfloor + 1$, otherwise $x_i = \lfloor \sqrt{n} \rfloor$ implies $x_{i+1} = \lfloor \sqrt{n} \rfloor$.

2.8   1979

13. From $3^x \equiv 1 \pmod 4$, we deduce $x = 2x'$. From $5^z \equiv 1 \pmod 3$, we deduce $z = 2z'$. It follows that

$$4^y = 5^{2z'} - 3^{2x'} = (5^{z'} - 3^{x'})(5^{z'} + 3^{x'}),$$

whence $5^{z'} - 3^{x'} = 2^a$ and $5^{z'} + 3^{x'} = 2^b$, with $a + b = 2y$. From $2 \cdot 3^{x'} = 2^b - 2^a$, it follows that $b > a$. We have $5^{z'} = 2^{a-1}(1 + 2^{b-a})$; whence $a = 1$. Then $b - a = 2y - 2$, $5^{z'} = 4^{y-1} + 1$, and $3^{x'} = 4^{y-1} - 1$; therefore y is even. From the binomial formula,

$$4^{y-1} - 1 = (3+1)^{y-1} - 1 = 3^{y-1} + \cdots + 3(y-1) = 3^{x'}.$$

If $y > 2$, then 3 divides $y - 1$, $y - 1 = 3y'$, and

$$4^{y-1} - 1 = 64^{y'} - 1 = 3^{x'} \equiv 0 \pmod{63},$$

contradiction. It follows that the given equation has the unique solution

$$x = y = z = 2.$$

14. We have

$$\frac{p\pi}{k} = \left\lfloor \frac{p}{k} \right\rfloor \cdot \pi + \left\{ \frac{p}{k} \right\} \cdot \pi$$

and then we can write

$$\sin \frac{p\pi}{k} = \sin\left(\left\lfloor \frac{p}{k} \right\rfloor \cdot \pi + \left\{ \frac{p}{k} \right\} \cdot \pi \right) = (-1)^{\lfloor \frac{p}{k} \rfloor} \sin \left\{ \frac{p}{k} \right\} \pi.$$

As $0 < \left\{ \frac{p}{k} \right\} \pi < \pi$, it follows $\sin\left\{\frac{p}{k}\right\}\pi > 0$, and it remains to show that $\lfloor \sqrt{p} \rfloor + \sum_{k=2}^{p-1} \lfloor \frac{p}{k} \rfloor$ is even. Denote

$$S_n = \sum_{k=2}^{n-1} \left\lfloor \frac{n}{k} \right\rfloor, n \geq 3.$$

As

$$\left\lfloor \frac{n}{n-1} \right\rfloor = 1, \left\lfloor \frac{n}{k} \right\rfloor - \left\lfloor \frac{n-1}{k} \right\rfloor = \begin{cases} 1, k \mid n \\ 0, k \nmid n \end{cases},$$

it is easy to see that $S_n - S_{n-1} = \tau(n) - 1$, where $\tau(n)$ is the number of positive divisors of n. Then $S_n = \tau(4) + \tau(5) + \cdots + \tau(n) - n + 4$. For $n = p_1^{\alpha_1} p_2^{\alpha_2} \ldots p_k^{\alpha_k}$, we have $\tau(n) = (\alpha_1 + 1)(\alpha_2 + 1) \ldots (\alpha_k + 1)$; therefore $\tau(n)$ is odd if and only if n is a square. It follows that

$$S_n \equiv \tau(2^2) + \tau(3^2) + \cdots + \tau(\lfloor \sqrt{n} \rfloor) - n \equiv \lfloor \sqrt{n} \rfloor - n - 1 \pmod{2}.$$

Putting $n = p$, we have $S_p + \lfloor \sqrt{p} \rfloor = 2\lfloor \sqrt{p} \rfloor - p - 1 \equiv 0 \pmod{2}$, since $p + 1$ is even. We deduce that $S_p + \lfloor \sqrt{p} \rfloor$ is even, as desired.

15. Answer: 2, 3, 4, 6, 12, 18, 24, and 30.

Lemma. Let $p_1 < p_2 < \cdots$ be the sequence of all prime numbers. Then

$$p_1 \cdots p_n > p_{n+1}^2$$

for all $n \geq 4$.

Proof. Induction on n. The inequality is true for $n = 4$. Suppose that it is true for $n = k$. Let $a = p_1 \ldots p_{k+1}, S = \sqrt{\frac{a}{4}}$. By the Bertrand's postulate, we know that there is a prime $q \in [S, 2S - 2]$. The induction hypothesis implies

$$S^2 = \frac{1}{4} p_1 \cdots p_{k+1} > \frac{1}{4} p_{k+1}^3 > p_{k+1}^2,$$

which means $q^2 > p_{k+1}^2$, but then $a = (2S)^2 > q^2 > p_{k+1}^2$.

Back to the problem, let p_k be the biggest prime that divides n, and let p_i be the smallest prime that doesn't divide n. The Bertrand's postulate implies that the ratio of consecutive primes is less than 2. We have $k \geq 4$.

If $i = 1$, then $4 < n$, but 4 isn't a prime.

If $i = 2$, then $9 < 14 \leq n$, but 9 isn't a prime.

If $i = k + 1$, then by the lemma, we know that $p_{k+1}^2 < n$, but p_{k+1}^2 isn't a prime.

If $3 \leq i \leq k - 1$, then we know that $p_i^2 < 2p_{i-1}p_k$, but p_i^2 isn't a prime.

So the only eligible n can be written as $n = 2^a 3^b 5^c$, with nonnegative integers a, b, and c. It is easy to see that $n < 49$. After checking all possibilities, we find the above values.

2.9 1980

16. We have to show that

$$\frac{2\cdot(3n)!}{n!\cdot(2n)!\cdot 3\cdot(3n-1)(3n-2)} = \frac{(3n-3)!}{n!\cdot(2n-2)!}$$

is an integer. It suffices to prove that every prime number p appears at the numerator with an exponent greater than the exponent at the denominator.

Case 1. $p \nmid (2n-2)(2n-1)$. Then the exponent of p in the denominator is equal to the exponent in $n! \cdot (2n-3)!$ and is less than the exponent in the denominator, since

$$\frac{(3n-3)!}{n!\cdot(2n-3)!} = \binom{3n-3}{n}.$$

Case 2. $p \nmid n(n-1)$. Then the exponent of p in the denominator is equal with the exponent in the expression $(n-1)! \cdot (2n-2)!$ and is less than the exponent of p in the numerator, since

$$\frac{(3n-3)!}{(n-1)!\cdot(2n-2)!} = \binom{3n-3}{n-1}.$$

Case 3. None of the above conditions hold, that is, $p|n(n-1)$ and $p|(2n-1)(2n-2)$; whence we deduce $p|\gcd(n(n-1),(2n-1)(2n-2)) = n-1$, since n is odd.

We can write $n - 1 = kp$, k positive integer, so that

$$\frac{(3n-3)!}{n!\cdot(2n-1)!} = \frac{(3kp)!}{(kp+1)!\cdot(2kp+1)!}.$$

From $p \nmid kp + 1$, $p \nmid 2kp + 1$, it follows that the exponent of p in the expression $(kp)! \cdot (2kp)!$ is less than the exponent in the numerator, since

$$\frac{(3kp)!}{(kp)!\cdot(2kp)!} = \binom{3kp}{kp}$$

is an integer.
In conclusion,

$$\frac{2\cdot(3n)!}{n!\cdot(2n)!}$$

is divisible by $3(3n-1)(3n-2)$ for each odd positive integer n.

17. We will prove by induction that $b_{n+100} = b_n$ for all positive integers n.

For $n = 1$, we have $b_1 = 1$; therefore we must show that $b_{101} = 1$. This reduces to

$$\sum_{k=1}^{99} k^k \equiv 0 \pmod{10}.$$

We have

$$\sum_{k=1}^{99} k^k \equiv \sum_{r=1}^{9} r^r \left(1 + r^{10} + r^{20} + \cdots + r^{90}\right)$$

$$\equiv \sum_{r=1}^{9} r^r \left(4r^4 + 5r^2 + 1\right) \equiv 0 \pmod{10}.$$

It follows that $b_{101} = b_1$.

Suppose that there is a positive integer n such that $b_{n+100} = b_n$. We must sow that $b_{n+101} = b_{n+1}$, which means that $(n+101)^{n+101} \equiv (n+1)^{n+1} \pmod{10}$. But

$$(n+101)^{n+101} \equiv (n+1)^{n+101} \pmod{10}.$$

Let r be the last digit of $n + 1$. In the sequence $(r^n)_{n \geq 1}$ for $r \geq 2$, the last digit has period 2 for $r = 4$ and 9 and has period 4 for $r = 2, 3, 7$, and 8. For $r = 0, 1, 5$, and 6, the last digit remains unchanged. We deduce that

$$r^{p+4} \equiv r^p \pmod{10} \text{ for all nonnegative integers } p.$$

Since $n + 101 \equiv n + 1 \pmod 4$, it follows that $(n+1)^{n+101} \equiv (n+1)^{n+1} \pmod{10}$; hence $b_{n+101} = b_{n+1}$, which ends the proof by induction.

18. Denote $E(a,b) = a^3b - ab^3 = ab(a-b)(a+b)$. If $ab = 0$, then $E(ab) = 0 \equiv 0 \pmod{10}$. From $E(-a,b) = E(a,-b) = -E(a,b)$ and $E(-a,-b) = E(a,b)$, it follows that we can suppose wlog that a and b are positive integers.

Clearly, $E(a, b)$ is even, since if a and b are both odd; then $a - b$ and $a + b$ are even. It remains to show that $E(a, b) \equiv 0 \pmod 5$. If $ab \equiv 0 \pmod 5$, then this property is obvious. Suppose that $ab \not\equiv 0 \pmod 5$ and consider the sets

$$A = \{1, 4, 6, 9\}, B = \{2, 3, 7, 8\}.$$

For every three positive integers not divisible by 5, there are two whose remainders modulo 10 are in the same set, which implies that their sum or difference is divisible by 5.

2.10   1981

19. It is easy to prove that $a_p = 2^{p-1}p + 2^{p-1} - 1$. Since p is an odd prime, the Fermat's theorem implies $2^{p-1} - 1 \equiv 0 \pmod{p}$. It follows that $a_p > p$ is divisible by p; therefore it is not a prime.

20. For $n = \overline{a_1 a_2 \ldots a_p}$, $a_p \neq 9$, we have $n + 1 = \overline{a_1 a_2 \ldots a_{p-1} b_p}$, with $b_p = a_p + 1$, hence $s(n+1) - s(n) = 1$, so that $\gcd(s(n), s(n+1)) = 1$.
Suppose that $n = \overline{a_1 a_2 \ldots a_r \underbrace{99 \ldots 9}_{p-r}}$, $a_r < 9$. Then $n+1 = \overline{a_1 a_2 \ldots b_r \underbrace{00 \ldots 0}_{p-r}}$,
$b_r = a_r + 1 \neq 0$. We have $s(n) = a_1 + a_2 + \cdots + a_r + 9(p - r)$ and $s(n+1) = a_1 + a_2 + \cdots + a_r + 1$. There are infinitely many ways of choosing $r, a_1, \ldots, a_r$ such that $s(n+1) \equiv 0 \pmod{m}$, for example, $r = km - 1$ and $a_1 = a_2 = \cdots = a_r = 1$.
Let $p = \lfloor \frac{m+1}{3} \rfloor^2 + r$. Then $9(p-r) - 1 \equiv 0 \pmod{m}$, so that $s(n) \equiv s(n+1) \equiv 0 \pmod{m}$.

2.11   1982

21. Let p be a prime divisor of $(n+1)^m - n$ and $(n+1)^{m+3} - n$. Then

$$(n+1)^{m+3} \equiv (n+1)^m (n+1)^3 \equiv n(n+1)^3 \pmod{p} \Rightarrow p \mid n\left((n+1)^3 - 1\right).$$

If $p \mid n$, then $(n+1)^m \equiv n \pmod{p}$ implies $1 \equiv 0 \pmod{p}$, contradiction. Thus, p divides $n^3 + 3n^2 + 3n$; hence $p \mid n^2 + 3n + 3$. Now,

$$(n+1)^m \equiv n \pmod{p} \Rightarrow (n+1)^{m^2+3m} \equiv n^{m+3} \pmod{p},$$
$$(n+1)^{m+3} \equiv n \pmod{p} \Rightarrow (n+1)^{m^2+3m} \equiv n^m \pmod{p}.$$

Thus, $n^{m+3} \equiv n^m \pmod{p}$, which means $p \mid (n-1)(n^2+n+1)$.
Case 1. $p \mid n - 1$. From $p \mid n^2 + 3n + 3$, we deduce $p \mid 7$, so $p = 7$. Since $n \equiv 1 \pmod{7}$ and $(n+1)^m \equiv n \pmod{7}$, we have $(1+1)^m \equiv 1 \pmod{7}$, so that $3 \mid m$. Note that $(n, m) = (7k - 6, 3l)$, with k, l positive integers, is a solution.
Case 2. $p \mid n^2 + n + 1$. Then $p \mid n^2 + 3n + 3$ implies $p \mid \gcd(n^2 + n + 1, n^2 + 3n + 3) = 1$, contradiction.

22. From $(p-k)^p \equiv -k^p \pmod{p^2}$ and $0 < a_k < p^2$, we deduce $a_k \equiv -a_{p-k} \pmod{p^2}$ and $0 < a_k + a_{p-k} < 2p^2$, so that $a_k + a_{p-k} = p^2$. It follows that

$$2 \sum_{k=1}^{p-1} a_k = \sum_{k=1}^{p-1} (a_k + a_{p-k}) = \sum_{k=1}^{p-1} p^2 = p^2(p-1).$$

2.12   1983

23. Let $P(n)$ be the given proposition. Choosing numbers of the form $\underbrace{22\ldots2}_{k}\underbrace{11\ldots1}_{2^k-2k}$ with $2^k - k$ digits, we see that the sum of digits is $2k + 2^k - 2k = 2^k$, and the product of digits is 2^k. As $2^k - k$ is unbounded, we conclude that $P(n)$ is true for infinitely many n.

    Now we prove that there are infinitely many positive integers n for which $P(n)$ is false. It is easy to prove the following inequality:

    $$9 \cdot (2 \cdot 3 \cdot 4 \cdot \ldots \cdot 9)^k > 2^{2^{k-3}}$$

    for all $k > k_0$, where k_0 is some fixed positive integer. If $n = 9 \cdot (2 \cdot 3 \cdot 4 \cdot \ldots \cdot 9)^k$, $k >> k_0$, then $P(n)$ is false. Suppose that there is a number n for which the proposition is true and denote by x_i, $1 \leq i \leq 9$, the number of digits equal to i (clearly, n does not contain zeros). Then we have

    $$x_1 + 2x_2 + \cdots + 9x_9 = 1^{x_1} \cdot 2^{x_2} \cdot \ldots \cdot 9^{x_9},$$

    or equivalently,

    $$2^{x_2} \cdot 3^{x_3} \cdot \ldots \cdot 9^{x_9} = n + (2x_3 + \cdots + 8x_9),$$

    which implies

    $$2^{x_2} \cdot 3^{x_3} \cdot \ldots \cdot 9^{x_9} > 2^k \cdot 3^k \cdot \ldots \cdot 9^k.$$

    We deduce that there is an index i, $2 \leq i \leq 9$, for which $x_i > k$. But then i^k divides LHS, $i^k | n$, so that i^k divides their difference $x_2 + 2x_3 + \cdots + 8x_9$. It follows that

    $$x_2 + 2x_3 + \cdots + 8x_9 \geq i^k \geq 2^k.$$

    Now we have

    $$2^{x_2+x_3+\cdots+x_9} \leq 2^{x_2} \cdot 3^{x_3} \cdot \ldots \cdot 9^{x_9} = x_1 + 2x_2 + \cdots + 9x_9 < 9n.$$

    We can write $2^k < 8(x_2 + x_3 + \cdots + x_9)$, or, $2^{k-3} < x_2 + x_3 + \cdots + x_9$. But then

    $$9 \cdot (2 \cdot 3 \cdot \ldots \cdot 9)^k = 9n > 2^{x_2+x_3+\cdots+x_9} > 2^{2^{k-3}},$$

    contradiction.

2.13 1984

24. Assume that the decimal representation of n uses digits $x \neq y$ and denote $g(n) = \gcd(n, f(n))$. Then $g(n)$ divides $n + f(n) = 1111(x+y) = 11 \cdot 101(x+y)$. A number of type $\overline{xxxy}, \overline{xxyx}, \overline{xyxx}, \overline{xyyy}, \overline{xxyy}, \overline{xyyx}$ is not divisible by 101, so that $g(n) | 11(x+y)$, therefore $g(n) \leq 11 \cdot 17 = 187$.

If $n = \overline{xyxy}$, then denote $\overline{xy} = a, \overline{yx} = b, \gcd(a,b) = d$; hence

$$n = 101a, \ f(n) = 101b, g(n) = 101d.$$

Now we need to maximize d. From $n > f(n)$, we get $x > y$. Clearly,

$$d \mid a - b = 9(x - y).$$

If $3 \nmid d$, then $d | x - y$; hence $d \leq 9 - 1 = 8$.

If $3 | d$, then $3 | x + y$, so that $a \in \{21, 42, 51, 54, 63, 72, 75, 81, 84, 87, 93, 96\}$. Then

$$d \in \{3, 6, 9, 12\}.$$

The value $d = 12$ occurs for $a = 84, b = 48$. Then $n = 8484, f(n) = 4848$, and the required maximum value is $g(n) = 101 \cdot 12 = 1212$.

2.14 1986

25. If k has a divisor $d > \frac{n}{2}, d < n$, then $\frac{k}{d} < \frac{n}{2}$; otherwise $k = d \cdot \frac{k}{d} > \frac{n^2}{4}$, contradiction. In this case $n!$ contains the distinct factors d and $\frac{k}{d}$, so that $k | n!$. If all divisors of k are less than $\frac{n}{2}$, then in particular $k \leq \frac{n}{2}$; therefore $k | n!$.

2.15 1987

26. Answer: $n = 1, 3$.

If $n = 1$, then $(x, y, z, n) = (3, 2, 1, 1)$ is a solution.

If $n = 3$, then $(x, y, z, n) = (1, 1, 1, 3)$ is a solution.

Assume that $x \geq y \geq z$. Then $3x^3 \geq n(xyz)^2$; whence $x \geq \frac{ny^2z^2}{3}$. From the given equation it follows that $x^2 | y^3 + z^3$; hence

$$2y^3 \geq y^3 + z^3 \geq x^2 \geq \frac{n^2 y^4 z^4}{9}.$$

If $z > 1$, then $y \leq \frac{18}{16n^2}$, so $y = 1$, but $y \geq z$, contradiction.

If $z=1$, then $y^3+1 \geq \frac{n^2y^4}{9}$, so $9+\frac{9}{y^3} \geq n^2y$.

If $y=1$, then $x^3+2=nx^2$, so the only solution is $(x,y,z,n)=(1,1,1,3)$.

If $y>1$, then $10 \geq n^2y$, so (a) $n=1$, with solution $(x,y,z,n)=(3,2,1,1)$; (b) $n=2$, no solution; (c) $n=3$, $y>1$, no solution.

27. From $A = mnp - m - n - p + \frac{1}{m} + \frac{1}{n} + \frac{1}{p} - \frac{1}{mnp}$, it follows that A is an integer if and only if $B = \frac{1}{m} + \frac{1}{n} + \frac{1}{p} - \frac{1}{mnp}$ is an integer. If two numbers are equal, say $m=n$, then $m(n+2p-Bmp)=1$; whence $m=n=1$, p is arbitrary, $B=2$, and $A=0$. If the numbers are distinct, $m > n > p \geq 1$, then $n \geq 2$, $m \geq 3$. We deduce

$$0 \leq B < B + \frac{1}{mnp} = \frac{1}{m} + \frac{1}{n} + \frac{1}{p} \leq \frac{1}{3} + \frac{1}{2} + 1 = \frac{11}{6} < 2,$$

so that $B=0$ or $B=1$. If $B=0$, then $mn+np+pm=1$, with no solutions in positive integers. If $B=1$, then

$$\frac{1}{m} + \frac{1}{n} + \frac{1}{p} = 1 + \frac{1}{mnp} \quad (1).$$

For $p \geq 3$ we must have $n \geq 4$, $m \geq 5$; therefore

$$\frac{1}{m} + \frac{1}{n} + \frac{1}{p} \leq \frac{1}{5} + \frac{1}{4} + \frac{1}{3} = \frac{47}{60} < 1.$$

For $p=1$, the equality (1) becomes $\frac{1}{m} + \frac{1}{n} = \frac{1}{mn}$ with no solutions in positive integers. If $p=2$, condition (1) get us $\frac{1}{m} + \frac{1}{n} = \frac{1}{2} + \frac{1}{2mn}$, which is equivalent to $(m-2)(n-2)=3$; whence $m=5$, $n=3$, and $A=21$. Answer: $A=0$ for $(m,n,p)=(1,1,1)$ and $A=21$ for $(m,n,p)=(5,3,2)$.

28. If $x,y \in B$, $x-y=1$ or $x-y=2$, then B does not satisfy the given condition. Take for B an arithmetic progression with common difference equal to 3: if $\lfloor \frac{n}{3} \rfloor = k$, then $B = \{1,4,7,\ldots,3k-1\}$; if $n=3k+1$ or $n=3k+2$, then $B = \{1,4,7,\ldots,3k+1\}$. In all cases, $|B| = \lfloor \frac{n+2}{3} \rfloor$ and the maximum is at least $\lfloor \frac{n+2}{3} \rfloor$. We shall prove that the maximum is in fact equal to this number. Suppose that $B = \{x_1, x_2, \ldots, x_m\}$, $m > \lfloor \frac{n+2}{3} \rfloor$, $x_1 < x_2 < \cdots < x_n$. From $x_{i+1} - x_i \geq 3$, $1 \leq i \leq m-1$, we deduce

$$x_m = x_1 + (x_2 - x_1) + \cdots + (x_m - x_{m-1}) \geq 1 + 3(m-1) > 1 + 3\left(\frac{n+2}{3} - 1\right) = n,$$

contradiction.

29. Suppose that there is an integer $n \geq 2$ such that $n | 3^n - 2^n$. Let p be the smallest prime divisor of n. From $p | 3^n - 2^n$ we deduce $(2,p) = (3,p) = 1$. By the Fermat's theorem, $2^{p-1} \equiv 3^{p-1} \equiv 1 \pmod{p}$; whence $p | 3^{p-1} - 2^{p-1}$. Let A be the set of positive integers m such that $p | 3^m - 2^m$. We have $p-1 \in A$, $n \in A$. We shall prove the following:

(a) If $a, b \in A$, then $a + b \in A$.
(b) If $a, b \in A$ and $a > b$, then $a - b \in A$.

Indeed, the equality $3^{a+b} - 2^{a+b} = 3^a(3^b - 2^b) + 2^b(3^a - 2^a)$ proves (a). Denote $q = a - b > 0$. From $3^a - 2^a = 3^{b+q} - 2^{b+q} = 3^q(3^b - 2^b) + 2^b(3^q - 2^q)$ follows (b).

Applying successively the Euclidean algorithm and properties (a) and (b), we deduce that $\gcd(a, b) \in A$. It follows that $\gcd(p - 1, n) \in A$, but $\gcd(p - 1, n) = 1$, since $p|n$, so that $1 \in A$, contradiction.

30. From $(a + b + c)^2 = a^2 + b^2 + c^2 + 2(ab + bc + ca)$, it follows that

$$a + b + c \mid 2(ab + bc + ca).$$

From $(ab + bc + ca)^2 = a^2b^2 + b^2c^2 + c^2a^2 + 2abc(a + b + c)$, it follows that

$$a + b + c \mid a^2b^2 + b^2c^2 + c^2a^2.$$

We shall prove by induction that

$$a + b + c \mid a^{2^k} + b^{2^k} + c^{2^k} \text{ and } a + b + c \mid 2\left[(ab)^{2^k} + (bc)^{2^k} + (ca)^{2^k}\right],$$

for every nonnegative integer k. The above equalities prove the properties for $k = 0$ and $k = 1$. For $k \geq 2$ we can use the identities:

$$\left(a^{2^k} + b^{2^k} + c^{2^k}\right)^2 = a^{2^{k+1}} + b^{2^{k+1}} + c^{2^{k+1}} + 2\left[(ab)^{2^k} + (bc)^{2^k} + (ca)^{2^k}\right]$$

$$\left[(ab)^{2^k} + (bc)^{2^k} + (ca)^{2^k}\right]^2 = (ab)^{2^{k+1}} + (bc)^{2^{k+1}} + (ca)^{2^{k+1}} + 2(abc)^{2^k}\left(a^{2^k} + b^{2^k} + c^{2^k}\right).$$

2.16 1988

31. We have

$$\prod_{k=1}^{n} k^{2k-n-1} = \frac{1^2}{1^{n+1}} \cdot \frac{2^{2 \cdot 2}}{2^{n+1}} \cdot \frac{3^{3 \cdot 2}}{3^{n+1}} \cdot \ldots \cdot \frac{n^{n \cdot 2}}{n^{n+1}} = \frac{1}{(n!)^{n+1}} \left(n! \cdot \frac{n!}{1!} \cdot \frac{n!}{2!} \cdot \ldots \cdot \frac{n!}{(n-1)!}\right)^2 =$$

$$= \frac{(n!)^{2n}}{(n!)^{n+1}} \cdot \frac{1}{[1! \cdot 2! \cdot \ldots \cdot (n-1)!]^2} = \frac{(n!)^{n-1}}{[1! \cdot 2! \cdot \ldots \cdot (n-1)!]^2} =$$

$$= \frac{n!}{1!(n-1)!} \cdot \frac{n!}{2!(n-2)!} \cdot \ldots \cdot \frac{n!}{(n-1)!1!} = \binom{n}{1}\binom{n}{2} \ldots \binom{n}{n-1}.$$

32. Define $f(y,x) = y^2 - axy - x^2$. We have $f(x_2, x_1) = f(a,1) = -1$, and by induction we deduce that if $f(x_{n+1}, x_n) = \pm 1$, then $f(x_{n+2}, x_{n+1}) = \mp 1$, which proves that every pair (x_{n+1}, x_n) is a solution of the equation.

Now suppose that $(y,x) \in \mathbb{N} \times \mathbb{N}$ is a solution of the equation. If $y = 0$ or $x = 0$, then $x = 1$ or $y = 1$. We must eliminate this case by hypothesis, since the pair (y,x) is not of the form (x_{n+1}, x_n). If $(y,x) \in \mathbb{N}^* \times \mathbb{N}^*$, then from $y^2 - axy - x^2 = \pm 1$, we deduce $y(y - ax) = x^2 \pm 1 \geq 0$, so that $y \geq ax$, with equality only if $x = 1$, $y = a$. In this case $(y,x) = (x_2, x_1)$. Suppose that $y > ax$. The pair of positive integers $(y^{(1)}, x^{(1)}) = (x, y - ax)$ is also a solution of the equation, since $f(y,x) = \pm 1$ if and only if $f(x, y - ax) = \mp 1$. We have also $y + x > y^{(1)} + x^{(1)}$. From $(y^{(1)}, x^{(1)}) \in \mathbb{N}^* \times \mathbb{N}^*$, it follows that $y^{(1)} \geq ax^{(1)}$. If $y^{(1)} = ax^{(1)}$, then $y^{(1)} = x_2$, $x^{(1)} = x_1$, and $(y,x) = (x_3, x_2)$. If $y^{(1)} > ax^{(1)}$, then consider the solution $(y^{(2)}, x^{(2)}) = (x^{(1)}, y^{(1)} - ax^{(1)}) \in \mathbb{N}^* \times \mathbb{N}^*$, for which $y^{(1)} + x^{(1)} > y^{(2)} + x^{(2)}$ and $y^2 \geq ax^{(2)}$. If $y^{(2)} = ax^{(2)}$, then $(y,x) = (x_4, x_3)$.

Further by induction from every solution $(y^{(n)}, x^{(n)})$ for which $y^{(n)} > ax^{(n)}$, we get a solution $(y^{(n+1)}, x^{(n+1)}) = (x^{(n)}, y^{(n)} - ax^{(n)})$ with $y^{(n)} + x^{(n)} > y^{(n+1)} + x^{(n+1)}$. Since the strictly decreasing sequence of positive integers $(y^{(n)} + x^{(n)})$ is finite, there exists a positive integer k such that $y^{(k)} = ax^{(k)}$. By inverse recurrence, from $(y^{(k)}, x^{(k)}) = (x_2, x_1)$, we deduce $(y,x) = (x_{k+1}, x_k)$.

2.17 1989

33. (a) From $a_n = (n^2 + 2)(n^2 + 3)(n^2 - 6)$, we see that $2|a_2$, $3|a_3$, $5|a_4$. For $p \geq 7$, consider the function $\varphi: \mathbb{Z}_p^* \to \mathbb{Z}_p^*$ defined by $\varphi(x) = x^2$, which is a group homomorphism, with kernel $\{1, -1\}$ and image containing all quadratic residues mod p. It follows that $\mathbb{Z}_p^*$ contains $\frac{p-1}{2}$ quadratic residues and $\frac{p-1}{2}$ quadratic nonresidues and the quadratic nonresidues are of the form $-x$, with x a quadratic residue. If $p \nmid n^2 + 2$ and $p \nmid n^2 + 3$, then -2 and -3 are quadratic nonresidues; therefore 6 is a quadratic residue mod p, so that $p|n^2 - 6$.
(b) $a_n \not\equiv 0 \pmod 8$.
(c) We have $1989 = 9 \cdot 13 \cdot 17$. Divisibility $9|a_n$ occurs if and only if $n \equiv 0$, 3, 4, 5, 6 (mod 9). Divisibility $13|a_n$ occurs if and only if $n \equiv 4, 9 \pmod{13}$. Divisibility $17|a_n$ occurs if and only if $n \equiv 7, 10 \pmod{17}$. By the Chinese remainder theorem, we find the minimal solution $n = 95$.

34. Equalities $a = wn - vp$, $b = up - wm$, $c = vm - un$ form an linear system with unknowns m, n, p. The determinant of this system is

$$\Delta = \begin{vmatrix} 0 & w & -v \\ -w & 0 & u \\ v & -u & 0 \end{vmatrix} = 0.$$

Furthermore, since we cannot have $u = v = w = 0$, there are nonzero minors of second order, say,

$$\begin{vmatrix} 0 & w \\ -w & 0 \end{vmatrix} \neq 0.$$

It follows that the rank of the system is 2. It suffices to solve in integers the system

$$\begin{cases} a = wn - vp \\ b = -wm + up \end{cases}.$$

This system has integer solutions if and only if there exists $p \in \mathbb{Z}$ such that the system of congruences

$$\begin{cases} vp \equiv -a \pmod{w} \\ up \equiv b \pmod{w} \end{cases} \quad (1)$$

has solutions. By the hypothesis, there exist integers A, B, C, with $Au + Bv + Cw = 1$. Multiplying by a, we get

$$Aua + Bva + Cwa = a \Rightarrow Aua + Bva \equiv a \pmod{w}.$$

From $au + bv \equiv 0 \pmod{w}$, we deduce

$$-Abv + Bav \equiv a \pmod{w} \Rightarrow v(Ab - Ba) \equiv -a \pmod{w},$$

which satisfies the first relation from (1) with $p = Ab - Ba$. The second relation is also satisfied:

$$u(Ab - Ba) \equiv Aub - Bua \equiv Aub + Bvb \equiv b(Au + Bv) \equiv b \pmod{w}.$$

35. Two consecutive recurrence relations are

$$a_{n+1}a_{n-2} = 1989 + a_n a_{n-1}, \quad a_n a_{n-3} = 1989 + a_{n-1} a_{n-2};$$

whence we deduce

$$a_{n+1}a_{n-2} - a_n a_{n-1} = a_n a_{n-3} - a_{n-1} a_{n-2} \Leftrightarrow \frac{a_{n+1} + a_{n-1}}{a_n} = \frac{a_{n-1} + a_{n-3}}{a_{n-2}},$$

equality which is true for $n \geq 3$. Descending the indices we have

$$\frac{a_{n+1}+a_{n-1}}{a_n} = \frac{a_{n-1}+a_{n-3}}{a_{n-2}} = \cdots = \frac{a_3+a_1}{a_2} = 200 \text{ for even } n \text{ and}$$

$$\frac{a_{n+1}+a_{n-1}}{a_n} = \frac{a_{n-1}+a_{n-3}}{a_{n-2}} = \cdots = \frac{a_4+a_2}{a_3} = 11 \text{ for odd } n.$$

The conclusion follows now by ascending induction, separately for odd terms and for even terms.

36. Let $w_1, \ldots, w_n$ be the given weights and consider the integers

$$w_1, w_1+w_2, \ldots, w_1+w_2+\cdots+w_n.$$

If one of these numbers is divisible by n, then it is equal to n.

If none of these numbers is divisible by n, then there are two congruent modulo n (by the pigeonhole principle), so their difference is divisible by n, hence equal to n.

Note. The result remains valid if the total weight is $2n$.

2.18 1990

37. We shall prove that for any positive integer n, there exists a positive n-digit number a_n containing only 1 and 2. Apply induction on n and suppose that we have a positive n-digit integer a_n consisting only of 1 and 2, which is divisible by 2^n and define a_{n+1} as follows: if $2^{n+1}|a_n$, then append a 2 before a_n, else append a 1. We have $a_1 = 2, a_2 = 12, a_3 = 112, a_4 = 2112, a_5 = 22112, \ldots$.

38. We shall prove that if $n+1$ is a prime number and $p \leq n$ is a prime, then

$$\max\{\nu_p(2), \ldots, \nu_p(n)\} = \max\left\{\nu_p\binom{n}{1}, \ldots, \nu_p\binom{n}{n-1}\right\}.$$

Let $n = a_k p^k + a_{k-1} p^{k-1} + \cdots + a_1 p + a_0$, $1 \leq a_i \leq p-1$, be the representation of n in base p. Since $n+1$ is a prime and $p \leq n$, we have $p > a_0 + 1$. Since $p^k \leq n < p^{k+1}$, it follows that $\nu_p(n) < k+1$. Furthermore, for every i, $2 \leq i \leq n$, we have $\nu_p(i) < k+1$. From $\nu_p(p^k) = k$, it follows that $\max\{\nu_p(2), \ldots, \nu_p(n)\} = k$. On the other hand,

$$\nu_p\binom{n}{i} = \sum_{h=1}^{k}\left(\left\lfloor\frac{n}{p^h}\right\rfloor - \left\lfloor\frac{i}{p^h}\right\rfloor - \left\lfloor\frac{n-i}{p^h}\right\rfloor\right),$$

for every i, $1 \leq i \leq n-1$. We shall prove that $\nu_p\binom{n}{i} \leq k$, for every $i \in \{1, \ldots, n-1\}$ and that there is an index i_0 for which the maximum k is attained. For every h,

$$\left\lfloor \frac{n}{p^h} \right\rfloor - \left\lfloor \frac{i}{p^h} \right\rfloor - \left\lfloor \frac{n-i}{p^h} \right\rfloor \leq 1;$$

whence $\nu_p \binom{n}{i} \leq k$, with equality if and only if

$$\left\lfloor \frac{n}{p^h} \right\rfloor - \left\lfloor \frac{i}{p^h} \right\rfloor - \left\lfloor \frac{n-i}{p^h} \right\rfloor = 1 \text{ for } h \in \{1, 2, \ldots, k\} \quad (1).$$

For $i = (a_k - 1)p^k + a_{k-1}p^{k-1} + \ldots + a_1 p + (a_0 + 1)$, we have $n - i = p^k - 1$ and

$$n - i = (p-1)p^{k-1} + (p-1)p^{k-2} + \cdots + (p-1)p + (p-1)$$

is the representation in base p. From the equalities

$$\left\lfloor \frac{n}{p^h} \right\rfloor = a_k p^{k-h} + a_{k-1} p^{k-h-1} + \cdots + a_h,$$

$$\left\lfloor \frac{i}{p^h} \right\rfloor = (a_k - 1) p^{k-h} + a_{k-1} p^{k-h-1} + \cdots + a_h,$$

$$\left\lfloor \frac{n-i}{p^h} \right\rfloor = (p-1)p^{k-h-1} + (p-1)p^{k-h-2} + \cdots + (p-1)$$

it follows (1).

Conversely, if $n + 1$ is not a prime, choose a prime number p such that $p < n$ and $p \mid n + 1$. The representation of n in base p is $n = a_k p^k + a_{k-1} p^{k-1} + \cdots + a_1 p + + (p - 1)$ and, as above, $\max\{\nu_p(2), \ldots, \nu_p(n)\} = k$ and $\nu_p \binom{n}{i} < k$.

39. If $p = 2$, then $q = 13$. Suppose that $p \geq q$ is odd, then $\gcd(p, q - 1) = 1$. Let a, b be positive integers such that $ap - b(q - 1) = 1$. From $2^p \equiv -3^p \pmod{q}$, we deduce $2^{ap} \equiv -3^{ap} \pmod{q}$, since a is odd, and $2^{b(q-1)+1} \equiv -3^{b(q-1)+1} \pmod{q}$. From the Fermat's theorem, we have $2^{q-1} \equiv 3^{q-1} \equiv 1 \pmod{q}$; whence it follows $2 \equiv -3 \pmod{q}$, contradiction with $q > 5$.

40. If $s_2(m)$ is the sum of digits of m in binary representation, then $\nu_2(m!) = m - s_2(m)$, so that we are looking for the smallest m with $s_2(m) = 1990$, and it is obvious that this is $m = 2^{1990} - 1$.

2.19 1991

41. Let n be a positive integer such that

$$\frac{a+1}{b} + \frac{b}{a} = n \quad (1).$$

If $a = b$ or $a = 1$ or $b = 1$, then $n = 3$. Suppose that there are triples (a, b, n) with $n \neq 3$ which verify (1); then $a > 1$, $b > 1$. Choose such a triple with the smallest $a + b$. If $a < b$, then the triple $(a, na - b, n)$ also satisfies (1) and $0 < na - b < b$:

$$na - b = \frac{a^2 + a}{b} < b + 1 \Rightarrow 0 < na - b \leq b.$$

If $na - b = b$, then $b^2 = a(a+1)$, which is impossible. If $a > b$, then the triple $(nb - a - 1, b, n)$ verifies (1) and

$$0 < nb - a - 1 = \frac{b^2}{a} < a,$$

which contradicts the minimality of the triple.

42. If $\gcd(a_m, a_n) = a_{\gcd(m,n)}$, then $n|m \Rightarrow a_n | a_m$. We shall prove that if there is a sequence (b_n) having the required property, then

$$\prod_{d|n, d<n} b_d = \mathrm{lcm}(a_d | \ d < n, d|\ n) \quad (1).$$

Denote by M_n the respective lcm. If $f|n$ and $f < n$, then $a_f | \prod_{d|n, d<n} b_d$, since $a_f = \prod_{d|f} b_d$. We deduce $M_n | \prod_{d|n, d<n} b_d$. In order to prove that $\prod_{d|n, d<n} b_d | M_n$, we shall use the following property:

$$k < l, k \nmid l \Rightarrow \gcd(b_k, b_l) = 1 \quad (2).$$

Indeed, from $k < l$ and $k \nmid l$, it follows that $\gcd(k, l) < k$. If $d | \gcd(k, l)$, then $d|k$, $d < k$ and

$$b_k = \frac{a_k}{\prod_{d|k, d<k} b_d} \text{ divides } \frac{a_k}{\prod_{d|(k,l)} b_d} = \frac{a_k}{a_{\gcd(k,l)}} = \frac{a_k}{\gcd(a_k, a_l)}.$$

Analogously, $b_l | \frac{a_l}{\gcd(a_k, a_l)}$, therefore $\gcd(b_k, b_l)$ divides $\gcd\left(\frac{a_k}{\gcd(a_k, a_l)}, \frac{a_l}{\gcd(a_k, a_l)}\right) = 1$.

The proof of (1) is finished if we show that every prime factor of $\prod_{d|n, d<n} b_d$ appears in M_n with a greater exponent. Suppose that p is a prime, $p^r | \prod_{d|n, d<n} b_d$ and p appears with the exponents $r_1, \ldots, r_s$ in $b_{d_1}, b_{d_2}, \ldots, b_{d_s}$ with $d_1 < d_2 < \cdots < d_s$ and $\sum_1^s r_i = r$. It follows that $d_1 | d_2, d_2 | d_3, \ldots, d_{s-1} | d_s$. Indeed, if there is an index $i \in \{1, \ldots, s-1\}$ such that $d_i \nmid d_{i+1}$, then $\gcd(b_{d_i}, b_{d_{i+1}}) = 1$, in contradiction with the fact that $p | b_{d_i}$ and $p | b_{d_{i+1}}$. Since $b_{d_1} b_{d_2} \cdots b_{d_s} | a_{d_s}$, it follows that $p^r | a_{d_s}$, which proves (1).

43. The sequence (a_n) can be defined by $a_n = k b_n + c_n$ for $n \geq 1$, where (b_n), (c_n) are the sequences defined by

$$b_1 = b_2 = 0, b_{n+1} = b_n + b_{n-1} + 1, n \geq 2 \quad (1)$$

$$c_1 = c_2 = 1, c_{n+1} = c_n + c_{n-1}, n \geq 2 \quad (2).$$

From this we deduce that $(b_n + 1)$ and c_n are the Fibonacci sequence (f_n), so that

$$a_n = k(f_n - 1) + f_n \quad (3).$$

From $1991 = 11 \cdot 181$, it follows that it suffices to see in what conditions $11 | a_{1991}$ or $181 | a_{1991}$, since 11 and 181 are prime numbers. We know that

$$2^{n-1} f_n = \binom{n}{1} + \binom{n}{3} \cdot 5 + \binom{n}{5} \cdot 5^2 + \cdots + \binom{n}{2t+1} \cdot 5^t + \cdots \quad (4).$$

If $n = p > 5$ is a prime number, from $2^{p-1} \equiv 1 \pmod{p}$ and (4), we deduce

$$f_p \equiv 5^{\frac{p-1}{2}} \pmod{p} \quad (5).$$

For $n = p + 1$ in (4), we have $2 f_{p+1} \equiv \left(1 + 5^{\frac{p-1}{2}}\right) \pmod{p}$; hence

$$f_{p+1} \equiv \frac{p+1}{2}\left(1 + 5^{\frac{p-1}{2}}\right) \pmod{p} \quad (6).$$

From $5^{p-1} \equiv 1 \pmod{p}$, it follows that $5^{\frac{p-1}{2}} \equiv \pm 1 \pmod{p}$. If $p = 11$ or $p = 181$, then

$$5^{\frac{p-1}{2}} \equiv 1 \pmod{p} \Rightarrow f_p \equiv 1 \pmod{p} \text{ and } f_{p+1} \equiv 1 \pmod{p}.$$

This shows that the sequence $(f_n \pmod{p})$ is periodic with period $p-1$ and $f_n \equiv f_r \pmod{p}$, where r is the remainder of division of n by $p-1$. Then $f_{1991} \equiv 1 \pmod{11}$ and from (3), it follows that

$$a_{1991} \equiv 1 \pmod{11} \Rightarrow \gcd(11, a_{1991}) = 1.$$

We search k such that $181 | a_{1991}$. We have $5^{90} \equiv 1 \pmod{181}$ and as above $f_{1991} \equiv f_r \pmod{181}$, where $r = 11$ is the remainder of division of 1991 by 180. Then

$$f_{1991} \equiv f_{11} \equiv 89 \pmod{181} \Rightarrow a_{1991} \equiv 88k + 89 \equiv 88(k+1) + 1 \pmod{181}.$$

The condition $a_{1991} \equiv 0 \pmod{181}$ implies $k + 1 = 109$; therefore $k = 108$ is the smallest positive integer solution of the congruence.

44. Compute the values for $n \leq 8$. We have

$$f(1) = 1, \ f(2) = 3, \ f(3) = 4, \ f(4) = 5, \ f(5) = 6, \ f(6) = 6, \ f(7) = 8, \ f(8) = 7.$$

If we ever reach $1 < f_s(n) \leq 8$, we either will have $f_t(n) = 6$ for some $t \geq s$, and then $f_{t+k}(n) = 6$ for all $k \geq 0$ or we will have $f_t(n) = 7$ for some $t \geq s$, and then

$$f_{t+k}(n) + f_{t+k+1}(n) = 15, \text{ for all } k \geq 0.$$

Now, for $n \geq 9$, if n is a prime, we will have $f(n) = 1 + n$. We also have $f(9) = 7$, so we check that $f(n) \leq 1 + n$ for all $n \leq 9$. If n is not a prime, then $n = ab$, with $a, b \geq 2$, $\gcd(a, b) = 1$ and

$$f(ab) = f(a) + f(b) - 1 \leq (a+1) + (b+1) - 1 \leq ab + 1.$$

Thus we have proved by induction that $f(n) \leq 1 + n$ for all n. This result can be improved:

$$f(ab) \leq a + b + 1 \leq ab - 2,$$

since it is equivalent to $(a-1)(b-1) \geq 4$, true for $n \geq 9$ (if $a = 2$ or $b = 2$, then the other is at least 5 and if $a, b \geq 3$, the inequality is true).

Finally, if n is a prime, we have $f(n) = 1 + n$, which is not a prime; therefore $f(1+n) \leq (1+n) - 2$; hence $f(f(n)) \leq n - 1$, while if n is composite, we have $f(n) \leq n - 2$ and so $f(f(n)) \leq 1 + f(n) \leq n - 1$. This means $f(f(n)) \leq n - 1$ for $n \geq 9$; therefore after sufficiently many steps, we will reach $f_s(n) \leq 8$ for some s, and the above analysis applies.

2.20 1992

45. The first condition is equivalent to $p|(2x_0 - 1)^2 + 11$, which means that -11 is a quadratic residue mod p, and the second is equivalent to $p|(2y_0 - 1)^2 + 99$, which means that $-99 = 3^2(-11)$ is a quadratic residue mod p.

46. We shall prove that for every $n \geq 1$, x_{n+2} is the square of the element from the first row and the first column of the matrix

$$A^n = \begin{pmatrix} a^2 + 1 & a \\ a & 1 \end{pmatrix}^n.$$

We have $A = B^2$, where $B = \begin{pmatrix} a & 1 \\ 1 & 0 \end{pmatrix}$. Consider the linear recurrence: $y_0 = 0$, $y_1 = 1$ and $y_{n+2} = ay_{n+1} + y_n$ for $n \geq 0$. The first terms of this sequence are $y_2 = a$, $y_3 = a^2 + 1$, and $y_4 = a^3 + 2a$, and by induction one can prove that

$$\begin{pmatrix} a & 1 \\ 1 & 0 \end{pmatrix} \begin{pmatrix} y_{n+1} & y_n \\ y_n & y_{n-1} \end{pmatrix} = \begin{pmatrix} y_{n+2} & y_{n+1} \\ y_{n+1} & y_n \end{pmatrix}.$$

From this, we deduce by induction that for every $n \geq 1$,

$$\begin{pmatrix} y_{n+1} & y_n \\ y_n & y_{n-1} \end{pmatrix} = B^n. \quad (1)$$

Consequently, we have to prove that $x_{n+2} = y_{2n+1}^2$ for every $n \geq 1$. We see that $x_2 = y_1^2 = 1$, $x_3 = y_3^2 = (a^2 + 1)^2$. It suffices to prove that the sequence $\left(y_{2n-3}^2\right)_{n \geq 2}$ verifies the recurrence relation of the sequence (x_n), namely,

$$y_{2n+1}^2 = (a^4 + 4a^2 + 2)y_{2n-1}^2 - y_{2n-3}^2 - 2a^2. \quad (2)$$

Calculating determinants in (1), we get

$$y_{n+1}y_{n-1} - y_n^2 = (-1)^n.$$

Write this equality for $2n$:

$$y_{2n+1}y_{2n-1} - y_{2n}^2 = 1. \quad (3)$$

The recurrence relation for (y_n) gives us

$$y_{2n} = \frac{y_{2n+1} - y_{2n-1}}{a}. \quad (4)$$

Replacing (4) in (3), we get

2.20 1992

$$y_{2n+1}^2 + y_{2n-1}^2 - (a^2 + 2)y_{2n+1}y_{2n-1} + a^2 = 0. \quad (5)$$

According to (3) from (5), we obtain

$$y_{2n+1}^2 + y_{2n-1}^2 = (a^2 + 2)y_{2n}^2 + 2. \quad (6)$$

Writing (2) as $y_{2n+1}^2 + y_{2n-1}^2 = (a^4 + 4a^2 + 3)y_{2n-1}^2 - y_{2n-3}^2 - 2a^2$, it follows that we have to prove the equality

$$(a^2 + 2)^2 y_{2n-1}^2 - (y_{2n-1}^2 + y_{2n-3}^2) - 2a^2 = (a^2 + 2)y_{2n}^2 + 2. \quad (7)$$

We use (6) and (7) to express $y_{2n-1}^2 + y_{2n-3}^2$, and then (3) becomes

$$(a^2 + 2)^2 y_{2n-1}^2 - (a^2 + 2)y_{2n-2}^2 = (a^2 + 2)y_{2n}^2 + 2(a^2 + 2). \quad (8)$$

Simplify by $a^2 + 2$:

$$(a^2 + 2)y_{2n-1}^2 - y_{2n-2}^2 = y_{2n}^2 + 2. \quad (9)$$

Replacing $a^2 y_{2n-1}^2 = (y_{2n} - y_{2n-2})^2$ in (9), we have

$$y_{2n-1}^2 = y_{2n} y_{2n-2} + 1, \quad (10)$$

which is proved if we calculate the determinant of the matrix B^{2n-1}.

47. Induction on n proves that

$$a_{n+1} = n!\left(a_1 + \frac{1}{1!} + \frac{1}{2!} + \cdots + \frac{1}{n!}\right),$$

$$b_{n+1} = n!\left(b_1 - \frac{1}{1!} - \frac{1}{2!} - \cdots - \frac{1}{n!}\right).$$

The sequence

$$e_n = \frac{1}{1!} + \frac{1}{2!} + \cdots + \frac{1}{n!}, n \geq 1,$$

is strictly increasing and convergent to $e - 1$; moreover, $1 \leq e_n < 2$. Let (n, m) a pair of positive integers such that $a_{n+1} = b_{m+1}$. From

$$n!(a_1 + e_n) = m!(b_1 - e_m),$$

it follows

$$\frac{n!}{m!} = \frac{b_1 - e_m}{a_1 + e_n}.$$

The following implication holds

$$\frac{b_1 - 1}{a_1 + 2} < \frac{b_1 - e_m}{a_1 + e_n} < \frac{b_1}{a_1} \Longrightarrow \frac{b_1 - 1}{a_1 + 2} < \frac{n!}{m!} < \frac{b_1}{a_1}.$$

The required result is a consequence of the above inequalities and of a known property: "For every pair (M_1, M_2) of positive real numbers, the set of pairs (n, m) of positive integers such that $M_1 < \frac{n!}{m!} < M_2$ is finite."

48. Answer: no.

In any set of consecutive integers, we can ignore seven even numbers. We shall prove that at least one of the remaining seven integers is not divisible by 3, 5, 7, or 11. Indeed, at most three of these odd integers are divisible by 3, at most two are divisible by 5, at most one is divisible by 7, and at most one is divisible by 11. It follows that each of these numbers can be divisible by 3, 5, 7, or 11 only if each of these primes divides its maximum number of terms, and no term is divisible by a product of these primes. In order to get three terms divisible by 3, they must be the first term, the middle term and the last term. Since the difference between any two of the remaining terms is less than 10, only one of these can be divisible by 5, which ends the proof.

49. The equation is equivalent to $(2x)^2 = (2y + p)^2 - p^2$; therefore any solution (x, y) belongs to a Pythagorean triple. It follows that there are positive integers m, n such that $2x = 2mn$, $p = m^2 - n^2$, $2y + p = m^2 + n^2$. But $p = (m - n)(m + n)$ is an odd prime; therefore $m - n = 1$, $m + n = p$; whence it follows

$$m = \frac{p+1}{2}, n = \frac{p-1}{2}.$$

Thus, the given equation has the unique solution

$$x = mn = \frac{p^2 - 1}{4}, y = \frac{m^2 + n^2 - (m^2 - n^2)}{2} = n^2 = \left(\frac{p-1}{2}\right)^2.$$

50. From the given equation, we get

$$x = \frac{5y \pm \sqrt{21y^2 - 20}}{2}.$$

Thus $21y^2 - 20 = s^2$; whence $s^2 - 21y^2 = -20$. This equation has the fundamental solution $(1, 1)$. To get all solutions, we solve the equation $s^2 - 21y^2 = 1$, which has the fundamental solution $(s_0, y_0) = (55, 12)$. All solutions can be found by

$$s_1 = 55 + 12 \cdot 21 = 307, y_1 = 12 + 55 = 67,$$
$$s_p = 55s_{p-1} + 252y_{p-1}, y_p = 12s_{p-1} + 55y_{p-1},$$
$$x_p = \frac{5y_p \pm \sqrt{21y_p^2 - 20}}{2}.$$

2.21 1993

51. We have

$$2F(p) = \sum_{k=1}^{\frac{p-1}{2}} \left(k^{120} + k^{120}\right) \equiv \sum_{k=1}^{\frac{p-1}{2}} k^{120} + \sum_{k=1}^{\frac{p-1}{2}} (p-k)^{120}$$
$$\equiv \sum_{k=1}^{p-1} k^{120} \pmod{p}.$$

Let r be a primitive root mod p, that is, $ord_p(r) = p - 1$. Then the sequence $r, r^2, \ldots, r^{p-1}$ is a permutation of $1, 2, \ldots, p-1$; therefore

$$F(p) = \frac{1}{2}\sum_{k=1}^{p-1} k^{120} \equiv \frac{1}{2}\sum_{i=1}^{p-1} r^{120i} \equiv \frac{r^{120}\left(r^{120(p-1)} - 1\right)}{2(r^{120} - 1)} \pmod{p}.$$

If $p - 1 | 120$, then $r^{120} \equiv 1 \pmod{p}$, so that

$$F(p) \equiv \frac{p-1}{2} \pmod{p},\ f(p) = \frac{1}{2} - \frac{p-1}{2p} = \frac{1}{2p}.$$

If $p - 1 \nmid 120$, $r^{120} \not\equiv 1 \pmod{p}$, $r^{120(p-1)} \equiv 1 \pmod{p}$, so that $p | F(p)$ and $f(p) = \frac{1}{2}$.

52. Answer: $(x, y) = (0, -1), (0, 1)$.

For $|y| < 2$ we have $x = 0$.

Suppose that $x > 0, y > 2$. Then $4 \lfloor y^2 - 1$. Replace $x = 2z$ to obtain

$$32z^4 = (y-1)(y+1).$$

Case 1. $y \equiv -1 \pmod{4}$. Then $y - 1 = 2t^4$, $y + 1 = 16s^4$, with $\gcd(s, t) = 1$. It follows that $8s^4 = t^4 + 1 \equiv 2 \pmod{8}$, contradiction.

Case 2. $y \equiv 1 \pmod{4}$. Then $y - 1 = 16s^4$, $y + 1 = 2t^4$, with $\gcd(s, t) = 1$; hence $t^4 - 1^4 = 2(2s^2)^2$. By infinite descent, it is easy to prove that the equation

$$x^4 - y^4 = 2z^2$$

has no integer solutions.

53. After squaring both sides, we obtain

$$(p^2 + 7pq + q^2)(p^2 + 14pq + q^2) = a^2, a \in \mathbb{N}^*.$$

If $p \neq q$, we can suppose that $p > q$, due to symmetry of the expressions. Observe that

$$\gcd(p^2 + 7pq + q^2, p^2 + 14pq + q^2) = \gcd(p^2 + 7pq + q^2, 7pq) =$$
$$= \gcd(p^2 + q^2) \in \{1, 7, mp, nq\}, m, n \in \mathbb{N}^*.$$

If gcd $=7$, then $p^2 + q^2 \equiv 0 \pmod{7}$ implies $p|7$, $q|7$, contradiction.
If gcd $\in \{mp, nq\}$, then $p|q$ or $q|p$, contradiction.
We deduce that gcd $=1$, therefore both factors are perfect squares,

$$p^2 + 7pq + q^2 = u^2, p^2 + 14pq + q^2 = v^2, u, v \in \mathbb{N}^*.$$

54. Suppose that $x^2 + y^2 + z^2 = 1993$ and $x + y + z = a^2$, with a positive integer. The Cauchy-Schwarz inequality implies

$$a^2 = x + y + z \leq \sqrt{3(x^2 + y^2 + z^2)} = \sqrt{3 \cdot 1993} < 78.$$

We deduce $a^2 \in \{1, 4, 9, \ldots, 64\}$. From $x^2 \equiv x \pmod{2}$ and $x^2 + y^2 + z^2 \equiv 1 \pmod{2}$, it follows that $x + y + z \equiv 1 \pmod{2}$; hence $a^2 \in \{1, 9, 25, 49\}$.

Since $(x + y + z)^2 > x^2 + y^2 + z^2$ and $25^2 = 625 < 1993$, it follows that $a^2 = 49$. Now we have the diophantine system of equations in positive integers:

$$x^2 + y^2 + z^2 = 1993, x + y + z = 49.$$

We shall prove that this system has no solution. Indeed,

$$y + z = 49 - x, (49 - x)^2 = (y + z)^2 > y^2 + z^2 = 1993 - x^2,$$

hence x satisfies the inequality $x^2 - 49x + 204 > 0$. It follows that

$$x > \frac{49 + \sqrt{1585}}{2} > \frac{49 + 39}{2} = 44,$$

which is impossible, since $45^2 = 2025$, or

$$x < \frac{49 - \sqrt{1585}}{2} < \frac{49 - 39}{2} = 5.$$

But $x \leq 4$, $y \leq 4$, and $z \leq 4$ imply $x + y + z \leq 12 < 49$, contradiction.

55. The first term $16 = 2^4$ has 5 divisors. The next number with 5 divisors is $3^4 = 81$, and the next number with 10 divisors is $48 = 2^4 \cdot 3$, so that the minimum common difference is $r = 48 - 16 = 32$. The general term $a_k = 2^4(1 + 2k)$ has a number of divisors divisible by 5, since $1 + 2k$ is odd:

$$\tau(a_k) = \tau(2^4)\tau(1 + 2k) = 5\tau(1 + 2k).$$

56. We have

$$m - 1 = \frac{4^p - 4}{3} = 4 \cdot \frac{(2^{p-1} + 1)(2^{p-1} - 1)}{3}$$

and we see that $2^{p-1} - 1$ is divisible by 3, and the Fermat's theorem shows that it is divisible by p; hence it is divisible by $3p$. It follows that $2p | m - 1$.

We deduce that $2^{m-1} - 1$ is divisible by $2^{2p} - 1$, which is divisible by m.

Generalization. Let a be an even positive integer, and let p be an odd prime number such that $\gcd(a^2 - 1, p) = 1$. Then $a^{n-1} - 1$ is divisible by $n = \frac{a^{2p} - 1}{a^2 - 1}$.

57. We see that $f(n) = k\varphi(n)$. It is easy to show that there exists a positive integer N such that all prime divisors of x_i, $i > N$, are primes $p \leq p_0$, where p_0 is the maximal prime divisor of k.

If $k = 2$, then $x_{i+1} = x_i$ for all $i > N$, so that the sequence is bounded.

If $k = 3$, then $x_i = 2^{m_i} 3^{n_i}$ for $i > N$ and the sequence is bounded.

If $k > 3$, then

$$k \prod_{p \leq p_0} \left(1 - \frac{1}{p}\right) > 1,$$

therefore the sequence is not bounded.

2.22 1994

58. If $z > 0$, then $7^x \equiv 3^y + 4 \pmod{20}$, which implies $x \equiv y \equiv 1 \pmod 4$. Hence $5^z \equiv 5 \pmod{16}$, so that $z \equiv 1 \pmod 4$. Examining the equation mod 13, we find that $y \equiv 1 \pmod 3$ and $x \equiv 1 \pmod{12}$. Therefore, $y \equiv 1 \pmod 6$ and $5^z \equiv 5 \pmod 7$ and hence $z \equiv 1 \pmod 6$. Taking mod 9, if $y > 1$, we have no solution. If $y = 1$, $z > 1$, we arrive at a contradiction by taking mod 25.

59. If $m-1 = \overline{a_1\ldots a_d}$, then $(10^n - 1)m = 10^n \cdot \overline{a_1\ldots a_d} + \underbrace{9\ldots 9}_{n} - \overline{a_1\ldots a_d}$ and the sum of digits is $9n$. More generally, if a, b, and n are positive integers, $b \geq 2$ and $a < b^n$, and then the sum of digits in base b of the number $(b^n - 1)a$ is $(b-1)n$.

60. Answer: $p = 3, 7$.
    Put $p = 2q + 1$, and then $2^{2q} - 1 = (2^q - 1)(2^q + 1)$. Since $\gcd(2^q - 1, 2^q + 1) = 1$, we have two cases:
    Case 1. $2^q - 1 = px^2$, $2^q + 1 = y^2$. From $(y+1)(y-1) = 2^q$, it follows $y + 1 = 2^m$, $y - 1 = 2^n$; hence $2^m - 2^n = 2$, which implies $m = 2, n = 1$, so that $q = 3, p = 7$.
    Case 2. $2^q - 1 = x^2$, $2^q + 1 = py^2$. For $q = 1$, we have $p = 3$. For $q \geq 2$, since $2^q - 1 \equiv 3 \pmod{4}$, it cannot be a square.

61. The sequence $(a_n)_{n \geq 1}$ is strictly increasing, since
$$a_{n+1} - a_n = 2 \cdot 3^n - 2^n > 0.$$

We shall prove the following double inequality:
$$a_m a_{2n-m} < a_n^2 < a_m a_{2n-m+1}.$$

This proves that there is no k such that $a_n^2 = a_m a_k$. The first inequality follows from
$$(3^n - 2^n)^2 - (3^m - 2^m)(3^{2n-m} - 2^{2n-m}) = (3^{n-m} - 2^{n-m})^2 > 0.$$

The second is also true, since
$$(3^m - 2^m)(3^{2n-m+1} - 2^{2n-m+1}) - (3^n - 2^n)^2 > 0 \Leftrightarrow$$
$$\Leftrightarrow 2 \cdot 3^{2n-m+1}(3^{m-1} - 2^{m-1}) + 2^{2n} + 2 \cdot 6^n - 3^m \cdot 2^{2n-m+1} > 0 \Leftrightarrow$$
$$(2 \cdot 3^{2n-m+1} - 2^{2n-m+1})(3^{m-1} - 2^{m-1}) + 2^{2n-m+1} \cdot 3^{m-1} + 2 \cdot 6^n - 3^m \cdot 2^{2n-m+1} > 0$$
$$\Leftrightarrow (3^{m-1} - 2^{m-1})(2 \cdot 3^{2n-m+1} - 2^{2n-m+1}) + 2^{n+1} \cdot 3^{m-1}(3^{n-m+1} - 2^{n-m+1}) > 0.$$

Remark. If $(b_n)_{n \geq 1}$ is a sequence of positive integers such that $\gcd(b_m, b_n) = b_{\gcd(m,n)}$ for every pair (m, n) of positive integers and $b_n^2 < b_1 b_{2n}$ for every positive integer n, then the sequence does not contain three terms in geometric progression.

62. Clearly, $ab \neq 0$. If (a, b) is a solution, then $(\pm a, \pm b)$, $(\pm b, \pm a)$ are solutions. Consider the solutions (a, b) in positive integers with $a \geq b$.

If (a, b) is a solution with $a > b > 1$, then $tab = a^2 + b^2 + 3 > a^2$; whence $tb > a$. Also, $a(tb - a) = b^2 + 3 \leq (a-1)^2 + 3 < a^2$; whence $tb - a < a$. Then $(tb - a, b)$ or $(b, tb - a)$ is a solution and $tb - a + b < a + b$. At the end of this process, we get a solution (r, s), $r \geq s$, with $r = s$ or $s = 1$.

If $r = s$, then $r^2 | 2r^2 + 3$, so that $r = 1$.

If $s = 1$, then $r | r^2 + 4$, so that $r \in \{1, 2, 4\}$.

It follows that the ending solutions are $(1, 1)$, $(2, 1)$, and $(4, 1)$ for which the respective values of t are 5, 4, and 5.

The solutions are as follows:

$$(a_0, b_0) = (1, 1), (a_{n+1}, b_{n+1}) = (5a_n - b_n, a_n) \text{ and}$$

$$(a'_0, b'_0) = (2, 1), (a'_{n+1}, b'_{n+1}) = (4a'_n - b'_n, a'_n).$$

63. Answer: 337.

From the condition

$$\frac{1^2 + 2^2 + \cdots + n^2}{n} = m^2,$$

we get the equation in positive integers

$$(n+1)(2n+1) = 6m^2.$$

Since $\gcd(n+1, 2n+1) = 1$ and $2n+1$ is odd, we must have

$$n + 1 = 2u^2, 2n + 1 = 3v^2,$$

so that

$$(2u)^2 - 3v^2 = 1.$$

The general solution in integers of this equation is

$$2u = \frac{(2+\sqrt{3})^k + (2-\sqrt{3})^k}{2},$$

where k is odd. The smallest $u > 1$ is

$$u = \frac{(2+\sqrt{3})^3 + (2-\sqrt{3})^3}{4} = \frac{64 - 12}{4} = 13,$$

therefore $n = 2 \cdot 13^2 - 1 = 337$.

64. (a) If $N = a^2 + 2$, then $(x, y, z, t) = (a^3 + 2a, a, 1, 1)$ is a solution.
    (b) If $a^2 + b^2 + c^2 + d^2 \equiv 0 \pmod{4}$, then a, b, c, and d are all even or all odd. Assume that for $N = 4^k(8t + 7)$, the equation has a solution (a, b, c, d). After k steps, we deduce

    $$a'^2 + b'^2 + c'^2 + d'^2 = (8t + 7)(2^{4k}a'b'c'd' + 1) \equiv 7 \pmod{8},$$

    contradiction.

2.23 1995

65. The answer is $p = 41$. We have

    $$2 = |3^1 - 2^0|, 3 = |3^0 - 2^2|, 5 = |3^2 - 2^2|, 7 = |3^2 - 2^1|, 11 = |3^3 - 2^4|, 13 = |3^1 - 2^4|,$$
    $$17 = |3^4 - 2^6|, 19 = |3^3 - 2^3|, 23 = |3^3 - 2^2|, 29 = |3^1 - 2^5|, 31 = |3^0 - 2^5|,$$
    $$37 = |3^3 - 2^6|.$$

    Case 1. $3^a - 2^b = -41$. Then $b > 2$, so that $3^a \equiv 7 \pmod 8$, which is impossible, since the remainders modulo 8 of 3^a are only 1 or 3.

    Case 2. $3^a - 2^b = 41$. It is easy to see that $a > 0$ and $b > 1$, so that $2^b \equiv 3^a - 41 \equiv 1 \pmod 3$ and $3^a \equiv 2^b + 41 \equiv 1 \pmod 4$. Thus, a, b are even, say $a = 2m, b = 2n$; whence $41 = 3^a - 2^b = (3^m - 2^n)(3^m + 2^n)$; therefore $3^m - 2^n = 1, 3^m + 2^n = 41$. But then $3^m = 21$, impossible.

    This shows that for $p = 41$ such a representation is impossible.

66. Answer: $(5, 3, 2)$ and circular permutations.

    Suppose, without loss of generality, that $x \geq y \geq z > 1$. From the given condition, it follows that x, y, z are pairwise coprime. Since x, y, z divide $xy + yz + zx - 1$, we have $xyz \leq xy + yz + zx - 1$. This condition is equivalent to

    $$(x-1)(y-1)(z-1) \leq (x-1) + (y-1) + (z-1) + 1,$$

    whence it is easy to deduce the result.

67. A composite number which satisfies condition (a) is a Carmichael number. The equivalence of conditions (a) and (b) is Korselt's criterion (1899).

    An example of such number is $n = 561 = 3 \cdot 11 \cdot 17$.

68. Answer: $(7, 6, 2), (9, 5, 2), (15, 4, 2), (8, 3, 3)$, and $(5, 4, 3)$ and all permutations.

    Assume that $x \geq y \geq z$. The inequality $2 \leq \left(1 + \frac{1}{z}\right)^3$ implies $z \leq 3$.

    Case 1. $z = 1$. Then $\left(1 + \frac{1}{x}\right)\left(1 + \frac{1}{y}\right) = 1$, which is impossible.

Case 2. $z = 2$. Then $\left(1 + \frac{1}{x}\right)\left(1 + \frac{1}{y}\right) = \frac{4}{3}$; therefore $\frac{4}{3} \leq \left(1 + \frac{1}{y}\right)^2$, which implies $y < 7$. Since $1 + \frac{1}{x} > 1$, we must have $y > 3$. Easy verifications yield the solutions $(7, 6, 2)$, $(9, 5, 2)$, and $(15, 4, 2)$.

Case 3. $z = 3$. Then $\left(1 + \frac{1}{x}\right)\left(1 + \frac{1}{y}\right) = \frac{3}{2}$; whence we get analogously $y < 5$. From $y \geq 3$, direct analysis leads to the solutions $(8, 3, 3)$ and $(5, 4, 3)$.

69. It is easy to see that $4(ab)^{1995} = (a^3 + b^3)^n \geq \left(2\sqrt{(ab)^3}\right)^n$ and since $a, b \geq 2$, it follows that $3n \leq 2 \cdot 1995$. For $\gcd(a, b) = d$, $a = dx$, $b = dy$, the equation becomes

$$d^{3n}(x^3 + y^3)^n = 4d^{3990}(xy)^{1995} \text{ or } (x^3 + y^3)^n = 4d^{3990-3n}(xy)^{1995}.$$

As $\gcd(x, y) = 1$ and $y \mid (x^3 + y^3)^n$, therefore $y = 1$ and similarly, $x = 1$. Then

$$2^n = 4d^{3990-3n}.$$

Now we are looking for positive integers n such that $3990 - 3n \mid n - 2$. But every common divisor of the numbers $3990 - 3n$ and $n - 2$ is a divisor of

$$3990 - 3n + 3(n - 2) = 3984 = 2^4 \cdot 3 \cdot 83.$$

Checking all possibilities, we find that all solutions (a, b, n) are

$$(2, 2, 998), \left(2^5, 2^5, 1247\right), \left(2^{55}, 2^{55}, 1322\right), \left(2^{221}, 2^{221}, 1328\right).$$

70. Clearly, $(0, 0, 0)$ is a solution. Assume that $n > 0$, $2^k \leq n < 2^{k+1}$, with $k \geq 0$. Then $(n + p + q, n)$ nice implies $n + p + q \equiv -1 \pmod{2^{k+1}}$. Let $i = \nu_2(n + 1)$. Then $0 \leq i \leq 2k + 1$.

Case 1. $i = 2k + 1$. Since $n < 2^{k+1}$, we have $n = 2^{k+1} - 1$. Since $2^{f(n)} = 2^{k+1} > p$ and $2^{k+1} \leq 2^{f(p)} < q$, we have $n + p + q < 3 \cdot 2^{k+1} - 1$. Hence $n + p + q \equiv -1 \pmod{2^{k+1}}$ implies $n + p + q = 2^{k+1} - 1$ or $2^{k+2} - 1$. In the first case, we have $(p, q) = (0, 0)$. In the second case, $p + q = 2^{k+1}$. If p is even, then $q = 0$, but $2^{f(n)} \not> 2^{k+1}$. If $p = 2^{k+1} - 1$, then $q = 1$; otherwise $p = 2^{k+1} - (2^t j + 1)$, $q = 2^t j + 1$, j odd. But $2^{f(p)} = 2^t > q$, contradiction. The solutions are $(n, p, q) = (2^{k+1} - 1, 0, 0), (2^{k+1} - 1, 2^k - 1, 1)$.

Case 2. $i = 0$. Then n is even, so $2^{f(n)} = 1 > p$. Thus, we have $p = 0$, so $2^{f(p)} = 1 > q$, hence $q = 0$ and $n + p + q$ is even. No solution.

Case 3. $2k + 1 > i > 0$. Let $n = 2^{k+1} - (2v + 1)2^i - 1$. Now, we have $2^{f(n)} = 2^i > p$ and $2^{f(p)} \geq 2^i > q$. Thus,

$$n + p + q < 2^{k+1} - (2v + 1)2^i - 1 + 2^i + 2^i = 2^{k+1} - (2v - 1)2^i - 1.$$

Since $2^{k+1} - 1 \leq n + p + q$, we must have

$$2^{k+1} - 1 < 2^{k+1} - (2v-1)2^i - 1 \Rightarrow (2v-1)2^i < 0 \Rightarrow v = 0.$$

So, we have $n = 2^{k+1} - 2^i - 1$. Then since $n + p + q \equiv -1 \pmod{2^{k+1}}$ and $2^{f(n)} = 2^i > p \geq q$, we must have $p + q = 2^i$. If p is even, then $q = 0$ and so $p = 2^i$, but $2^{f(n)} = 2^i \not> p$.

If $p = 2^i - 1$, then $q = 1$, otherwise $p = 2^i - (2^t j + 1)$, $q = 2^t j + 1$, where j is odd. But $2^{f(p)} = 2^t > q = 2^t j + 1$, contradiction. Thus, we have solutions

$$(n, p, q) = (2^{k+1} - 2^i - 1, 2^i - 1, 1).$$

2.24 1996

71. There is a sequence $a_1, a_2, \ldots, a_{2 \cdot 1996}$ of pairwise coprime numbers that are congruent to 1 mod 2016! (we can choose prime numbers, by Dirichlet's theorem). Then the set

$$X = \{a_1 a_2, a_3 a_4, \ldots, a_{2 \cdot 1996 - 1} a_{2 \cdot 1996}\}$$

satisfies the conditions.

72. The following equality is true

$$\sum_{k=2}^{n} a_k = \sum_{k=2}^{n} \left(\frac{1}{p_1} + \frac{1}{p_2} + \cdots + \frac{1}{p_k} \right) = \sum_{\substack{p \leq n \\ p \text{ prime}}} \frac{1}{p} \left[\frac{n}{p} \right].$$

For every positive integer n, we have

$$\sum_{\substack{p \leq n \\ p \text{ prime}}} \frac{1}{p} \left[\frac{n}{p} \right] \leq \sum_{\substack{p \leq n \\ p \text{ prime}}} \frac{1}{p} \cdot \frac{n}{p} = n \sum_{\substack{p \leq n \\ p \text{ prime}}} \frac{1}{p^2} \leq n \left[\frac{1}{4} + \sum_{k \geq 1} \frac{1}{(2k+1)^2} \right] <$$

$$< n \left[\frac{1}{4} + \frac{1}{4} \sum_{k \geq 1} \frac{1}{k(k+1)} \right] = \frac{n}{2}.$$

By the GM–AM inequality,

$$a_2 a_3 \ldots a_n < \left(\frac{a_2 + a_3 + \cdots + a_n}{n-1}\right)^{n-1} < \frac{1}{2^{n-1}}\left(1 + \frac{1}{n-1}\right)^{n-1} < \frac{3}{2^{n-1}}.$$

Adding these inequalities, we obtain

$$\sum_{n \geq 2} a_2 \ldots a_n < \frac{1}{2} + \frac{1}{6} + \frac{1}{12} + \frac{1}{60} + 3\left(\frac{1}{2^5} + \frac{1}{2^6} + \cdots\right) = \frac{30 + 10 + 5 + 1}{60} +$$
$$+ \frac{3}{2^5}\left(1 + \frac{1}{2} + \cdots\right) = \frac{46}{60} + \frac{6}{32} = \frac{229}{240} < 1.$$

73. Answer: $(0, 1, 2), (3, 0, 3), (4, 2, 5)$.

If $x = 0$, then $3^y = (z+1)(z-1)$, so that $z + 1 = 3^u$, $z - 1 = 3^{y-u}$; whence $3^u - 3^{y-u} = 2$, with unique solution $u = y = 1$; therefore $z = 2$.

If $y = 0$, then $2^x = (z+1)(z-1)$, so that $z + 1 = 2^v$, $z - 1 = 2^{x-v}$; whence $2^v - 2^{x-v} = 2$, with unique solution $x = v + 1 = 3$; therefore $z = 3$.

Suppose that $x \geq 1, y \geq 1$. Then x odd implies $z^2 \equiv 2 \pmod 3$, contradiction and y odd implies $z^2 \equiv 3 \pmod 4$, contradiction. It follows that x, y are both even, $x = 2s, y = 2t$, and $4^s + 9^t = z^2$. Then $4^s = (z + 3^t)(z - 3^t)$; whence $2^w - 2^{2s-w} = 2 \cdot 3^t$, with unique solution $t = 1, w = 3, 2s - w = 1$, so that $x = 4$, $y = 2, z = 5$.

2.25 1997

74. Answer: $m = 588$.

Denote $b_n = 4a_n + 3$. Then

$$b_0 = 7, b_1 = 1351;$$
$$b_{n+1}b_{n-1} - b_n^2 = 16m.$$

The solution of this recurrence relation sis $b_n = ax^n + by^n$, where a, b are constants and x, y are solutions of an equation of the form $x^2 - 2cx + 1 = 0$. Then

$$x, y = c \pm \sqrt{c^2 - 1}, b_{n+1}b_{n-1} - b_n^2 = ab(x^2 + y^2 - 2) = 4ab(c^2 - 1) = 16m.$$

It follows that

$$a, b = \frac{7}{2} \pm \frac{1351 - 7c}{2\sqrt{c^2 - 1}}, 8m = 2ab(c^2 - 1) = 9457c - 912625,$$

whence we deduce $c \equiv 1 \pmod 8$. If c is an integer, then (a) and (b) are satisfied and

$$b_{n+1} = cb_n - b_{n-1}.$$

For the condition (c), if

$$c_n^2 = \frac{(a_n+1)(2a_n+1)}{6},$$

then

$$(4a_n+3)^2 = b_n^2 = \frac{(2+\sqrt{3})^{k(n)} + (2-\sqrt{3})^{k(n)}}{2}, n \in \mathbb{N}.$$

We have $k(0) = 2$, $k(1) = 6$, and

$$c = \frac{(2+\sqrt{3})^l + (2-\sqrt{3})^l}{2}, l \in \mathbb{N}.$$

From $c \equiv 1 \pmod{8}$, it follows that $l = 4k$ and

$$c = \frac{(2+\sqrt{3})^{4k} + (2-\sqrt{3})^{4k}}{2}.$$

Then we have

$$m = \frac{9457\left[(2+\sqrt{3})^{4k} + (2-\sqrt{3})^{4k}\right] - 1{,}825{,}250}{16}.$$

If $k = 1$, then $m = 588$.

75. If $y = 1$, then $(p, x) = (2, 1)$.

If $(y, p) = (2, 3)$, then $x = 2$.

For $y > 1$, $p \neq 3$, by Zsigmondy's theorem, $y^p + 1$ has a prime factor that does not divide $y + 1$. Since $y^p + 1$ is divisible by $y + 1$, it follows that $y^p + 1$ has at least two prime factors, contradiction.

76. From $4 \notin A$ we deduce that $p > 2$. Since p is odd, from $p^2 = a^2 + 2b^2$, we deduce that a is odd, b is even, and $(a, b) = 1$. The equality $(p - a)(p + a) = 2b^2$ implies

$$p - a = 2^m c, p + a = 2^n d, (*)$$

$m, n \geq 1$ and $c, d, m + n$ odd. Adding these equalities, we get

$$2p = 2^m c + 2^n d = 2^{\min\{m,n\}} e,$$

where e is odd. We deduce that $\min\{m, n\} = 1$.

Case 1. $m = 1, n = 2r, r \geq 1$. From $p^2 - a^2 = (p-a)(p+a) = 2c \cdot 2^{2r}d = 2b^2$, we deduce that $2^{2r}cd = b^2$. It is easy to see that $(c,d) = 1$; whence it follows that c, d are squares, $c = u^2, d = v^2$. Then $2p = 2(u^2 + 2^{2r-1}v^2)$ and $p = u^2 + 2(2^{r-1}v^2)^2$.

Case 2. $m = 2s, n = 1, s \geq 1$. From $p^2 - a^2 = 2^{2s}c \cdot 2d = 2b^2$, we obtain analogously, $c = u^2, d = v^2$, and $p = v^2 + 2(2^{s-1}u)^2$.

77. Let L be the set of ordered triples (a, b, c) with $0 \leq a, b, c \leq 5$. Then L is an ordered set under the relation: $(a, b, c) \leq (a_1, b_1, c_1) \Leftrightarrow a \leq a_1, b \leq b_1, c \leq c_1$.

It is obvious that $p^a q^b r^c | p^{a_1} q^{b_1} r^{c_1}$ if and only if $(a, b, c) \leq (a_1, b_1, c_1)$. Now we have to solve the following problem: find the number of elements of a maximal subset $W \subset L$ such that any two elements of W are not comparable, and then add to W an extraelement x.

In this way, the set $W \cup \{x\}$ has the required property and it is minimal with this property. We call such a set W an antichain. A subset $V_k \subset L$ such that for any $(a, b, c) \in V$ the sum $a + b + c = k$ is obviously an antichain of L. Geometrically, V_k is the intersection of the set of laticial points from the cube

$$C = \{(a, b, c) | \ 0 \leq a, b, c \leq 5\}$$

with the plane P_k given by the equation $a + b + c = k$, where $1 \leq k \leq 14$.

For instance, V_1 contains three points $e_1 = (1, 0, 0)$, $e_2 = (0, 1, 0)$, $e_3 = (0, 0, 1)$, which are the unitary vectors of a base in $\mathbb{R}^3$. The antichain V_2 contains six points, which can be obtained as follows: for any $(a, b, c) \in V_1$,

$$(a, b, c) + e_1 = (a + 1, b, c), (a, b, c) + e_2$$
$$= (a, b+1, c), (a, b, c) + e_3 = (a, b, c+1)$$

are the points of V_2. The set V_8 has 27 elements: 3 triples of the form $(0, b, c)$, 4 triples of the form $(1, b, c)$, 5 triples of the form $(2, b, c)$, 6 triples of the form $(3, b, c)$, 5 triples of the form $(4, b, c)$, and 4 triples of the form $(5, b, c)$. The total number is

$$3 + 4 + 5 + 6 + 5 + 4 = 27.$$

There exists a partition of the set L with 27 chains such that every chain has a unique representative from V_8.

In this way, every set with $27 + 1 = 28$ elements contains at least 2 elements which belong to a chain, and hence, these two elements are comparable. Therefore, $n = 28$.

78. The pair $(1, P(1))$ verifies the conditions: $1|P(P(1))$ and $P(1)|P(1)$. Suppose that only a finite number of pairs (x, y) verify the conditions and choose a pair with $x \leq y$ and y maximal. We shall prove that the pair $\left(y, \frac{P(y)}{x}\right)$ also verifies the conditions.

Since $x|P(y)$, the number $\frac{P(y)}{x}$ is an integer. It is obvious that $\frac{P(y)}{x} | P(y)$; therefore we have to prove only that $y| P\left(\frac{P(y)}{x}\right)$.

From $y|P(x)$, i.e., $y|x^n + a_1 x^{n-1} + \cdots + a_1 x + 1$, it follows that x and y are coprime; hence the congruence $xz \equiv 1 \pmod{y}$ has a solution z. From $P(y) \equiv 1 \pmod{y}$, we get

$$P\left(\frac{P(y)}{x}\right) \equiv P(zP(y)) \equiv P(z) \pmod{y}.$$

Since P is a reciprocal polynomial, we have

$$x^n P\left(\frac{1}{x}\right) = P(x).$$

From $y|P(x)$ follows $x^n P(\frac{1}{x}) \equiv 0 \pmod{y}$, hence $P(z) \equiv 0 \pmod{y}$. In this way the pair $\left(y, \frac{P(y)}{x}\right)$ verifies the condition. Moreover, $P(y) \geq y^n + 1 > y^2 \geq xy$ and therefore $\frac{P(y)}{x} > y$, which contradicts the maximality of y in the pair (x, y).

79. Observe that $\gcd(a^2 + a - 1, a^3 + a^2 - 1) = 1$ and more generally, every two consecutive numbers from the given set are coprime. Suppose that

$$a^{n_1+1} + a^{n_1} - 1, a^{n_2+1} + a^{n_2} - 1, \ldots, a^{n_k+1} + a^{n_k} - 1$$

are pairwisely coprime, and let N be their product. The numbers $a^0 = 1$, $a, a^2, \ldots, a^N$ are distinct integers, and then, by the pigeonhole principle, there exist exponents $i > j$ such that $a^i \equiv a^j \pmod{N}$. Then $N|a^j(a^{i-j} - 1)$. From the definition of N, it follows that N and a are coprime, whence it follows $\gcd(N, a^{i-j+1} + a^{i-j} - 1) = 1$.

80. Start with $x_1 = 3, y_1 = 4$; hence $x_2 = 5$. Since $p \geq 7$, there exists an $t \in \{1, \ldots, p-1\}$ such that $3t \equiv 5 \pmod{p}$; thus $x_2 \equiv 3t \pmod{p}$. Letting $y_2 \equiv 4t \pmod{p}$, we have

$$x_3 \equiv 5t \equiv 3t^2 \pmod{p}, \ldots, x_n \equiv 3t^{n-1} \pmod{p}.$$

Since p is a prime, by Fermat's theorem, $t^{p-1} \equiv 1 \pmod{p}$, hence we need to stop when $n = d$ and d is the smallest positive integer with $t^d \equiv 1 \pmod{p}$.

2.26 1998

81. (a) If $k = 1$, then $\binom{n}{j}$ is an arithmetic progression.

If $k = 2$, then $\binom{n}{j}, \binom{n}{j+1}$ is an arithmetic progression.

If $k = 3$, then $\binom{n}{j}, \binom{n}{j+1}, \binom{n}{j+2}$ is an arithmetic progression if and only if

$$\binom{n}{j} + \binom{n}{j+2} = 2\binom{n}{j+1},$$

or equivalently, $4j^2 - (4n - 8)j + n^2 - 5n + 2 = 0$; whence

$$j = \frac{n - 2 \pm \sqrt{n+2}}{2}.$$

It follows that j is an integer if and only if $n = t^2 - 2$, $t \in \mathbb{Z}$. We conclude that in this case there no exists j for all n.

(b) If $1 \leq k \leq 3$, then there exists j, as above.
If $k = 4$, then $n = t^2 - 2$, $t \in \mathbb{Z}$.
If $k > 4$, then such n does not exist.

82. It is obvious that $x = 1$ and $x = 2$ cannot give solutions; hence $x \geq 3$. The equality

$$x^{n+1} + 2^{n+1} + 1 = x(x^n + 2^n + 1) + 2^{n+1} - 2^n x - x + 1$$

implies that $x^n + 2^n + 1 | x^{n+1} + 2^{n+1} + 1$ if and only if $x^n + 2^n + 1 | 2^n(x - 2) + x - 1$. Then $x^n + 2^n + 1 \leq 2^n(x - 2) + x - 1$, which is equivalent to

$$\left(\frac{x}{2}\right)^n + 1 \leq x - 2 + \frac{x}{2} \cdot \frac{1}{2^{n-1}} - \frac{1}{2^{n-1}}.$$

Denote $\frac{x}{2} \geq 1 + y$. Then $y \geq \frac{1}{2}$ and the Bernoulli's inequality $(1 + y)^n \geq 1 + ny$ together with the above inequality get us

$$2 + ny \leq 2y + \frac{y}{2^{n-1}},$$

whence

$$n \leq 2 + \frac{1}{2^{n-1}} - \frac{2}{y} < 2 + \frac{1}{2^{n-1}},$$

so that $n \leq 2$. The case $n = 2$ is impossible, since $x^2 + 5 > 5x - 9$. If $n = 1$, the required condition is

$$\frac{x^2 + 5}{x + 3} = x - 3 + \frac{14}{x + 3} \in \mathbb{N},$$

therefore $x = 4$ or $x = 11$. The solutions (x, n) are $(4, 1)$ and $(11, 1)$.

83. The general solution of the diophantine equation in positive integers

$$\frac{1}{x} = \frac{1}{y} + \frac{1}{z}$$

is $x = hmn$, $y = hm(m + n)$, $z = hn(m + n)$, where h, m, n are arbitrary positive integers, $\gcd(m, n) = 1$, $\gcd(x, y, z) = h$. Indeed, take $y = tz$, then

$$\frac{x}{t} = \frac{y}{t(t+1)} = \frac{z}{t+1}.$$

Choose $t = \frac{m}{n}$, with m, n coprime positive integers, eliminate denominators, and observe that if (x, y, z) is a solution; then (hx, hy, hz) is a solution.

Now we prove that $h = \gcd(x, y, z)$. Indeed, let p be a common prime divisor of x, y, z. If $p \nmid h$, then $p|x$ implies $p|m$ or $p|n$. If $p|m$, then p divides $z - x = hn^2$, hence $p|n$, contradiction.

Finally, $hxyz = (h^2mn(m+n))^2$ and $h(y-x) = (hm)^2$ are both squares.

84. The following well-known results solve the problem:

(a) $\sum_{i=1}^{k} \left(\frac{1}{p_i} + \frac{1}{p_i^2} \right) \geq \ln \left[\prod_{i=1}^{k} \left(1 + \frac{1}{p_i} + \frac{1}{p_i^2} + \cdots \right) \right];$

(b) $\prod_{i=1}^{k} \left(1 + \frac{1}{p_i} + \frac{1}{p_i^2} + \cdots \right) \geq 1 + \frac{1}{2} + \cdots + \frac{1}{m};$

(c) $1 + \frac{1}{2} + \cdots + \frac{1}{m} \geq \ln m.$

2.27 1999

85. We prove that if $p > 13$ is prime, then there is a prime $q < p$ such that the least residue of p modulo q is not square-free, which is the required property. In fact, we shall prove a stronger statement: if $p > 13$ is odd, then there is a prime q such the least residue of p modulo q is either 4, 8, or 9.

Suppose that some odd $p > 13$ is a counterexample: there's no prime $q < p$ for which the remainder upon dividing p by q belongs to $\{4, 8, 9\}$.

If there is a prime $q > 3$ dividing $p - 4$, then 4 is the least residue of p modulo q, contradiction. Otherwise, $p - 4 = 3^k$ for some positive integer k.

Consider now $p - 8$. If it has any prime factor greater than 8, we have also a contradiction. Otherwise, all prime factors of $p - 8$ are 2, 3, 5, and 7. But 2 is impossible, since $p - 8$ is odd, and 3 is impossible, since $3 | p - 4$. Thus, $p - 8$ is divisible by 5 and 7 or both.

Next, consider $p - 9$. As above, it can have only prime factors 2, 3, 5, and 7. Since $p - 4$ is divisible by 3 and not by 5; $p - 9$ is divisible by neither 3 nor 5. If $7 | p - 9$, then $7 \nmid p - 8$. If $7 \nmid p - 9$, then $p - 9 = 2^n$. But powers of 2 can only leave remainders 1, 2, and 4 upon division by 7; therefore $7 \nmid p - 8$.

Recall that the only possible prime factors of $p - 8$ are 5 and 7, and we just ruled out 7. We conclude that $p - 8 = 5^m$. From $p - 4 = 3^k$, we obtain

$$3^k - 4 = 5^m.$$

This diophantine equation has only the solution $(k, m) = (2, 1)$, which corresponds to $p = 13$. Indeed, reducing mod 4, we see that k must be even, $k = 2t$. Now,

$$3^{2t} - 4 = (3^t - 2)(3^t + 2) = 5^m.$$

The factors in parentheses cannot be both divisible by 5, since them differ by 4, so one of them is 1, which implies $t = 1$, $k = 2$, as required.

86. If $p = 2$, then $2^2 + 3^2 = 13 \neq a^n$.

If $p > 2$, then the following implications hold:

$$5 | 2^p + 3^p \Longrightarrow 5 | a \Longrightarrow 5^n | a^n \Longrightarrow 5^n | 2^p + 3^p.$$

From the lifting the exponent lemma, we deduce $\nu_5(2^p + 3^p) = \nu_5(5) + \nu_5(p) = 2$ if and only if $p = 5$. But then $2^5 + 3^5 = 275 \neq a^n$.

87. (a) Denote by $s(n)$ the sum of digits of the positive integer n and by $r(n)$ the remainder of $s(n)$ modulo 11. If $\overline{xy}$ is the number obtained writing together the numbers x, y, then $s(\overline{xy}) = s(x) + s(y)$. In the following 10×10 matrix, the entries are values of $r(n)$ for $n = 10a + b$, $0 \leq a, b \leq 9$:

$$\begin{pmatrix} 0 & 1 & 2 & 3 & 4 & 5 & 6 & 7 & 8 & 9 \\ 1 & 2 & 3 & 4 & 5 & 6 & 7 & 8 & 9 & 10 \\ 2 & 3 & 4 & 5 & 6 & 7 & 8 & 9 & 10 & 0 \\ 3 & 4 & 5 & 6 & 7 & 8 & 9 & 10 & 0 & 1 \\ 4 & 5 & 6 & 7 & 8 & 9 & 10 & 0 & 1 & 2 \\ 5 & 6 & 7 & 8 & 9 & 10 & 0 & 1 & 2 & 3 \\ 6 & 7 & 8 & 9 & 10 & 0 & 1 & 2 & 3 & 4 \\ 7 & 8 & 9 & 10 & 0 & 1 & 2 & 3 & 4 & 5 \\ 8 & 9 & 10 & 0 & 1 & 2 & 3 & 4 & 5 & 6 \\ 9 & 10 & 0 & 1 & 2 & 3 & 4 & 5 & 6 & 7 \end{pmatrix}.$$

We notice that any consecutive rows contain all remainders modulo 11.

Consider the numbers $n, n+1, n+2, \ldots, n+38$. We have two cases:

Case 1. All numbers are of the form $100m + a$, with $0 \le a \le 99$, that is, $100m + b$, with $0 \le b \le 61$. The remainders $r(b), r(b+1), \ldots, r(b+38)$ cover entirely two consecutive rows. Since $r(100m + a) = r(m) + r(a)$, we can choose a such that $r(m) + r(a) \in \{0, 11\}$ (in fact, in this case, 39 can be replaced with 29).

Case 2. One of the numbers is divisible by 100. Suppose, for instance, $n + k = 100m$. Then we have in our list the numbers:

$$100(m-1) + 100 - k, 100(m-1) + 100 - k$$
$$+ 1, \ldots, 100(m-1) + 99;$$
$$100m + 1, 100m + 2, \ldots, 100m + 38 - k.$$

If $k \ge 20$, then the remainders $r(100 - k), r(100 - k + 1), \ldots, r(99)$ cover the last two rows of the matrix, so we can choose one such that $r(m - 1) + r(100 - k + i)$ is divisible by 11. If $k < 20$, then $39 - k \ge 20$, and we apply the same argument for the second part of the list.

(b) Any 38 consecutive numbers with the requested property have the following form:

$$100(m-1) + 81, 100(m-1) + 82, \ldots, 100(m-1) + 99,$$

followed by

$$100m, 100m + 1, \ldots, 100m + 18.$$

Since $r(81), r(82), \ldots, r(99)$ cover all remainders except 8 and $r(0), r(1), \ldots, r(18)$ cover all remainders except 10, we must have $r(m-1) = 3$ and $r(m) = 1$. This is possible only if $m - 1 = 10x + 9$, with $r(x) = 5$ and

$r(x+1) = 1$. Analogously, $x = 10y + 9$, with $r(y) = 7$ and $r(y+1) = 1$. For this, we must have $y = 10z + 9$, with $r(z) = 9$ and $r(z+1) = 1$ and $z = 10w + 9$, with $r(w) = 0$ and $r(w+1) = 1$. The least w with such property is $w = 0$; hence $m - 1 = 9999$ and the numbers are

$$999981, 999982, \ldots, 1000018.$$

88. We have

$$S_n = \frac{(2+\sqrt{3})^{2n+1} + (2-\sqrt{3})^{2n+1}}{4} = (k-1)^2 + k^2$$

with k a positive integer. The quadratic eq. $2k^2 - 2k + 1 - S_n = 0$ has the discriminant

$$\Delta = 4(2S_n - 1) = \left[\frac{(1+\sqrt{3})^{2n+1} + (1-\sqrt{3})^{2n+1}}{2^n}\right]^2.$$

The solution of the equation is

$$k = \frac{2^{n+1} + (1+\sqrt{3})^{2n+1} + (1-\sqrt{3})^{2n+1}}{2^{n+2}},$$

so that it suffices to prove that k is an integer. Consider the sequence of integers $a_n = (1+\sqrt{3})^n + (1-\sqrt{3})^n$. We shall prove that $2^{\lfloor n/2 \rfloor}$ divides a_n. From $a_0 = a_1 = 2$, $a_2 = 8$ and $a_n = 2(a_{n-1} + a_{n-2})$ for $n \geq 3$, this property follows by induction.

89. (a) We will prove that $\{a_n\}$ is periodical. Put $a_i = 2^{t_i} b_i$, with b_i odd, for all i. From the definition, we deduce that a_i are even and belong to $(\frac{p}{2}, p)$ for all i. Since the interval contains only finitely many even elements, it follows that there are $i < j$ such that $a_i = a_j$; whence $a_{i+1} = a_{j+1}, a_{i+2} = a_{j+2}, \ldots$, since $a_i = p - b_{i-1}, a_j = p - b_{j-1}$ So $b_{i-1} = b_{j-1}$; hence $a_{i-1} | a_{j-1}$ or $a_{j-1} |$ a_{i-1}. But $\frac{1}{2} < \frac{a_i}{a_j} < 2$ for all i, j, so that $a_{i-1} = a_{j-1}$. We have $a_0 = a_{j-i}$, $a_1 = a_{j-i+1}, \ldots$; hence the sequence is periodical with period $j - i$.

(b) Let T be the principal period of the sequence. Then $a_1, \ldots, a_T$ are distinct even numbers in the interval $(\frac{p}{2}, p)$. We have

$$a_1 \ldots a_T = 2^{t_1 + \cdots + t_T} b_1 \ldots b_T.$$

Clearly, $a_i \equiv -b_{i-1} \pmod{p}$ for all $i > 0$; therefore $a_1 \ldots a_T \equiv (-1)^T b_1 \ldots b_T \pmod{p}$, so that $2^{2(t_1 + \cdots + t_T)} \equiv 1 \pmod{p}$. But $2^h \not\equiv 1 \pmod{p}$ for all $h = 1, 2, \ldots, p - 2$; hence

$$t_1 + \cdots + t_T \geq \frac{p-1}{2}. \quad (1)$$

The exponent of 2 in the product of all numbers from $(\frac{p}{2}, p)$ is

$$\sum_{i \geq 1}\left(\left\lfloor\frac{p}{2^i}\right\rfloor - \left\lfloor\frac{p}{2^{i+1}}\right\rfloor\right) = \frac{p-1}{2}$$

and $a_1, \ldots, a_T$ are even distinct numbers from $(\frac{p}{2}, p)$; hence the exponent of 2 in the product $a_1 \ldots a_T$ is $\leq \frac{p-1}{2}$ (2). From (1) and (2), we deduce

$$t_1 + \cdots + t_T = \frac{p-1}{2},$$

so that $a_1, \ldots, a_T$ are all even numbers in $(\frac{p}{2}, p)$.
It follows that $T = \lfloor\frac{p-1}{4}\rfloor$.

2.28 2000

90. A Pythagorean triple (x,y,z) is *primitive* if $(x,y,z) = 1$. If a Pythagorean triple contains 2^{n+1}, then there is a primitive Pythagorean triple containing 2^k, with $0 \leq k \leq n+1$.
    Case 1. $z = 2^k$. No solution, since $x^2 + y^2 \equiv 2 \pmod 4$, but $z^2 = 2^{2k} \equiv 0, 1 \pmod 4$.
    Case 2. $x = 2^k$ or $y = 2^k$. If, say, $x = 2^k$, then y, z are odd and

$$x^2 = 2^{2k} = z^2 - y^2 = (z-y)(z+y),$$

whence it follows that there are positive integers m, n such that $z - y = 2^n$, $z + y = 2^m$. We deduce $y = 2^{m-1} - 2^{n-1}$, $z = 2^{m-1} + 2^{n-1}$; therefore $m = 2k - 1$, $n = 1$. For every k such that $2 \leq k \leq n+1$, there is exactly one primitive Pythagorean triple containing 2^k, so that 2^{n+1} is in exactly n Pythagorean triples.

91. If $y = 0$, then $(x, y, z) = (x, 0, 0)$ is the unique solution.
    If $y > 0$, then for $x \neq 1$, $y \neq 2$, by Zsigmondy's theorem, there exists a prime p such that $p | (x+1)^{y+1} + 1$, but $p \nmid x + 2$, contradiction. It follows that is this case there is only the solution $(x, y, z) = (1, 2, 1)$.

92. Answer: $(1,1,4011973)$, $(7,7,81877)$, $(1,2,1022464)$, $(2,4,255616)$, $(4,8,63904)$, $(8,16,15976)$, $(16,32,3994)$.

2.28 2000

The given equation is equivalent to

$$1997(13y^2 + 1996x^2) = x^2y^2z.$$

It follows that the prime number 1997 divides x^2y^2z. If $1997 \nmid z$, then

$$1997 \mid x^2y^2 \implies 1997^2 \mid x^2y^2z = 1997(13y^2 + 1996x^2) \implies 1997 \mid (13y^2 + 1996x^2).$$

We deduce $13y^2 \equiv x^2 \pmod{1997}$; therefore 13 is a quadratic residue modulo 1997. But the quadratic reciprocity law implies

$$\left(\frac{13}{1997}\right) = \left(\frac{1997}{13}\right) = \left(\frac{8}{13}\right) = \left(\frac{2}{13}\right)^3 = \left(\frac{2}{13}\right) = -1,$$

contradiction.

Hence $1997 \mid z$, $z = 1997w$ and $13y^2 + 1996x^2 = x^2y^2w$ (*). We see that $x^2 \mid 13y^2$ and $y^2 \mid 1996x^2$, so there exist positive integers h, k such that $13y^2 = kx^2$, $1996x^2 = hy^2$. We have $h, k > 1$; otherwise $13y^2 = x^2$ or $499 \cdot (2x)^2 = y^2$, which is not possible. Multiplying these equalities and simplifying by x^2y^2, we have

$$hk = 2^2 \cdot 13 \cdot 499;$$

therefore $k \in \{2, 4, 13, 26, 52, 499, 998, 1996, 6487, 12974\}$. Since an equality $pm^2 = qn^2$, with p and q prime numbers and m and n integers is not possible, we can put away $k = 2$ and $k = 499$.

Case 1. $k = 4$. Then $13y^2 = 4x^2$; whence $13 \mid x$, $y^2 = \frac{4}{13}x^2$. Replacing this in (*), we get $wx^2 = 2^2 \cdot 5^3 \cdot 13$. No solution.

Case 2. $k = 13$. Then $y^2 = x^2$; whence $y = x$, so replacing in (*), we have $wx^2 = 2009$. The only squares that divide 2009 are 49 and 1. If $x = 7$, then $w = 41$; hence $z = 1997 \cdot 41 = 81877$. If $x = 1$, then $w = 2009$; hence $z = 4011973$.

Case 3. $k = 26$. Then $13y^2 = 26x^2$, or equivalently, $y^2 = 2x^2$, impossible.

Case 4. $k = 52$. Then $13y^2 = 52x^2$, so $y = 2x$. Replacing in (*), we have $wx^2 = 512$, so $x^2 \in \{1, 4, 16, 64, 256\}$ and we find, respectively,

$$z \in \{1022464, 255616, 63904, 15976, 3994\}.$$

Case 5. $k = 998$. Then $13y^2 = 998x^2$; hence $13 \mid x$. Replacing in (*), we get $wx^2 = 39$, impossible.

Case 6. $k = 1996$. Then $13y^2 = 1996x^2$, hence $13 \mid x$. Replacing in (*), we have $wx^2 = 26$, impossible.

Case 7. $k = 6487$. Then $13y^2 = 6487x^2$ implies $y^2 = 499x^2$, impossible.

Case 8. $k = 12974$. Then $13y^2 = 12974x^2$ implies $y^2 = 998x^2$. Replacing in (*), we have $wx^2 = 15$, impossible.

93. For fixed k, choose $m > k$ such that $n + \binom{m}{3}$ is odd. This is always possible: if n is odd, then take $m \equiv 0 \pmod 4$; if n is even, then take $m \equiv 3 \pmod 4$. We have $n + \binom{m}{3} = 2a + 1$, with a a positive integer. The identity

$$2a + 1 = \binom{a}{3} - \binom{a+1}{3} - \binom{a+2}{3} + \binom{a+3}{3}$$

implies

$$n = \binom{a}{3} - \binom{a+1}{3} - \binom{a+2}{3} + \binom{a+3}{3} - \binom{m}{3}.$$

This is the required representation, since for large m we have

$$a = \frac{n - 1 + \binom{m}{3}}{2} > m.$$

94. Choose an even number a and consider the identity

$$(a^3 + 1)^4 = (2a^3)^3 + (2a)^3 + \left[(a^3 - 1)^2\right]^2.$$

Since a is even, $a^3 - 1$ and $a^3 + 1$ are odd; therefore

$$x = 2a^3, y = 2a, z = (a^3 - 1)^2, t = a^3 + 1$$

have no common divisor greater than 1.

95. Since a is odd, $\gcd(a, 2^k) = 1$, for any $k \geq 0$. By Euler's theorem, $a^{\varphi(2^k)} \equiv 1 \pmod{2^k}$. Since $\varphi(2^k) = 2^{k-1}$ and we are looking for the least exponent n such that $a^n \equiv 1 \pmod{2^{2000}}$, it follows that n is a divisor of $2^{1999} = \varphi(2^{2000})$. If $a \equiv 1 \pmod{2^{2000}}$, it follows that $n = 1$. We shall omit this case. Consider the decomposition

$$a^{2^n} - 1 = (a-1)(a+1)(a^2+1)\left(a^{2^2}+1\right)\cdots\left(a^{2^{n-1}}+1\right).$$

Assume that $a = 2^s b + 1$, where $2 \leq s \leq 1999$ and b is odd. Then a has the binary representation

$$a = 1\ldots 1\underbrace{00\ldots 1}_{s\text{ digits}}.$$

It can be proved by induction on k that the even number $a^{2^k} + 1$ is not divisible by 2^2. Then, by the above decomposition, $a^{2^m} - 1$ is divisible by 2^{s+m} and is not divisible by 2^{s+m+1}; hence the required number is 2^{2000-s}.

Assume that $a \equiv -1 \pmod{2^s}$ and $a \not\equiv -1 \pmod{2^{s+1}}$, where $s \geq 2$. Equivalently, a has the binary representation

$$a = 1\ldots 0\underbrace{11\ldots 1}_{s\text{ digits}}.$$

As above, $a - 1$ is divisible by 2, is not divisible by 2^2, and $a^{2^k} + 1$ is divisible by 2 and is not divisible by 2^2, for $k \geq 1$. It follows that $a^{2^m} - 1$ is divisible by 2^{s+m} and is not divisible by 2^{s+m+1}. In this case, the required exponent is $n = 2^{1999-s}$, when $s < 1999$, and $n = 2$, when $s \geq 1999$.

2.29 2001

96. If (a, b, c) is a Pythagorean triple, that is, $a^2 + b^2 = c^2$; then $x = ac$, $y = bc$, $z = ab$ is a solution, and conversely, every solution can be obtained in this manner. It follows that $xyz \geq (abc)^2 \geq (3 \cdot 4 \cdot 5)^2 = 3600$.

97. If $x = 2^k m$, with k, m positive integers and m odd, then $x \in S$ if and only if k is even. Call k the 2-exponent of x and m the odd part of x. Consider $x, y \in S$. If x, y have different odd parts, then $y \neq 2x$. This means that maximize $|S|$ is maximize

$$|S \cap \{\text{integers with odd part } m\}|$$

for each m and take the union of maximal subsets for each m. Then

$$\{\text{integers with odd part } m\} = \{2^0 m, 2^1 m, 2^2 m, \ldots\}.$$

We have a solution here for $y = 2x$ if and only if we take two elements with consecutive 2-exponents. Induction shows that the best way to do this is to take all with even 2-exponents.

Thus, each $|S \cap \{\text{integers with odd part } m\}|$ is maximized; hence $|S|$ is maximized when S contains the numbers with even 2-exponents. The number of elements of $\{1, 2, \ldots, p\}$ with 2-exponent k is $\left\lfloor \frac{p}{2^k} \right\rfloor - \left\lfloor \frac{p}{2^{k+1}} \right\rfloor$. Adding over all k gives

$$\sum_{k\geq 0}\left(\left\lfloor\frac{p}{2^{2k}}\right\rfloor-\left\lfloor\frac{p}{2^{2k+1}}\right\rfloor\right)=\sum_{k\geq 0}(-1)^k\left\lfloor\frac{p}{2^k}\right\rfloor.$$

When $p=2001$, we can end the summation at $k=10$. The final result is

$$n=\sum_{k=0}^{10}(-1)^k\left\lfloor\frac{2001}{2^k}\right\rfloor.$$

98. The diophantine equation $u^2+v^2+1=3uv$ has infinitely many solutions. Each pair (u,v) of two consecutive terms of the sequence given by

$$u_1=u_2=1,\, u_{n+1}=3u_n-u_{n-1},\, n\geq 2$$

is a solution. Then

$$(x,y,z,n)=\left(u,3v-u,3v^3,3v^2+3\right)$$

is a solution of the given equation.

99. Answer: 2001.

It suffices to find the exponent of 37 in the representation of the number

$$\underbrace{99\ldots 9}_{3\cdot 37^{2000}\text{ digits}}=10^{3\cdot 37^{2000}}-1=1000^{37^{2000}}-1,$$

since 9 and 37 are coprime. We shall prove by induction on k that the exponent of 37 in the representation of the number $1000^{37^k}-1$ is $k+1$. For $k=0$ we have

$$1000^{37^0}-1=3^3\cdot 37;$$

therefore the exponent is $0+1=1$.

Suppose that the assertion is valid for some k and prove it for $k+1$. From

$$1000^{37^{k+1}}-1=\left(1000^{37^k}-1\right)\left(1+1000^{37^k}+1000^{2\cdot 37^k}+\cdots+1000^{36\cdot 37^k}\right),$$

we deduce that it suffices to show that the exponent of 37 of the number in the second parenthesis is 1. But $1000\equiv 1\pmod{37}$ implies $1000^{37^k}\equiv 1\pmod{37}$. If $1000^{37^k}=37q+1$, with q positive integer, then the second parenthesis equals

$$1+ (37q + 1) + (37q + 1)^2 + \cdots + (37q + 1)^{36} \equiv$$
$$\equiv 1 + (37q + 1) + (2 \cdot 37q + 1) + \cdots + (36 \cdot 37q + 1) \equiv 37^2 \cdot 18q + 37 \equiv$$
$$\equiv 37 \pmod{37^2};$$

therefore this number is divisible by 37 and is not divisible by 37^2, which ends the proof by induction.

If $k = 2000$, then the exponent of 37 in the given number is $2000 + 1 = 2001$.

100. We see that $m|n$ if and only if $a_m|a_n$.

Suppose that $r \nmid k$, then $a_{\gcd(k,r)}^2 = a_r a_{\gcd(r,s)}$ $(*)$, but $a_{\gcd(k,r)} < a_r < a_k$, which contradicts the fact that k is the smallest with the desired property.

It remains to prove the equality $(*)$. Choose a prime $p|a_k$ and prove that

$$2\nu_p(a_{\gcd(k,r)}) = \nu_p(a_r) + \nu_p(a_{\gcd(r,s)})$$

(it is obvious for primes $p \nmid a_k$). We know that $2\nu_p(a_k) = \nu_p(a_r) + \nu_p(a_s)$ and consider the cases $\nu_p(a_r) < \nu_p(a_k)$ and $\nu_p(a_r) > \nu_p(a_k)$. If we write explicitly $\nu_p(a_{\gcd(k,r)})$, $\nu_p(a_r)$, $\nu_p(a_{\gcd(r,s)})$, then the equality is verified.

101. Answer: $(2,2)$, $(3,3)$, $(2,13)$, $(13,2)$, $(3,7)$, $(7,3)$.

If $p = q$, then $5^p + 1 \equiv 5 + 1 \equiv 0 \pmod{p}$; whence $p \in \{2,3\}$.

Suppose that $p \neq q$. From $5^{2p} \equiv 1 \pmod{q}$, it follows that $\text{ord}_q(5)|2p$, so that $\text{ord}_q(5) = 1, 2, p$ or $2p$.

If $\text{ord}_q(5) = 1$ or 2, then either $5 \equiv 1 \pmod{q}$, or $5^2 \equiv 1 \pmod{q}$; thus $q = 2, 3$.

If $q = 2$, then $p|5^2 + 1$; therefore $p = 13$.

If $q = 3$, then $p|5^3 + 1$; therefore $p = 7$.

If $\text{ord}_q(5) \in \{p, 2p\}$ and $\text{ord}_p(5) \in \{q, 2q\}$, then $p|q - 1$ and $q|p - 1$, contradiction, since $p \neq q$.

102. We shall prove that the required pairs are $(p, p - 1)$, where $p \geq 3$ is a prime. Such a pair is a solution, by the Fermat's theorem. Let (m, n) be a solution and let $p|m$ be a prime. Since $m|a^n - 1$, we obtain $p|a^n - 1$, for every $a = 1, 2, \ldots, n$. We cannot have $n \geq p$, since $p|p^n - 1$ gives a contradiction.

Consider the polynomial $f = \prod_{i=1}^{n} (X - i) - (X^n - 1)$, and let g be its reduced polynomial modulo p. Then deg.$(g) \leq n - 1$ and, by the hypothesis, g has the roots $1, 2, \ldots, n \pmod{p}$. All these residue classes are distinct; hence, by the Lagrange's theorem, g is the identical zero polynomial. The leading coefficient of g is $\frac{n(n+1)}{2}$; therefore $p|n(n + 1)$. Since $p \geq n + 1$, it follows that $p = n + 1$. We deduce $n = p - 1$ and $m = p^k$. We shall prove that $k = 1$. Indeed, if $k \geq 2$, then we have $p^2|a^{p-1} - 1$ for all $a = 1, 2, \ldots, p - 1$; hence $p^2|(p - 1)^{p-1} - 1$, implying $p^2|p(p - 1)$, contradiction.

103. Clearly, $x \geq 0$. If y is odd, then RHS is even, no solution.

If y is even, then RHS $\equiv 1 \pmod 8$, so that x is even and LHS is a perfect square.

Case 1. $y \geq 0$. Then $y = 0, 2$ are solutions; hence $(x, y) = (0, 0)$ $(2, 2)$ are solutions. We see that $y = 1$ is not a solution. For $y > 2$, we have

$$(y^2)^2 < y^4 + 4y + 1 < (y^2 + 1)^2,$$

no solutions.

Case 2. $y < 0$. We see that $y = -1, -2$ are not solutions. If $y < -2$, the inequalities

$$(y^2 - 1)^2 < y^4 + 4y + 1 < (y^2)^2$$

show that there are no solutions.

104. Suppose that prime $p | a_k$ and consider $S_1 = \{a_{3k-1}, a_{3k+2}\}$. We have $|S_1| = 4$, and due to the recursive equation, all elements are congruent mod p, that is,

$$a_{3k+i} - a_{3k+i-1} \equiv a_{\lfloor \frac{3k+i}{3} \rfloor} \equiv a_k \equiv 0 \pmod{p}, 0 \leq i \leq 3.$$

Consider now $S_2 = \{a_{9k+i} | -3 \leq i \leq 8\}$. So $|S_2| = 12$ and the difference between consecutive elements from S_2 is an element of S_1. It follows that S_2 increases linearly mod p. Since $p \leq 13$, there are at most nonzero elements mod p. Therefore, $a_{9k-4} \equiv 0 \pmod{p}$ or there exists in S_2 an element divisible by p.

Thus, we have found the next element divisible by p, and we can continue this process to infinity.

2.30 2002

105. We shall prove that $m = n = 0$ is the unique solution. Clearly, $n = 1$ is not a solution. Considering the equality modulo 3, we conclude that n is even. Then 16 divides $3^n - 1$; therefore 4 divides n. The equality $(2^{4k} - 1)(3^{4k} - 1) = m^2$, with k a positive integer, is equivalent to

$$m^2 + (3^{2k} - 2^{2k})^2 = (6^{2k} - 1)^2.$$

Taking it mod 4, we see that there are no solutions.

106. Answer: $S = \{1995\}$.

We must have $k = d_8 d_9$.

If $p_1 = 2$, then $2 | k$ and $4 \nmid k$; hence one of the numbers d_8, d_9 is even and the other is odd. But then $d_9 - d_8$ is odd, contradiction.

If $p_1 > 2$, then $p_1 = 3$; otherwise $k \geq 5 \cdot 7 \cdot 11 \cdot 13 = 5005$, contradiction. Similarly, $p_2 = 5$; otherwise $k \geq 3 \cdot 7 \cdot 11 \cdot 13 = 3003$, contradiction. Analogously, $p_3 = 7$ and $p_4 < 19$. From $3p_4 - 5 \cdot 7 = 22$, we deduce $p_4 = 19$, so that $k = 3 \cdot 5 \cdot 7 \cdot 19 = 1995$.

107. The numbers $a_n = 2np + (n^2)$ have the desired property, where (x) denotes the remainder of x upon division by p. Suppose that $a_k + a_l = a_m + a_n$, $k \leq l, m \leq n$. By the construction of a_i, we have $2p(k+l) \leq a_k + a_l < 2p(k+l+1)$. Hence we must have $k + l = m + n$; therefore $(k^2) + (l^2) = (m^2) + (n^2)$. Thus,

$$k + l \equiv m + n \pmod{p}, k^2 + l^2 \equiv m^2 + n^2 \pmod{p}.$$

Then $(k - l)^2 = 2(k^2 + l^2) - (k + l)^2 \equiv (m - n)^2 \pmod{p}$, so that $k - l \equiv \pm (m - n) \pmod{p}$, which leads to $\{k, l\} = \{m, n\}$; hence $k = m, l = n$.

108. The quadratic residues mod 11 are 0, 1, 4, 9, 5, and 3. The sum of two residues is 0 mod 11 if and only if both residues are 0. Then $11|x$, $11|y$, so that $121|x^2 + y^2$.

Now the problem is to count the number of numbers that can be written as $a^2 + b^2$ with $a, b \in \{0, 1, 2, \ldots, 90\}$. The answer is 3009.

The number of ordered pairs (x, y) which are solutions is $90^2 = 8100$.

109. We shall solve in integers the equation $x^n + (x + 1)^n = (x + 2)^n$.

If $n = 1$, then $x = 1$.

If $x = -1$, then n is an arbitrary even number.

If $n > 1, x \neq -1$, then from $(x+1)^n = (x+2)^n - x^n$, we deduce that x is odd. If $\nu_2(n) = 0$, then $\nu_2((x+2)^n - x^n) = 1$, and we have no solution, since $\nu_2((x+1)^n) < > 1$. If $\nu_2(n) > 0$, then

$$\nu_2((x+2)^n - x^n) = \nu_2\left((x+2)^2 - x^2\right) - 1 + \nu_2(n)$$
$$= \nu_2(x+1) + 1 + \nu_2(n).$$

Hence $\nu_2((x+1)^n) = n\nu_2(x+1) = \nu_2(x+1) + 1 + \nu_2(n)$, so that

$$n = 1 + \frac{1 + \nu_2(n)}{\nu_2(x+1)} \leq 2 + \nu_2(n).$$

We deduce $n \leq 4$, and applying the Fermat's last theorem for the particular cases $n = 3$ and $n = 4$, we obtain $n = 2, y = 1$. The equation $x^2 + (x+1)^2 = (x+2)^2$ has the unique positive solution $x = 3$.

110. Among $2004! + 2, 2004! + 3, \ldots, 2004! + 2004$, there is no prime number. Denote by $p(n)$ the number of primes in the set $n, n + 1, \ldots, n + 2001$. Then we have

$$|p(n+1) - p(n)| \leq 1, p(1) \geq 150, p(2004! + 2) = 0,$$

whence the conclusion.

111. If $x = y = z$, there is 1 triple.

If $x = 0$, $yz \neq 0$, then there are triples $(x, y, z) = (0, k, p - k)$, with $1 \leq k \leq p - 1$. Counting also the symmetric cases, there are $3(p - 1)$ triples.

If $xyz \neq 0$, then $y \equiv bx \pmod{p}$, $z \equiv bcx \pmod{p}$, where b, c are positive integers not divisible by p. Then

$$(x + y + z)^2 \equiv axyz \pmod{p} \iff (bc + b + 1)^2 \equiv ab^2 cx \pmod{p};$$

Whence we get $x \equiv (bc + b + 1)^2 (ab^2 c)^{-1} \equiv (ac)^{-1}(1 + b^{-1} + c)^2 \pmod{p}$. (The exponent -1 denotes the inverse modulo p.) For $1 \leq c \leq p - 2$ and $b \equiv (c + 1)^{-1} \pmod{p}$, we have $x \equiv 0 \pmod{p}$, no solutions. In this case, for each fixed c, there are $p - 2$ triples. For $c = p - 1$, we have $x \equiv (ab^2)^{-1} \pmod{p}$, so that there are $p - 1$ triples.

The total number of triples is

$$1 + 3(p - 1) + (p - 2)^2 + (p - 1) = p^2 + 1.$$

112. More generally, there is n such that p^n contains m consecutive zeros, for each positive integer m.

For $p \neq 2, 5$, take a solution $n > 0$ of the congruence $p^n \equiv 1 \pmod{10^{m+1}}$, which is solvable, since $\gcd(p, 10) = 1$.

For $p = 2$, choose an integer such that $10^{t-m-1} > 2^t$ and a solution $n > t$ of the congruence $2^n \equiv 2^t \pmod{10^t}$, equivalent to $2^{n-t} \equiv 1 \pmod{5^t}$, which is solvable, since $\gcd(2, 5) = 1$. Since $n > t$, we have $2^n = 2^t + k \cdot 10^t$, $k \geq 1$; thus 2^n has at least $t - \lfloor t \log_{10} 2 \rfloor \geq m$ consecutive zeros.

The case $p = 5$ is considered similarly.

113. For $n = 2, 3, 4, 6, 8, 33, \ldots$, we have $\pi(n) | n$. We shall prove that this fact occurs infinitely often. Choose a positive integer $m > 1$. Since

$$\lim_{k \to \infty} \frac{\pi(mk)}{mk} = 0,$$

there is a maximal k such that $\frac{\pi(mk)}{mk} \geq \frac{1}{m}$ ($k = 1$ satisfies this inequality). If the equality holds, we are done; otherwise $\pi(mk) > k$. Since k is maximal, we have

$$\frac{\pi(mk + m)}{mk + m} < \frac{1}{m};$$

thus, $k \geq \pi(mk + m) \geq \pi(mk) > k$, contradiction.

114. From

$$\binom{p^n}{p} = p^{n-1} \binom{p^n - 1}{p - 1}$$

and

$$\binom{p^n-1}{p-1} - 1 \equiv -1^{p-1} \cdot \frac{(p-1)!}{(p-1)!} - 1 \equiv 1 - 1 \equiv 0 \pmod{p},$$

the conclusion follows.

115. We will prove that there are infinitely many pairs (x, y) of positive integers satisfying the given condition.

The pair $(x, y) = (1, 1)$ satisfies the conditions. Suppose that there is a pair (x, y), $x < y$, of positive integers which satisfies the conditions. Then the pair (y, x_1), $x_1 = \frac{y^2+m}{x}$, also satisfies the conditions.

Let $d = \gcd(x_1, y)$. If $d > 1$, then $d | x_1 y^2 + m$, so that $d | x$. This is a contradiction, since $\gcd(x, y) = 1$. Now it remains to show that $x_1 | y^2 + m$ and $y | x_1^2 + m$. The first relation is obvious, for the second observe that it is equivalent to

$$y | (xx_1)^2 + mx^2 = y^4 + 2my^2 + m^2 + mx^2,$$

which is true, since $y | m(x^2 + m)$ by hypothesis (the equivalence is a consequence of $\gcd(x, y) = 1$). Hence (y, x_1) satisfies all conditions.

116. Consider the sequence $(b_n)_{n \geq 0}$ defined by $b_0 = -1$, $b_1 = 1$, $b_{n+1} = 4b_n - b_{n-1}$, for all $n \geq 1$. We shall prove by induction that.

(i) $2a_n - 1 = b_n^2$; (ii) $2b_n b_{n-1} = a_n + a_{n-1} - 4$, for all positive integer n.

We have $b_0^2 = 2a_0 - 1 = 1$, $b_1^2 = 2a_1 - 1$, $2b_0 b_1 = -2 = a_0 + a_1 - 4$. If the equalities (i) and (ii) are true for n, then

$$b_{n+1}^2 = (4b_n - b_{n-1})^2 = 16b_n^2 - 8b_n b_{n-1} + b_{n-1}^2 =$$
$$= 16(2a_n - 1) - 4(a_n + a_{n-1} - 4) + (2a_n - 1) =$$
$$= 28a_n - 2a_{n-1} - 1 = 2(14a_n - a_{n-1}) - 1 = 2a_{n+1} - 1$$

and

$$2b_{n+1}b_n = 2(4b_n - b_{n-1})b_n = 8b_n^2 - 2b_n b_{n-1} = 8(2a_n - 1) - a_n - a_{n-1} + 4 =$$
$$= 15a_n - a_{n-1} - 4 = 14a_n - a_{n-1} + a_n - 4 = a_{n+1} + a_n - 4.$$

117. We prove first that n is square-free. Assume that $n = q^b r$, where q is a prime divisor of n, $b \geq 2$ and r is a positive integer coprime to q. If $r \geq 2$, then the number $a = \varphi(q^b)r + 1 = (q^b - q^{b-1})r + 1$ has the property $a \equiv 1 \pmod{r}$, which implies $a^{a-1} \equiv 1 \pmod{r}$. For $b \geq 2$, the prime q is a divisor of $\varphi(q^b)$; hence $\gcd(q, a) = 1$. By Euler's theorem, we have $a^{\varphi(q^b)} \equiv 1 \pmod{q^b}$; therefore $a^{a-1} = a^{\varphi(q^b)r} \equiv 1 \pmod{q^b}$. By the Chinese remainders theorem, $a^{a-1} \equiv 1 \pmod{n}$, and since $\varphi(q^b)r + 1 < q^b r$, we have an $a \neq n - 1$ satisfying the conditions, contradiction.

If $n = 2qr$, where $q > 2$ is a prime and $r > 2$ is odd, $\gcd(q, r) = 1$, then q is not divisible by $r - 1$, and $a = (q - 1)r - 1$ satisfies $a \equiv 1 \pmod{2r}$, $\gcd(a, q) = 1$. By the Fermat's theorem, $a^{q-1} \equiv 1 \pmod{q}$ and $a^{a-1} \equiv 1 \pmod{q}$. Then we have a solution of the congruence $a^{a-1} \equiv 1 \pmod{n}$ with $1 < a < n$.

If $n = 2pqr$, where p and q are distinct odd primes, $r \geq 1$, $p | qr - 1$, $q | pr - 1$, then the number $a = (p - 1)(q - 1)r + 1$ has the required properties. Indeed, as p is not a divisor of $(q - 1)r - 1$ nor q is a divisor of $(p - 1)r - 1$, a is relatively prime to pq. By the Fermat's theorem and the Chinese remainder theorem, we get as in the first part that a satisfies $a^{a-1} \equiv 1 \pmod{n}$ and $1 < a < n$.

We conclude that $n = 2p$, with p prime.

118. Clearly, $d_1 = 1$. If n is odd, then all divisors are odd and $n = d_1^2 + d_2^2 + d_3^2 + d_4^2$ is even, contradiction. It follows that n is even and $d_2 = 2$. Suppose that n is divisible by 4, then $d_3 = 3$ or $d_3 = 4$. If $d_3 = 3$, then $d_4 = 4$ and $n = 30$ is divisible by 4, contradiction. If $d_3 = 4$, then $0 \equiv n \equiv 1^2 + 2^2 + 4^2 + d_4^2 \equiv 1, 2 \pmod{4}$, contradiction.

We deduce that $d_3 = p$ is an odd prime and $d_4 = 2p$ or $d_4 = q > p$ is an odd prime. In this case $2 \equiv n \equiv 1^2 + 2^2 + p^2 + q^2 \equiv 3 \pmod{4}$, contradiction. Finally, $d_4 = 2p$ and $n = 1^2 + 2^2 + p^2 + (2p)^2 = 5(1 + p^2)$. But $p | n$ implies $p = 5$ and $n = 130$.

119. The number $a^2 + ab + b^2$ divides $a(a^2 + ab + b^2) - ab(a + b) = a^3$ and similarly divides b^3, so that it divides $(a^3, b^3) = (a, b)^3$. Then

$$ab < a^2 + ab + b^2 \leq (a, b)^3 \leq |a - b|^3;$$

whence $|a - b| > \sqrt[3]{ab}$.

120. If m, n, $\frac{m^2 + n^2}{mn - 1} = q$ are positive integers, then $q = 5$.

Let $m_1 \geq n$ be a minimal solution of the equation

$$m^2 - qmn + n^2 + q = 0.$$

Then

$$m_2 = qmn - m_1 = \frac{n^2 + q}{m_1} \geq m_1 \geq n.$$

We have the following equivalences:

$$(m_1 - n)(m_2 - n) \geq 0 \Leftrightarrow m_1 m_2 - n(m_1 + m_2) + n^2 \geq 0 \Leftrightarrow$$
$$\Leftrightarrow n^2 + q - n(qn) + n^2 \geq 0 \Leftrightarrow (q - 2)n^2 \leq q \Leftrightarrow n^2 \leq 1 + \frac{2}{q - 2}.$$

So, a minimal solution exists if and only if $n = 1$. Then $q = 5$, $m_1 = 2$, $m_2 = 3$.

All solutions are consecutive terms of the sequence $a_n = 5a_{n-1} - a_{n-2}$ with first terms

$a_1 = 1, a_2 = 2, (1,2,9,43,206,\ldots)$ and $a_1 = 1, a_2 = 3, (1,3,14,67,321,\ldots)$.

2.31 2003

121. The equation is equivalent to

$$(n^2 + 3n + 1)^2 = m(m+2)\left[(m+1)(m+2)(m+3)^2\right]^2 + 1.$$

Set $a = n^2 + 3n + 1$, $b = (m+1)(m+2)(m+3)^2$. Then (a,b) is the solution of the Pell equation $a^2 - m(m+2)b^2 = 1$. The fundamental solution of this equation is $(m+1, 1)$, so that

$$a + b\sqrt{m(m+2)} = \left(m + 1 + \sqrt{m(m+2)}\right)^k, k \in \mathbb{N}^*.$$

By size considerations, it's easy to eliminate the case $k \geq 5$, since

$$n^2 + 3n + 1 \leq (m+3)^{\frac{5}{2}}.$$

For $k \leq 4$, we must only check small m. No solutions.

122. Answer: 0 and each positive integer value ≥ 2.
    If $x = p^3$, $y = p^5 - p$, then

$$\frac{x^3 - y}{1 + xy} = \frac{p^9 - p^5 + p}{p^8 - p^4 + 1} = p.$$

For $p \geq 2$, it follows $0 < x < y$; therefore each value $p \geq 2$ is admissible.
    If $P = 1$, then $x^3 - 1 = (x+1)y$, implies that $x + 1$ divides $x^3 - 1$; therefore $x + 1$ divides $x^3 + 1 - (x^3) = 2$; whence $x = 1$, $y = 0$, contradiction.
    If $x = 2$, $y = 8$; then $P = 0$.
    If $P \leq -1$, then $y + 1 > y - x^3 \geq 1 + xy$, so that $x < 1$, contradiction.

123. It is easy to see that $a_1 = 3$, $a_2 = 7$, $a_{n+1} = 3a_n - a_{n-1}$ for all $n \geq 2$ and

$$a_n = \left(\frac{3+\sqrt{5}}{2}\right)^n + \left(\frac{3-\sqrt{5}}{2}\right)^n = u^n + u^{-n}, n \geq 1.$$

Define $b_n = a_n + (-1)^n$. Then if 3^k is the maximum power of 3 dividing n, we will prove that $b_{3^k} | b_n$.

If n is odd, then $b_n = u^n + u^{-n} - 1$. With $x = u^{3^k}$, we have $b_n = x^r + x^{-r} - 1$, $n = = 3^k r$, $3 \nmid r$. We will prove that $x + x^{-1} - 1 \mid x^r + x^{-r} - 1$ for r odd, $3 \nmid r$.

If $r = 6t + 1$ (the case $r = 6t - 1$ is similar), we have $y_t = x^{6t+1} + x^{-(6t+1)} - 1$ and

$$y_t - y_{t-1} = (x + x^{-1} - 1)(x + x^{-1} + 1)\left(x^{3t-1} + x^{-(3t-1)}\right) \times$$
$$\times \left((x^{3t} + x^{-3t}) - (x^{3t-2} + x^{-(3t-2)})\right),$$

and $x^l + x^{-l}$ are all integers. The result follows inductively.

If n is even, the proof is similar.

124. Answer: $n = 35$.

The equalities $1 = 3^2 - 2^3$, $5 = 2^3 - 3^1$, $7 = 2^3 - 3^0$, $11 = 3^3 - 2^4$, $13 = 2^4 - 3^1$, $17 = 3^4 - 2^6$, $19 = 3^3 - 2^3$, $23 = 3^3 - 2^2$, $25 = 3^3 - 2^1$, $9 = 2^5 - 3^1$, $31 = 2^5 - 3^0$ show that $n \geq 35$. We will prove that the equation $|2^a - 3^b| = 35$ has no solution in nonnegative integers. It is easy to verify that there are no solution for $a < 4$.

Case 1. $2^a - 3^b = 35$. Considering this equation mod 16, we get $3^b \equiv 13$ (mod 16), without solutions, since the remainders 3^b (mod 16) are 1, 3, 9, 11.

Case 2. $3^b - 2^a = 35$. Then $-(-1)^a \equiv -1$ (mod 3) and $(-1)^b \equiv -1$ (mod 4), so that a is even and b odd. The eq. $3x^2 - y^2 = 35$ has no integer solutions, since 3 is not a square mod 7.

125. We shall prove by induction on n that for positive integers n, $b \geq 2$, the condition $n \mid b^n - 1$ implies $(b-1)n \mid b^n - 1$. If $n = 1$, the result is obvious.

Suppose that the claim holds for all numbers less than n, b and study two cases:

Case 1. $\gcd(n, b-1) = 1$. Then the conditions $n \mid b^n - 1$ and $b - 1 \mid b^n - 1$ imply $(b-1)n \mid b^n - 1$.

Case 2. $p \mid \gcd(n, b-1)$, where p is a prime number. Then $n = mp$, $b^n - 1 = (b^p)^m - 1$ and by the induction hypothesis, $(b^p - 1)m \mid b^n - 1$. Since $p \mid b - 1$, we get $b \equiv 1 \pmod{p}$ and $1 + b + \cdots + b^{p-1} \equiv 0 \pmod{p}$; hence

$$(b^p - 1)m = (b - 1)m(1 + b + \cdots + b^{p-1})$$

is divisible by $(b-1)mp = (b-1)n$, so that $(b-1)n$ divides $b^n - 1$, as required.

Taking $b = 10$, we get the desired result.

126. We shall prove first the properties:

a. If a and b are odd integers and $\gcd(a, b) = 1$, then $\gcd(2^a + 1, 2^b + 1) = 3$.

Proof. Denote $d = \gcd(2^a + 1, 2^b + 1)$; then $3 \mid d$. From

$$\gcd(2^{2a} - 1, 2^{2b} - 1) = 2^{\gcd(2a, 2b)} - 1 = 2^2 - 1 = 3,$$

it follows that $d \mid 3$; hence $d = 3$.

b. If a is odd and $3 \nmid a$, then $9 \nmid 2^a + 1$.
   Proof. For $a = 3k + 1$, we have $2^a + 1 = 9t - 1$, and for $a = 3k + 2$, we have $2^a + 1 = 9t - 3$.
   Now we shall prove the result by induction. Let $N = p_1 \ldots p_n, N' = \frac{N}{p_n}$.
   For $n = 1$, $2^p + 1$ has the divisors $1, 3, \frac{2^p+1}{3}, 2^p + 1$. Since $p > 3$, we have $3 \neq \frac{2^p+1}{3}$.
   Assume that we have shown it for $n - 1$, that is, $2^{N'} + 1$ has at least 4^{n-1} divisors. We have $2^{N'} + 1 \mid 2^N + 1, \frac{2^{p_n}+1}{3} \mid 2^N + 1$, and we can use properties (1) and (2) for $a = p_n$, $b = N'$ to get $\frac{2^{p_n}+1}{3} \mid \frac{2^N+1}{2^{N'}+1}$. On the other hand, we have $\left(\frac{2^{p_n}+1}{3}\right)^2 < \frac{2^N+1}{2^{N'}+1}$, which means that $\frac{2^N+1}{2^{N'}+1}$ has at least four divisors $d_1, \ldots, d_4$. For each $d \mid 2^{N'} + 1$, it follows $d_i d \mid 2^N + 1$ and all $d_i d$ are distinct, since gcd $(d_i, d) = 1$. We deduce that $2^N + 1$ has at least $4 \cdot 4^{n-1} = 4^n$ divisors.

127. For each k, there are $\left\lfloor \frac{N}{x_k} \right\rfloor$ multiples of x_k not greater than N. These multiples are all distinct, since lcm $(x_i, x_j) > N$, $1 \leq i < j \leq n$; hence

$$\sum_{i=1}^{n} \left\lfloor \frac{N}{x_i} \right\rfloor < N.$$

Now,

$$\sum_{i=1}^{n} \left\{ \frac{N}{x_i} \right\} < n < N,$$

since x_i are distinct. Adding these inequalities, we get

$$\sum_{i=1}^{N} \frac{N}{x_i} < 2N,$$

whence the result.

128. Answer: $(a, b, c) = (1, 1, 1), (1, 2, 3)$.
   Let $\gcd(a, b) = d$; then $a^2 b \mid a^3 + b^3 + c^3$ implies $d \mid c$ and $\gcd(a, b, c) = 1$ implies $d = 1$; hence a, b, c are pairwise coprime, so $(abc)^2 \mid a^3 + b^3 + c^3$. Then

$$(abc)^2 \leq a^3 + b^3 + c^3 \leq 3c^3,$$

which implies $3c \geq (ab)^2$. We have

$$c^2 \mid a^3 + b^3 \Rightarrow 2b^3 \geq a^3 + b^3 \geq c^2 \geq \left(\frac{a^2 b^2}{3}\right)^2 \Rightarrow 18 \geq a^4 b \geq a^5 \Rightarrow a = 1.$$

If $b = 1$, then $c^2 \mid a^3 + b^3$ implies $c = 1$.

If $b > 1$, then $\gcd(b, c) = 1$ implies $c \geq b + 1$, so

$$2c^3 \geq c^3 + b^3 + 1 = a^3 + b^3 + c^3 \geq (bc)^2 \Rightarrow c > \frac{b^2}{2}$$

and

$$1 + b^3 = a^3 + b^3 \geq c^2 \geq \left(\frac{b^2}{2}\right)^2 \Rightarrow b \leq 4.$$

From $c^2 | a^3 + b^3 = 1 + b^3$, we have $b = 2$, $c = 3$.

129. If $a = 1$, then $b = 1$, $p = 17$.

If $a > 1$, then $19 - 2 = 17$ divides $19^a - 2^a$; therefore $p = 17$. On the other hand, the Zsigmondy's theorem implies that there is a prime q which divides $19^a - 2^a$, but does not divide $19 - 2 = 17$, contradiction.

130. With the substitution $x = a - k$, $y = a$, $z = a + k$, the given equation becomes

$$a^2 - 3k^2 = 1,$$

which is a Pell equation. It has solutions $(a_0, k_0) = (1, 0)$, $(a_1, k_1) = (2, 1)$; hence the general solution is given by $a_{n+2} = 4a_{n+1} - a_n$, $k_{n+2} = 4k_{n+1} - k_n$.

131. If $2 \notin M$, then take $A = \{p\}$, with $p \in M$. Since $p - 1$ is even, it follows that $2 \in M$, contradiction. It follows that $2 \in M$.

If M is finite, $M = \{2, p_2, \ldots, p_k\}$, $k \geq 3$, then define $A = M - \{p_2\}$ and denote by P the product of elements of M. We have

$$2p_3 \ldots p_k - 1 = \frac{P}{p_2} - 1 = p_2^\alpha \Rightarrow P = p_2^{\alpha+1} + p_2,$$

and if we consider $A = \{p_3, \ldots, p_k\}$, then we obtain

$$p_3 \ldots p_k - 1 = \frac{P}{2p_2} - 1 = 2^\beta p_2^\gamma \Rightarrow P = 2p_2(2^\beta p_2^\gamma + 1).$$

We deduce

$$p_2^{\alpha+1} + p_2 = 2p_2(2^\beta p_2^\gamma + 1) \Rightarrow p_2^\alpha + 1 = 2^{\beta+1} p_2^\gamma + 2 \Rightarrow 1 \equiv 2 \pmod{p_2},$$

contradiction; hence M is infinite.

Suppose that there is a prime $q \notin M$ and let $M = \{2, p_2, p_3, \ldots, p_k, \ldots\}$. By the Pigeonhole principle, it follows that at least two of the numbers

$$2-1, 2p_2-1, \ldots, 2p_2\cdots p_{q+1}-1$$

are congruent modulo q; let them $2p_2\cdots p_i - 1, 2p_2\cdots p_j - 1$, $1 \le i < j \le q+1$. Then

$$2p_2\cdots p_i(p_{i+1}\cdots p_j - 1) \equiv 0 \pmod{q} \Rightarrow p_{i+1}\cdots p_j - 1$$
$$\equiv 0 \pmod{q} \Rightarrow q \in \mathcal{M},$$

contradiction. It follows that $\mathcal{M}$ contains all primes.

132. Denote $f(x) = x + d(x)$ for each positive integer x. We shall prove by induction on k that there are numbers $x_1 < x_2 < \cdots < x_k$ such that x_k begins with the digit 1; it has at least two digits and $f(x_1) = f(x_2) = \ldots f(x_k)$.

For $k = 1$ take $x_1 = 11$.

Suppose that the statement holds for k, and let $x_1 < x_2 < \ldots < x_k$ be the numbers for which the statement holds. Let n be the number of digits of x_k. Obviously we have

$$f(x_1 + 10^n a) = f(x_2 + 10^n a) = \cdots = f(x_k + 10^n a)$$

for all positive integers a; since if x_i is increased by $10^n a$, then $d(x_i)$ is increased by $d(a)$. Let $b = \overline{99\ldots 9} = 10^n - 1$. We have $f(b) > f(x_k)$, since $b > x_k$, $d(b) > d(x_k)$. Moreover, if $\alpha = f(b) - f(x_k)$, then from $b - x_k \ge 8 \cdot 10^{n-1} > 9n$, and from the fact that $d(b) > d(x_k)$, it follows that $\alpha > 9n$.

But $f(b+1) - f(b) = 2 - 9n$, thus $f(b+1) - f(x_k) = \alpha + 2 - 9n > 0$. Let us consider a positive integer t such that $9t \ge \alpha + 2 - 9n > 9(t-1)$, and let us denote by y_i the number $\overline{99\ldots 9}\cdot 10^n + x_i = (10^t - 1)\cdot 10^n + x_i$ and by c the number $(10^t - 1)\cdot 10^n + b$. It is easy to see that $f(y_1) = f(y_2) = \ldots f(y_k)$ and

$$-9 < f(c+1) - f(y_k) = \alpha + 2 - 9n - 9t \le 0.$$

If $f(c+1) - f(y_k) = -2l$, $l \in \{0,1,2,3,4\}$, then $f(c+l+1) - f(y_k) = f(c+1) - -f(y_k) + 2l = 0$; thus $y_{k+1}c + l + 1$ satisfies the requirements.

If $f(c+1) - f(y_k) = -2l + 1$, $l \in \{1,2,3,4\}$, then $f(c+1+l+4) - f(y_k) = f(c+1) - -f(y_k) + 2(l+4) = -2l + 1 + 2l + 8 = 9$. Therefore if we put another 9 in the front of each y_i-s and in the front of c, we obtain the numbers $z_i = \overline{9y_i}$ and $d = \overline{9c}$, for which we have $f(z_1) = f(z_2) = \cdots = f(z_k)$ and for which

$$f(d+1+l+4) - f(z_k) = f(c+1+l+4) - 9 - f(y_k) = 0.$$

Finally we observe that the first digit of $z_{k+1} = d+1+l+4$ is 1 and also the first digit of y_{k+1}, and obviously these numbers have at least two digits; thus the statement is proved.

133. We have

$$a^{25} - a = a(a-1)(a+1)(a^2-a+1)(a^2+1)(a^2+a+1)(a^4-a^2+1) \times$$
$$\times (a^4+1)(a^8-a^4+1).$$

For $a = 2$, $2^{25} - 2 = 2 \cdot 3^2 \cdot 5 \cdot 7 \cdot 13 \cdot 17 \cdot 241$.
For $a = 3$, $3^{25} - 3 = 2^5 \cdot 3 \cdot 5 \cdot 7 \cdot 13 \cdot 41 \cdot 73 \cdot 6481$.

It follows that n divides $2 \cdot 3 \cdot 5 \cdot 7 \cdot 13 = 2730$. But it is easy to see, by the Fermat theorem, that $a^{25} - a$ is divisible by 2, 3, 5, 7, and 13, for example,

$$a^{25} - a = a \cdot (a^{12})^2 - a \equiv a - a \equiv 0 \pmod{13}.$$

We deduce that $n = 2730$.

134. Answer: $(1,2,3)$, $(2,3,4)$, $(2,9,16)$, as well as multiplying these triples by $2^a 3^b$.

Cancelling common powers of 2 and 3, we can assume that x, y, and z are coprime. If two are divisible by 3, then the third is also divisible by 3; hence at most one is divisible by 3.

If none of x, y, z is divisible by 3, then they are powers of 2. But then we can suppose that $x = 1$; hence $y = z = 1$, contradiction.

If exactly one of the numbers is divisible by 3, then at least one is even. If exactly one is even, then it is the middle one, and we can take $x = 1$, $y = 2^a$, $z = 3^b$, so

$$3^b + 1 = 2^{a+1}.$$

But this implies $a = b = 1$; hence $(x,y,z) = (1,2,3)$.

If two numbers are even, then the middle one is odd, hence a power of 3. Then $(x,y,z) = (2^a, 3^b, 2^c)$, so that $2 \cdot 3^b = 2^a + 2^c$. But then $a = 1$ and we get the solutions $(2,3,4)$ and $(2,9,16)$.

135. If $p = 8k - 1$ is a prime and $p \mid 2^n + 1$, then n cannot be even, since $2 \nmid \frac{p-1}{2}$. Hence $n + 1$ is even, $2^{n+1} \equiv -2 \pmod{p}$, so that -2 is a quadratic residue mod p. Then

$$1 = \left(\frac{-2}{p}\right) = \left(\frac{-1}{p}\right)\left(\frac{2}{p}\right) = (-1)^{\frac{p-1}{2}}(-1)^{\frac{p^2-1}{8}} = (-1)^{\frac{(p-1)(p+5)}{8}}$$
$$= (-1)^{(4k-1)(2k+1)} = -1,$$

contradiction.

2.32 2004

136. Answer: no.

A Heronian triangle has side lengths

$$a = n(m^2 + k^2), b = m(n^2 + k^2), c = (m+n)(mn - k^2)$$

and area $A = kmn(m+n)(mn - k^2)$, where k, m, and n are positive integers such that $mn > k^2$. Since $2004 = 2 \cdot 2 \cdot 3 \cdot 167$ has only four factors, it follows that we cannot have $A = 2004$.

137. Answer: $n = 1, 2, 3, 8$.

Let $\mathbb{P}$ be the set of all prime numbers, and let $A(n) = \tau(\text{lcm}(1, 2, \ldots, n))$ be the number of positive divisors of $\text{lcm}(1, 2, \ldots, n)$. Denote by $p \triangleright m$ the exponent of prime number p in the canonical representation of m. Observe that

$$A(n) = \tau\left(\prod_{p \in \mathbb{P}} p^{\max(p \triangleright 1, \ldots, p \triangleright n)}\right) = \tau\left(\prod_{p \in \mathbb{P}} p^{\lfloor \log_p n \rfloor}\right)$$
$$= \prod_{p \in \mathbb{P}} (\lfloor \log_p n \rfloor + 1).$$

It follows that $A(n)$ is a power of 2 if and only if all the numbers $\lfloor \log_p n \rfloor + 1$, $p \in \mathbb{P}$, are powers of 2. Let $\lfloor \log_2 n \rfloor + 1 = 2^k$, $\lfloor \log_3 n \rfloor + 1 = 2^l$. As $\log_2 n \geq \log_3 n$, we get $k \geq l$.

Case 1. $k = l$. Then $\lfloor \log_2 n \rfloor = \lfloor \log_3 n \rfloor$ (*) holds for $n = 1, 3$ and does not hold for $n = 2$ and $4 \leq n < 8$. If $n \geq 8$, then

$$\log_2 n - 3 = \log_2 n - \log_2 8 = \log_2 3 (\log_3 n - \log_3 8)$$
$$> \log_3 n - \log_3 8;$$

hence $\log_2 n - \log_3 n > 3 - \log_3 8 > 3 - 2 = 1$; therefore $\lfloor \log_2 n \rfloor > \lfloor \log_3 n \rfloor$. It follows that (*) holds if and only if $n = 1, 3$.

Case 2. $k > l$. Then $\lfloor \log_2 n \rfloor + 1 \geq 2(\lfloor \log_3 n \rfloor + 1)$, that is,

$$\lfloor \log_2 n \rfloor \geq 2\lfloor \log_3 n \rfloor + 1 \ (**).$$

It is easy to see that this holds for $n = 2, 8$ and does not hold for $4 \leq n \leq 7$ and $9 \leq n < 27$. If $n \geq 27$, then

$$\log_2 n - \log_2 27 = \log_2 3(\log_3 n - \log_3 27) < 2(\log_3 n - 3)$$
$$= 2\log_3 n - 6;$$

hence $2\log_3 n - \log_2 n > 6 - \log_2 27 > 6 - 5 = 1$. We deduce that $\lfloor\log_2 n\rfloor < \lfloor 2\log_3 n\rfloor \le 2\lfloor\log_3 n\rfloor + 1$, for all $n \ge 27$, so that $(**)$ holds if and only if $n = 2, 8$.

The only possible values of n are 1, 2, 3, and 8, for which we get

$$A(1) = 2^0, A(2) = 2^1, A(3) = 2^2, A(8) = 2^5.$$

138. Answer: no.

Case 1. $19 | n + k$, for some $0 \le k \le 17$. Then the equality

$$\prod_{x \in B} x = \prod_{x \in C} x$$

is impossible, since exactly one of these numbers is divisible by 19.

Case 2. $n = 19m + 1$. Let a, b, c be the products of elements of the respective sets. Then $b = c$ implies $a = b^2$. By the Wilson's theorem, $a \equiv 18! \equiv -1 \pmod{19}$. But 19 is a prime number of the form $4t + 3$, so that -1 is not a square modulo 19.

139. Induction on n: Let a_n be a positive integer with the desired property. Then $n = 1$ works, since we can take $a_1 = 5$.

Suppose that a_n exists and construct a_{n+1}. Choose a digit k and append it to the left of a_n. Then $5^n | 10^n k$ and $5^n | a_n$, so that $a_n = 5^n b_n$ with b_n positive integer. Define $a_{n+1} = 5^n(2^n k + b_n)$ and choose k such that $5 | 2^n k + b_n$. This fact is always possible, since $\{1, 3, 5, 7, 9\}$ is a complete residue set modulo 5; therefore $\{2^n k\}$, k odd digit, is also a complete residue set modulo 5, so that we can always choose k such that $2^n k \equiv -b_n \pmod{5}$.

140. If $x + y = 2^k, 0 \le k \le n$, then $1 + xy = 2^{n-k}$; whence $1 + x(2^k - x) = 2^{n-k}$. We get triples

$$(1, 2^k - 1, 2k), (2^k - 1, 1, 2k), (2^k + 1, 2^k - 1, 3k + 1), (2^k - 1, 2^k + 1, 3k + 1).$$

141. If $q^m - 1$ has a prime factor that does not divide $q - 1$, then since $q^m - 1$ is divisible by $q - 1$; we deduce that $q^m - 1$ has at least two distinct prime factors, by the Zsigmondy's Theorem. The exceptional cases are $q = 2$ and $m = 6$, and $m = 2$, $q + 1$ is a power of 2.

The former does not work. For the latter, $p = 2$, since it must be even. Then

$$2^k = q^2 - 1 = 2^y(2^y - 2);$$

whence $q = 3, k = 3$.

142. Suppose that $m = u + v + w$, where u, v, and w are good integers whose product is a perfect square of an odd integer. It follows that $uvw \equiv 1 \pmod 4$, therefore two or none of the numbers are congruent to 3 mod 4. In both cases $m = u + v + w \equiv 3 \pmod 4$.

Conversely, every $m \equiv 3 \pmod 4$ has infinitely many representations of the desired type. We shall represent $m = 4k + 3$ as

$$4k + 3 = xy + yz + zx, \ x, y, z \text{ odd}.$$

The product of the summands is an odd square. If $x = 1 + 2l$, $y = 1 - 2l$, then z must be $2l^2 + 2k + 1$. The summands xy, yz, and zx are distinct, except for finitely many l, so it remains to prove that for infinitely many integers l, $|xy|$, $|yz|$, and $|zx|$ are not perfect squares. Observe that $|xy| = 4l^2 - 1$ is not a perfect square for $l \neq 0$.

Let $p, q > m$ be different prime numbers. By the Chinese remainder theorem, the system of congruences $1 + 2l \equiv p \pmod{p^2}$, $1 - 2l \equiv q \pmod{q^2}$ has infinitely many solutions l. For any solution l, the number $z = 2l^2 + 2k + 1$ is divisible by neither p nor q; hence $|xz|$ and $|yz|$ are divisible by p, but not by p^2, respectively, by q, but not by q^2. It follows that xz and yz are also good numbers.

143. (a) There are arbitrary long arithmetic progressions of perfect powers. Indeed, we can choose, for example, $a_1 = 2^2$, $a_2 = 3^2$. If for some $n \geq 1$, the sequence $a_k = mk + d$, $1 \leq k \leq n$ is an arithmetic progression of perfect powers, that is, $a_k = b_k^{c_k}$, $c_k \geq 2$; then we can find an arithmetic progression $a'_k = m'k + d'$ such that a'_k is a perfect power for $1 \leq k \leq n + 1$. Let $m' = a_{n+1}^r m$, $d' = a_{n+1}^r d$, with $r = \text{lcm}(c_1, \ldots, c_n)$. Then $a'_k = a_k a_{n+1}^r = \left(b_k a_{n+1}^{\frac{r}{c_k}}\right)^{c_k}$ is a perfect power for $1 \leq k \leq n$, and $a'_{n+1} = a_n^{r+1}$ is also a perfect power.

(b) If there is such an arithmetic progression (a_n), then

$$\sum_{n \geq 1} \frac{1}{a_n} = \infty.$$

On the other hand,

$$\sum_{n \geq 1} \frac{1}{a_n} \leq 1 + \sum_{a \geq 2} \sum_{b \geq 2} \frac{1}{a^b} = 1 + \sum_{a \geq 2} \frac{1}{a(a-1)} = 2,$$

contradiction.

144. Suppose that $a^2 + ab + b^2 = k(ab - 1)$ with k nonnegative integer. If $a = b$, then k is an integer for $a \in \{0, 2\}$; whence $k \in \{0, 4\}$. If $b = 0$, then $k = -a^2$; whence $k = 0$. We look for solutions $a > b > 0$. The inequalities

$$k > \frac{a+b}{b} \Longleftrightarrow \frac{a^2 + ab + b^2}{ab - 1} > \frac{a+b}{b} \Longleftrightarrow b^3 + a + b > 0 \text{ and}$$

$$k < \frac{2b+a}{b} \Longleftrightarrow \frac{a^2 + ab + b^2}{ab - 1} < \frac{2b+a}{b} \Longleftrightarrow a > b + \frac{3b}{b^2 - 1}, b > 1,$$

show that if (a, b) is a solution, then $(b, (k - 1)b - a)$ is also a solution. If $3b < b^2 - 1$ then for each solution (a, b) with $a > b \geq 4$, we have a solution (b, c) with $b > c > 0$. By the descent method, each solution corresponds to a solution with $b \leq 3$.

If $b = 1$, then $k = a + 2 - \frac{3}{a-1}$, $a \in \{2, 4\}$, $k = 7$.
If $b = 2$, then $4k = 2a + 5 + \frac{21}{2a-1}$, $a \in \{4, 11\}$, $k \in \{4, 7\}$.
If $b = 3$, then $9k = 3a + 10 + \frac{91}{3a-1}$, with no solutions.
In conclusion, $k \in \{0, 4, 7\}$.

145. Consider the characteristic polynomial $f = X^4 - cX^2 - bX - a$ associated with the given recursion formula. We can check that all the roots of f are simple: if α is a root of both f and $f' = 4x^3 - 2cX - b$, then α is a root of $g = 4f - Xf' = -2cX^2 - 3bX - 4a$, a root of $h = cf + 2Xg = -6bX^2 - (2c^2 + 8a)X - bc$ and a root of $l = 3bg - ch = (2c^3 + 8ac - 9b^2)X + bc^2 - 12ab$. Since $l(\alpha) = 0$ and $2c^3 + 8ac - 9b^2$ is odd, it follows that α is rational, and, because f is monic, we deduce that α is an integer; whence $f'(\alpha)$ is odd, contradiction.

The above result shows that there are coefficients a_1, a_2, a_3, a_4, uniquely determined by x_1, x_2, x_3, x_4, such that

$$x_n = a_1\alpha_1^n + a_2\alpha_2^n + a_3\alpha_3^n + a_4\alpha_4^n,$$

where $\alpha_1, \alpha_2, \alpha_3,$ and α_4 are the roots of f. We notice that

$$x_0 = 4 = \alpha_1^0 + \alpha_2^0 + \alpha_3^0 + \alpha_4^0,$$
$$x_1 = 0 = \alpha_1 + \alpha_2 + \alpha_3 + \alpha_4,$$
$$x_2 = 2c = \alpha_1^2 + \alpha_2^2 + \alpha_3^2 + \alpha_4^2 = -2\sum \alpha_1\alpha_2,$$
$$x_3 = 3b = \alpha_1^3 + \alpha_2^3 + \alpha_3^3 + \alpha_4^3 = 3\sum \alpha_1\alpha_2\alpha_3,$$

which proves that $a_1 = a_2 = a_3 = a_4 = 1$.

Remark now that if p is a prime; then $p | \binom{p}{k}$ for each $k = 1, 2, \ldots, p-1$; hence

$$(X+Y)^p = X^p + Y^p + pQ(X,Y),$$

where Q is a symmetric integer polynomial in two variables This easily extends to

$$(X+Y+Z+T)^p = X^p + Y^p + Z^p + T^p + pR(X,Y,Z,T),$$

where R is a symmetric polynomial in four variables.

In order to prove the required conclusion, we shall use induction on m. For $m=1$ we notice that

$$x_p = \alpha_1^p + \alpha_2^p + \alpha_3^p + \alpha_4^p = -pR(\alpha_1, \alpha_2, \alpha_3, \alpha_4),$$

and since R can be represented as an integer symmetric polynomial in symmetric fundamental polynomials, $R(\alpha_1, \alpha_2, \alpha_3, \alpha_4)$ is an integer.

Assume now that x_{p^m} is divisible by p. Then

$$\begin{aligned} x_{p^{m+1}} &= \alpha_1^{p^{m+1}} + \alpha_2^{p^{m+1}} + \alpha_3^{p^{m+1}} + \alpha_4^{p^{m+1}} \\ &= pR\left(\alpha_1^{p^m}, \alpha_2^{p^m}, \alpha_3^{p^m}, \alpha_4^{p^m}\right) - \\ &\quad - \left(\alpha_1^{p^m} + \alpha_2^{p^m} + \alpha_3^{p^m} + \alpha_4^{p^m}\right)^p = pS(\alpha_1, \alpha_2, \alpha_3, \alpha_4) - x_{p^m}^p \end{aligned}$$

is also divisible by p, because S is also a symmetric integer polynomial; hence $S(\alpha_1, \alpha_2, \alpha_3, \alpha_4)$ is an integer.

146. If $\omega = e^{\frac{2\pi i}{3}}$, then

$$3\sum_{k=0}^{m}\binom{3m}{3k}(3n-1)^k = \left(1+\sqrt[3]{3n-1}\right)^{3m} + \left(1+\omega\sqrt[3]{3n-1}\right)^{3m} + $$
$$+\left(1+\omega^2\sqrt[3]{3n-1}\right)^{3m}.$$

The right side of the above equality is the sum of the $3m-$ power of the roots x_1, x_2, x_3 of the polynomial

$$(X-1)^3 - (3n-1) = X^3 - 3x^2 + 3X - 3n.$$

Let $s_k = x_1^k + x_2^k + x_3^k$. Then $s_0 = s_1 = s_2 = 3$ and $s_{k+3} = 3s_{k+2} - 3s_{k+1} + 3ns_k$. It follows by induction that each s_k is an integer divisible by $3^{\lfloor \frac{k}{3} \rfloor + 1}$. A repeated application of the above recurrence relation yields

$$s_{k+7} = 63ns_{k+2} - 9(n^2 - 3n - 3)s_{k+1} + 27n(2n+1)s_k.$$

Since $s_3 = 9n$, it follows inductively that s_{6k+3} is divisible by $3^{2k+2}n$ for all nonnegative integers k and the conclusion follows.

147. If m is odd, then $n|(n-1)^m - 1$ implies $n|2$; hence $n \leq 2$.

Take now $m = 2^t q$, $t \geq 1$, q odd. If $n = 2^u(2v+1)$ is m-good, then $2v+1$ divides $(2v-1)^m - 1$; whence $2v+1 | 2^m - 1$. If $a = 8v+5$, then gcd $(a, n) = 1$, so

$$2^u | (a^q)^{2^t} - 1 = (a^q - 1)(a^q + 1)(a^{2q} + 1) \cdots \left(a^{2^{t-1}q} + 1\right).$$

But $a^q \equiv 5 \pmod 8$ implies that the exponent of the factor 2 in the last product is $t + 2$; therefore $u \leq t+2$; whence $n \leq 4 \cdot 2^t(2v+1) \leq 4m(2^m - 1)$.

148. If $x^{p-1} \not\equiv 1 \pmod{p^2}$, then there is a prime divisor q of x such that $q^{p-1} \not\equiv 1 \pmod{p^2}$; otherwise we can multiply some numbers which are all $1 \pmod{p^2}$ to obtain x^{p-1}, contradiction.

For $p = 5$, we have $2^{5-1} \not\equiv 1 \pmod{25}$, $3^{5-1} \not\equiv 1 \pmod{25}$.
For $p = 7$, we have $2^{7-1} \not\equiv 1 \pmod{49}$, $3^{7-1} \not\equiv 1 \pmod{49}$.
Suppose that $p \geq 11$ and call a prime q good if $q^{p-1} \not\equiv 1 \pmod{p^2}$. Observe that

$$(p-1)^{p-1} \equiv p+1 \not\equiv 1 \pmod{p^2}, (p+1)^{p+1} \equiv -p+1 \not\equiv 1 \pmod{p^2}.$$

We conclude that both $p-1$ and $p+1$ have a good prime divisor. If these prime divisors are odd, then we are finished. Otherwise, 2 is good. From

$$(2p-1)^{p-1} \equiv -2p(p-1) + 1 \not\equiv 1 \pmod{p^2},$$
$$(2p+1)^{p-1} \equiv 2p(p-1) + 1 \not\equiv 1 \pmod{p^2},$$

we deduce that $2p - 1$ or $2p + 1$ has a good prime divisor, since

$$(2p-1)(2p+1) = 4p^2 - 1 \equiv 0 \pmod 3.$$

149. Answer: $n = 2004$.

More generally, if $a < b$ are positive integers, then $n = 2b - a - 1$ is the greatest n for which $4^a + 4^b + 4^n$ is a quadrate. Indeed, if $n = 2b - a - 1$, then

$$4^a + 4^b + 4^n = \left[2^a\left(1 + 2^{2(b-a)-1}\right)\right]^2.$$

Suppose that $n \geq 2b - a$. Then $4^a + 4^b + 4^n = (2^a)^2(1 + 4^{b-a} + 4^{n-a})$ and

$$(2^{n-a})^2 < 1 + 4^{b-a} + 4^{n-a} < (1 + 2^{n-a})^2,$$

so that $4^a + 4^b + 4^n$ cannot be a square.

2.33 2005

150. Answer: $a = 9, p = 2, b = 6$.

Suppose that $a < 9$. If $p|a$, then test the cases $a < 9, p|a$. If $p \nmid a$, then $\frac{a^{p-1}-1}{p}$ is an integer, by the Fermat's theorem, and it is coprime with a. Then it follows that a is a square; hence $a = 4$. Denoting $x = \frac{b}{2}$, we have

$$4^{p-1} - 1 = px^2 \Leftrightarrow (2^{p-1} + 1)(2^{p-1} - 1) = px^2$$

(since $p \nmid 4$ implies $p \neq 2$). The factors of the LHS are coprime; therefore one is y^2 and the other is pz^2. If $2^{p-1} + 1 = y^2$, then $y + 1$ and $y - 1$ are both powers of 2, which is impossible. If $2^{p-1} - 1 = y^2$, then $y^2 \equiv 3 \pmod 4$, contradiction.

151. If $n = 2$, then $f \leq \frac{2}{3}$, with equality for $(x_1, x_2) = (1, 2)$. Let $d = \gcd(x_1, x_2)$, $x_1 = da, x_2 = db$. Then $f = \frac{2}{a+b}$. Using the fact that x_1 and x_2 are distinct, it follows that $a + b \geq 3$, consequently $f \leq \frac{2}{3}$.

If $n \geq 3$, note that $\gcd(x_i, S_i) \leq x_i$; hence $f \leq 1$. The equality holds if x_i divides $x_1 + \cdots + x_n$ for every i. If the equality holds for $(x_1, \ldots, x_{n-1})$, then it holds also for $(x_1, \ldots, x_{n-1}, x_1 + \cdots + x_{n-1})$. Since the equality holds for $(1, 2, 3)$, then we can construct a n-tuple which enjoys the condition, for every $n \geq 3$.

152. Due to the homogeneity, we can suppose that $(a, b, c) = 1$. Multiplying by abc we get $a^2c + b^2a + c^2b = 3abc$. If $abc = \pm 1$, then the statement is obvious. Otherwise, let p be a prime divisor of abc. From the new equation, it follows that p divides exactly two of the numbers a, b, and c. By symmetry, suppose that p divides a, b, but not divides c. Denote $m = v_p(a)$. $n = v_p(b)$.

If $n \leq 2m - 1$, then p^{n+1} divides $a^2c, b^2a, 3abc$, so that p^{n+1} divides c^2b; therefore p divides c, contradiction.

If $n \geq 2m + 1$, then p^{2n+1} divides $b^2a, c^2b, 3abc$, so that p^{2n+1} divides a^2c, therefore p divides c, contradiction.

We deduce $n = 2m$ and

$$abc = \prod_{p|abc} p^{3m} = \left(\prod_{p|abc} p^m\right)^3.$$

153. Answer: $(x, y) = (2, 6)$ or $(3, 6)$.

From $x^y = (x+y)^x > x^x$, it follows $y > x$. Take $y = rx$, with $r > 1$ rational number. Then $(x + rx)^x = x^{rx}$ implies $1 + r = x^{r-1}$. If $r - 1 = \frac{p}{q}$, with p, q positive integers, then $x^{r-1} = \sqrt[q]{x^p}$ is either integer or irrational. But the left hand side is rational; therefore r is a positive integer.

If $r = 2$, then $x = 3$ and $y = 2x = 6$.

If $r = 3$, then $x = 2$ and $y = 3x = 6$.

If $r \geq 4$, then $1 + r < x^{r-1}$, for every $x \geq 2$; therefore no more solutions exist.

154. We can rewrite the equation as

$$(8x+1)^2 = (4x)^2 + (8y+1)^2;$$

whence we deduce that there are integers a, b such that

$$8x+1 = a^2 + b^2, \; 4x = 2ab, \; 8y+1 = a^2 - b^2.$$

Subtracting the third equality from the first, we get $8(x-y) = 2b^2$; therefore b is even and

$$x - y = \left(\frac{b}{2}\right)^2$$

is a square, as desired.

155. Answer: 502, 1003.

Denote by g the inverse transformation. Since $2005 + 1 \not\equiv 0 \pmod{3}$, we deduce

$$g(2005) = \frac{2005+1}{2} = 1003.$$

Since $1003 + 1 \not\equiv 0 \pmod{3}$, we deduce

$$g(1003) = \frac{1003+1}{2} = 502.$$

Since $502 \equiv 4 \pmod{6}$, the inverse transformation stops (more generally, numbers $6k$, $6k+4$ cannot be similar to a smaller number).

156. If A contains $1, a, b$, then A contains $a+1$ (property 2), and similarly, A contains all positive integers greater than a. Then A contains $2a, 3a, \ldots, (a-1)a$; therefore A contains $2, 3, \ldots, a-1$. It follows that $A = \mathbb{N}^*$.

157. It is easy to see that the quadruples

$$\left(2^n, 2^2, 2^n, 2^{n+1}\right), \left(2^{n-1}, 2^{n-1}, 2^{n-1}, 5 \cdot 2^{n-1}\right), \left(2^{n-1}, 3 \cdot 2^{n-1}, 3 \cdot 2^{n-1}, 3 \cdot 2^{n-1}\right)$$

are solutions of the given equation. We shall prove that these quadruples and their permutations are all solutions.

Let (a, b, c, d) be any solution. Then $a^2 + b^2 + c^2 + d^2 \equiv 0 \pmod{4}$; therefore

$$a^2 \equiv b^2 \equiv c^2 \equiv d^2 \equiv 0 \pmod{4} \text{ or } a^2 \equiv b^2 \equiv c^2 \equiv d^2 \equiv 1 \pmod{4},$$

since $x^2 \equiv 0, 1 \pmod{4}$ for every integer x. In the latter case a, b, c, d are odd; hence $a^2 \equiv b^2 \equiv c^2 \equiv d^2 \equiv 1 \pmod{8}$, and we can set

$$a^2 = 1 + 8a_1, b^2 = 1 + 8b_1, c^2 = 1 + 8c_1, d^2 = 1 + 8d_1;$$

where a_1, b_1, c_1, d_1 are nonnegative integers. From the equation

$$1 + 2(a_1 + b_1 + c_1 + d_1) = 7 \cdot 4^{n-1},$$

we see that $n = 1$ and $a_1 + b_1 + c_1 + d_1 = 3$. Up to permutations, (a_1, b_1, c_1, d_1) is one of the quadruples $(0, 0, 0, 3)$, $(0, 0, 1, 2)$, $(0, 1, 1, 1)$; whence $(a, b, c, d) = (1, 1, 1, 5)$ or $(1, 3, 3, 3)$ (up to permutations).

Suppose that a, b, c, and d are all even. Then

$$a = 2^\alpha t, b = 2^\beta u, c = 2^\gamma v, d = 2^\delta w,$$

for some odd positive integers t, u, v, w and positive integers $\alpha, \beta, \gamma, \delta$, for which we assume wlog $\alpha \leq \beta \leq \gamma \leq \delta$. Then

$$2^{2\alpha}\left(t^2 + 2^{2(\beta-\alpha)}u^2 + 2^{2(\gamma-\alpha)}v^2 + 2^{2(\delta-\alpha)}w^2\right) = 7 \cdot 2^{2n}.$$

If $\beta > \alpha$, then $\alpha = n$ and $t^2 + 2^{2(\beta-\alpha)}u^2 + 2^{2(\gamma-\alpha)}v^2 + 2^{2(\delta-\alpha)}w^2 = 7$, which is clearly impossible. Thus, $\alpha = \beta$ and

$$2^{2\alpha}\left(t^2 + u^2 + 2^{2(\gamma-\alpha)}v^2 + 2^{2(\delta-\alpha)}w^2\right) = 7 \cdot 2^{2n}.$$

Since $t^2 + u^2$ is twice an odd integer, we must have $\gamma = \alpha$ (otherwise the total exponent of 2 would be odd on the left and even on the right), and finally,

$$2^{2\alpha}\left(t^2 + u^2 + v^2 + 2^{2(\delta-\alpha)}w^2\right) = 7 \cdot 2^{2n}.$$

If $\delta = \alpha$, then $2^{2\alpha}(t^2 + u^2 + v^2 + w^2) = 7 \cdot 2^{2n}$, and hence $\alpha < n$ and

$$t^2 + u^2 + v^2 + w^2 = 7 \cdot 2^{2(n-\alpha)}.$$

By the previous results, $\alpha = n - 1$ and $(t, u, v, w) = (1, 1, 1, 5)$ or $(1, 3, 3, 3)$ (up to permutations), and thus,

$$(a, b, c, d) = \left(2^{n-1}, 2^{n-1}, 2^{n-1}, 5 \cdot 2^{n-1}\right) \text{ or } \left(2^{n-1}, 3 \cdot 2^{n-1}, 3 \cdot 2^{n-1}, 3 \cdot 2^{n-1}\right)$$

(up to permutations). If $\delta > \alpha$, then $t^2 + u^2 + v^2 + 2^{2(\delta-\alpha)}w^2$ is odd and

$$t^2 + u^2 + v^2 + 2^{2(\delta-\alpha)}w^2 = 7 \cdot 2^{2(n-\alpha)}.$$

The only possibility is $n = \alpha$, $\delta = \alpha + 1$, $t = u = v = w = 1$; therefore

$$(a, b, c, d) = \left(2^n, 2^n, 2^n, 2^{n+1}\right).$$

158. Consider the sets $A_n = \{9(n-1), 9(n-1)+1, \ldots, 9n-1\}$, and let S_n be the sum of nice numbers in A_n and $x_n = 9(n-1)$ be the smallest number in A_n. We shall prove that $S_n = 3x_n + 12$ for all positive integers n. Clearly, exactly one number in the set $\{x_n, x_n+1, x_n+2\}$ is nice.

Case 1. x_n is nice. Since x_n ends with 00, we see that in A_n only $x_n + 5$ (which ends in 12), $x_n + 7$ (which ends in 21) and x_n are nice; therefore $S_n = 3x_n + 12$.

Case 2. $x_n + 1$. Since $x_n + 1$ ends with 01, we see that in A_n only $x_n + 3$ (which ends with 10), $x_n + 8$ (which ends with 22) and $x_n + 1$ are nice; therefore $S_n = 3x_n + 12$.

Case 3. $x_n + 2$ is nice. Since $x_n + 2$ ends with 02, we see that in A_n only $x_n + 4$ (which ends with 11), $x_n + 6$ (which ends with 20) and $x_n + 2$ are nice; therefore $S_n = 3x_n + 12$.

We conclude that $S_{n+1} = S_n + 27$ for all positive integers n. All the sets A_n have 3 nice integers, excepting A_1, which has only two. The first 2005 nice integers are all those in $A_1, \ldots, A_{668}$ and the last two in A_{669}, so that their sum is

$$12 + (12 + 27) + \cdots + (12 + 667 \cdot 27) + 6013 + 6015 = 6035050.$$

159. For $n_0 = 2$, $a = (m-1)^2(m+3)$, $b = m(m-1)(m+3)$, $c = m^2(m+3)$, both condition are satisfied:

$$m^3 < a < b < c < (m+1)^3,$$

$$abc = [m(m-1)(m+3)]^3, m \geq 2.$$

160. For $m = an$, the given equation reduces to $a\tau(m) = m$. Choose a prime $p > 3$, take $a = p^{p-1}$, and suppose that there is a positive integer m such that $p^{p-1}\tau(m) = m$. Then $p^{p-1}|m$, so that $m = p^\alpha k$, $\alpha \geq p-1$, $p \nmid k$. If $k = p_1^{\alpha_1} \ldots p_r^{\alpha_r}$ is the decomposition of k into primes; then $\tau(k) = (\alpha_1 + 1)\ldots(\alpha_r + 1)$, $\tau(m) = (\alpha+1)\tau(k)$, and the equation becomes

$$p^{\alpha-p+1}k = (\alpha+1)\tau(k) \ (*).$$

Observe that $\alpha \neq p-1$; otherwise the RHS is divisible by p and the LHS is not; hence $\alpha \geq p$ and

$$p^{\alpha-p+1} \geq \frac{p(\alpha+1)}{p+1}.$$

For $\alpha \geq p$, $\alpha + 1$ cannot be divisible by $p^{\alpha-p+1}$, so that $p|\tau(k)$, that is, there is at least one i for which $p|\alpha_i + 1$, consequently $\alpha_i \geq p-1$ for some i. We deduce

$$k \geq \frac{p_i^{\alpha_i}\tau(k)}{\alpha_i+1} \geq \frac{2^{p-1}\tau(k)}{p},$$

$$p^{\alpha-p+1}k \geq \frac{p(\alpha+1)}{p+1} \cdot \frac{2^{p-1}\tau(k)}{p} > (\alpha+1)\tau(k),$$

which contradicts (∗).

161. Answer: 3, 4, 6, p^{2k}, $2p^{2k}$, for $p \equiv 11 \pmod{12}$.

If $\alpha \geq 2$ and $2^\alpha | n$, then $n = 4$ or $4 | \varphi(n)$.

If $n = p^\alpha$, $2p^\alpha$, p odd, then $\varphi(n) = (p-1)p^{\alpha-1} \equiv 2 \pmod{12}$, which implies $p = 3$ ($\alpha = 1$) or $p \equiv 2 \pmod 3$ (α even). As $p \equiv 3 \pmod 4$, we find the above solutions.

162. Let n be a positive integer. For each k, $1 \leq k \leq n$, the number $\gcd(k, n)$ is a divisor of n. If $d | n$ and $\gcd(k, n) = \frac{n}{d}$, then $k = \frac{nl}{d}$ for some positive integer l and

$$\gcd(k, n) = \frac{\gcd(l, d)n}{d},$$

which implies $\gcd(l, d) = 1$ and $l \leq d$. It follows that $\gcd(k, n) = \frac{n}{d}$ for exactly $\varphi(d)$ positive integers $k \leq n$. Therefore

$$\psi(n) = \sum_{k=1}^{n} \gcd(k, n) = \sum_{d|n} \frac{\varphi(d)n}{d} = n \sum_{d|n} \frac{\varphi(d)}{d}.$$

(a) If $\gcd(m, n) = 1$, then each divisor f of mn can be uniquely expressed as $f = de$, where $d|n$, $e|m$. Now we have

$$\psi(mn) = mn \sum_{f|mn} \frac{\varphi(f)}{f} = mn \sum_{d|n,\, e|m} \frac{\varphi(de)}{de} = mn \sum_{d|n,\, e|m} \frac{\varphi(d)}{d} \cdot \frac{\varphi(e)}{e} =$$

$$= \left(n \sum_{d|n} \frac{\varphi(d)}{d}\right)\left(m \sum_{e|m} \frac{\varphi(e)}{e}\right) = \psi(m)\psi(n).$$

(b) If p is a prime, k is a positive integer and $n = p^k$; then

$$\frac{\psi(n)}{n} = \sum_{i=0}^{k} \frac{\varphi(p^i)}{p^i} = 1 + \frac{k(p-1)}{p}.$$

Setting $p = 2$, $k = 2(a-1)$, we have $\psi(n) = an$.

163. Denote
$$\sigma = \sum_{m=1}^{n^2-1} \{\sqrt{m}\}.$$

Then $A = 2n^3 - \frac{n}{2} - 3\sigma$. The function $f(x) = \sqrt{x}$ is concave; thus

$$\int_m^{m+1} f(x)dx > \frac{\sqrt{m} + \sqrt{m+1}}{2}, \quad (*)$$

$$\int_{m-\frac{1}{2}}^{m+\frac{1}{2}} f(x)dx < \sqrt{m}. \quad (**)$$

From $(*)$ we conclude that

$$\sigma < \int_0^{n^2-1} f(x)dx + \frac{\sqrt{n^2-1}}{2} = \frac{2\sqrt{(n^2-1)^3}}{3} + \frac{\sqrt{n^2-1}}{2} < \frac{2n^3}{3} - \frac{n}{2}.$$

From $(**)$ we conclude that

$$\sigma > \int_{\frac{1}{2}}^{n^2-\frac{1}{2}} f(x)dx = \frac{2}{3}\left(n^2 - \frac{1}{2}\right)^{\frac{3}{2}} - \frac{2}{3}\left(\frac{1}{2}\right)^{\frac{3}{2}} > \frac{2n^3}{3} - \frac{n}{2} - \frac{1}{3}.$$

(the last inequality follows from the Bernoulli's inequality).
Thus $n < A < n+1$; therefore $\lfloor A \rfloor = n$; hence $k = 1$.

164. The equation is equivalent to $(x - N)(y - N) = N^2$; hence it has $\tau(N^2)$ solutions. Since $2005 = 5 \cdot 401$, it follows that $\tau(N^2) = (2 \cdot 2 + 1)(2 \cdot 100 + 1)$, so that

$$N = p^2 q^{100} = (pq^{50})^2$$

is a square (p, q are arbitrary prime numbers).

165. (a) Let r_i be the quadratic residues mod p. The given sum is equal to

$$\sum_{i=1}^{\frac{p-1}{2}} \left(2\left\{\frac{r_i}{p}\right\} - \left\{\frac{2r_i}{p}\right\}\right).$$

Every term is 1 if and only if $r_i > \frac{p-1}{2}$, else it is 0; hence the sum is equal to the number of quadratic residues greater than $\frac{p-1}{2}$. Since -1 is a quadratic residue, it follows that the number of quadratic residues greater than $\frac{p-1}{2}$ is equal to the number of quadratic residues less than $\frac{p-1}{2}$; hence

$$S = \frac{1}{2} \cdot \frac{p-1}{2} = \frac{p-1}{4}.$$

(b) 2 is a quadratic residue; hence

$$S = \sum_{i=1}^{\frac{p-1}{2}} \left(2\left\{\frac{r_i}{p}\right\} - \left\{\frac{2r_i}{p}\right\}\right) = \sum_{i=1}^{\frac{p-1}{2}} \frac{r_i}{p} = \sum_{i=1}^{\frac{p-1}{2}} \frac{i^2}{p} - T,$$

since the set of $2r_i$ coincides with the set of r_i mod p. It follows that

$$T = \sum_{i=1}^{\frac{p-1}{2}} \frac{i^2}{p} - S = \frac{p^2 - 1}{24} - \frac{p-1}{4} = \frac{(p-1)(p-5)}{24}.$$

2.34 2006

166. The substitutions $x = a + 1$, $y = b + 1$, $z = c + 1$ lead to the equation

$$abc = a + b + c.$$

Suppose that $c = 0$; then $b = -a$ and we have the solution $(1 + a, 1 - a, 1)$. For $a \geq b \geq c > 0$, it follows $c^2 a \leq 3a$, so that $c = 1$, with solution $(4, 3, 2)$.

167. Let $k = \lfloor \sqrt{p} \rfloor$, $a = p - k^2 \leq 2k + 1$. Hence $a \in S$. Also, $|k - a| < \sqrt{p}$. For $b = p^2 - (k - a)^2$, we have $1 < a < b$ and $a|b$.

168. Note that

$$f_5(x, y) = (x + y)^5 - x^5 - y^5 = 5xy(x + y)(x^2 + xy + y^2),$$

$$f_7(x, y) = (x + y)^7 - x^7 - y^7 = 7xy(x + y)(x^2 + xy + y^2)^2.$$

It is obvious that a prime $p \neq 5, 7$ divides f_5 if and only if p divides f_7, since $p|a^2$ is equivalent to $p|a$ for any nonzero integer a. Now,

(a) 29 is nice, since it is a prime $\neq 2, 5$.
(b) $2006 = 2 \cdot 17 \cdot 59$ is nice.
(c) every prime $p > 7$ is nice.

126   2 Solutions

169. Answer: $n = 1$.

Let p be the smallest prime divisor of n. If p divides a or $a + 1$, then p divides their difference $a + 1 - a = 1$, contradiction. Then there exists b, $1 \leq b \leq p - 2$ such that $ab \equiv 1 \pmod{p}$. Multiplying $(a + 1)^n \equiv a^n \pmod{p}$ by b^n, we get $(b + 1)^n \equiv 1 \pmod{p}$. But $(b + 1)^{p-1} \equiv 1 \pmod{p}$, by the Fermat's theorem. From $\gcd(n, p - 1) = = 1$ (p is the smallest prime divisor of n), we deduce $b + 1 \equiv 1 \pmod{p}$, so that p divides b, contradiction.

170. We show by induction that there is $k \in \mathbb{N}$ such that $2^k - m$ has at least n odd prime divisors. If m is even, we can write $n = 2^e s$ with odd s and take $k > e$, so that we have $2^k - m = 2^e(2^{k-e} - s)$. Then it suffices to prove the assertion for m odd.

Suppose that m is odd, take $k \in \mathbb{N}$ such that $2^k - m > 1$, and choose a prime divisor p of $2^k - m$. Assume that we have $k \in \mathbb{N}$ such that $2^k - m$ has n odd prime divisors $p_1, p_2, \ldots, p_n$. For any i the value of $2^j \pmod{p_i}$ is periodic for j, which implies that there are $e_i, f_i \in \mathbb{N}$ such that $2^j \equiv m \pmod{p_i}$ if and only if $j = e_i t + f_i$ for some $t \in \mathbb{N}$. Since $p_i > 2$, each e_i is greater than 1. Thus, we can take f'_i such that $f_i \not\equiv f'_i \pmod{e_i}$. By the Chinese remainder theorem, we can take k' such that $k' \equiv f'_i \pmod{e_i}$ and we have $p_i | 2^{k'} - m$ for $1 \leq i \leq n$. We can also select k' such that $2^{k'} - m > 1$. Then we can take an odd prime divisor of $2^{k'} - m$, which is different from $p_1, p_2, \ldots, p_n$. Choose j such that $2^j \equiv m \pmod{p_{n+1}}$, where $j = e_{n+1} t + f_{n+1}$ for some positive integers e_{n+1}, f_{n+1}. By the Chinese remainder theorem again, we can take K such that $K \equiv f'_i \pmod{e_i}$, and we have $p_i | 2^K - m$ for $1 \leq i \leq n + 1$; therefore $2^K - m$ has at least $n + 1$ prime factors.

171. Answer: $(x, y) = (1, 1), (3, 2), (4, 5)$.

We shall use the notation $p^n \| t$, which means that $p^n | t$ and $p^{n+1} \nmid t$. Note that if p is a prime, $p^i \| a$, $p^j \| b$; then $p^{i+j} \| ab$. We need two lemmas.

Lemma 1. For any positive integer n, we have $2^{n+2} \| 3^{2^n} - 1$.

Proof. Induction on n. For $n = 1$, $2^{1+2} \| 3^{2^1} - 1$ holds. Suppose that Lemma 1 is true for $n \geq 1$ and prove that it is true for $n + 1$. Note that

$$3^{2^{n+1}} - 1 = \left(3^{2^n} - 1\right)\left(3^{2^n} + 1\right).$$

From the induction hypothesis, we know that $2^{n+2} \| 3^{2^n} - 1$. But $n + 2 \geq 3$, so that $4 | 2^{n+2}$; therefore $4 | 3^{2^n} - 1$. As a consequence, we get that $4 \nmid 3^{2^n} + 1$, and since $3^{2^n} + 1$ is even, it follows that $2 \| 3^{2^n} + 1$. Using the above property, we get

$$2^{n+3} \| \left(3^{2^n} - 1\right)\left(3^{2^n} + 1\right) = 3^{2^{n+1}} - 1.$$

Lemma 2. For any positive integer $n \geq 3$, we have $2^n > n + 2$.

Proof. Induction over n. For $n = 3$, the inequality $2^3 > 3 + 2$ is true. Suppose that Lemma 2 is true for some $n \geq 3$ and prove it for $n + 1$. We have

$$2^{n+1} = 2 \cdot 2^n > 2(n+2) = 2n + 4 > n + 3.$$

Back to the equation, we rewrite it as

$$2^x y = 2\left(3^{x-1} + 3^{x-2} + \cdots + 1\right).$$

Let n, α be nonnegative integers such that $x = 2^n \alpha$, with α odd.

Case 1. $n = 0$. Since $S = 3^{x-1} + 3^{x-2} + \cdots + 1$ is a sum of x odd terms and x is also odd, it follows that S is odd. Therefore $2 \parallel 2^x y$; whence $x = 1$, which implies $y = 1$.

Case 2. $n \geq 1$. The equation becomes

$$2^{2^n \alpha} y = \left(3^{2^n}\right)^{\alpha} - 1 = \left(3^{2^n} - 1\right)\left(\left(3^{2^n}\right)^{\alpha-1} + \left(3^{2^n}\right)^{\alpha-2} + \cdots + 1\right).$$

Since $T = \left(3^{2^n}\right)^{\alpha-1} + \left(3^{2^n}\right)^{\alpha-2} + \cdots + 1$ is a sum of α odd terms and α is also odd, it follows that T is odd. Combine this with Lemma 1 to obtain $2^{n+2} \parallel$ RHS. Let t be a positive integer such that $2^t \parallel 2^{2^n \alpha} y$. The equality holds if and only if $t = n + 2$. Clearly, $t \geq 2^n \alpha$.

If $n \geq 3$, then from Lemma 1 we get $t \geq 2^n \alpha \geq 2^n > n + 2$, so that the equality is impossible. If $n = 2$, we must have $t = 4$. But $t \geq 4\alpha \geq 4$; whence $\alpha = 1$; therefore

$$x = 2^n \alpha = 4, y = 5.$$

If $n = 1$ we must have $t = 3$. But $t \geq 2\alpha$, therefore $\alpha = 1, x = 2^n \alpha = 2, y = 2$.

In conclusion, the solutions of the equation are $(1, 1)$, $(2, 2)$, and $(4, 5)$.

172. If $b \neq a$, then $b > a$. Let $p > b$ be a prime number and let $n = (a+1)(p-1) + 1$. Observe that $n \equiv 1 \pmod{p-1}$ and $n \equiv -a \pmod{p}$. It follows that for any integer r,

$$r^n = r \cdot \left(r^{p-1}\right)^{a+1} \equiv r \pmod{p}.$$

Now, $a^n + n \equiv a - a \equiv 0 \pmod{p}$; hence $p \mid b^n + n$. But $b^n + n \equiv b - a \pmod{p}$, which contradicts $p > b$. It follows that $a = b$.

173. Define the polynomial $P(x) = (x+a)(x+b)(x+c) - (x-d)(x-e)(x-f)$
$= Sx^2 + Qx + R$, where $Q = ab + bc + ca - de - ef - fd$, $R = abc + def$. Since $S \mid Q, R$ it follows that $S \mid P(x)$ for every $x \in \mathbb{Z}$. We deduce that

$$S \mid P(d) = (d+a)(d+b)(d+c).$$

Since $S > d+a$, $d+b$, $d+c$, we deduce that S cannot divide any factor; therefore S is composite.

174. We will prove first that among the numbers $n = 2^k$, $k \in \mathbb{N}$, there are infinitely many satisfying the inequality $w(n) < w(n+1)$. If there is a value $n = 2^k$ for which the converse inequality $w(n) \geq w(n+1)$ holds, then

$$1 = w(2^k) \geq w(2^k + 1) \geq 1,$$

so that $w(2^k + 1) = 1$ and $2^k + 1 = p^m$, where p is a prime number and m a positive integer. If m is even, $m = 2l$; then

$$2^k = p^{2l} - 1 = (p^l - 1)(p^l + 1);$$

therefore both numbers $p^l - 1$ and $p^l + 1$ are powers of 2, which is possible only when $p^l + 1 = 4$, $p^l - 1 = 2$, that is, for $p = 3$, $l = 2$, $k = 3$. If m is odd, then $m = 1$; otherwise $2^k = p^m - 1 = (p-1)(p^{m-1} + \cdots + p + 1)$ implies that the number

$$p^{m-1} + \cdots + p + 1 > 1$$

is odd and a power of 2, contradiction. The equality $2^k + 1 = p$ implies that k is a power of 2. It follows that if $k \neq 3$ is not a power of 2, then $w(2^k) < w(2^k + 1)$.

Suppose that the assertion from the enounce is false. Then among the numbers k satisfying $w(2^k) < w(2^k + 1)$, only finitely many satisfy also the inequality

$$w(2^k + 1) < w(2^k + 2).$$

Then there is $k_0 = 2^{q_0} > 5$ such that for all k, $k_0 + 1 < k < 2k_0$, we have

$$w(2^k + 1) \geq w(2^k + 2) = w(2^{k-1} + 1) + 1;$$

whence we deduce

$$w(2^{2k_0 - 1} + 1) \geq w(2^{2k_0 - 2} + 1) + 1 \geq \ldots \geq (k_0 - 1) + w(2^{k_0} + 1) \geq k_0.$$

If $p_1 < p_2 < p_3 < \ldots$ are all prime numbers, then

$$2^{2k_0-1} + 1 \geq p_1 p_2 \ldots p_{k_0} = (2 \cdot 3 \cdot 5 \cdot 7 \cdot 11)(p_6 \ldots p_{k_0}) >$$
$$> 4^5 \cdot 4^{k_0-5} = 2^{2k_0} > 2^{2k_0-1} + 1,$$

contradiction.

175. Answer: all prime numbers.

We will show that n has the desired property if and only if it is prime.

For $n = 2$ we can take only $a = 1$. For $n > 2$ even, we have $4|n!$, but $a^n + 1 \equiv 1, 2 \pmod{4}$, contradiction. If n is odd, then $(n! - 1)^n + 1 \equiv (-1)^n + 1 \equiv 0 \pmod{n!}$. Suppose that n is composite and let d be a prime divisor of n. Then

$$\left(\frac{n!}{d} - 1\right)^n + 1 = \sum_{k=1}^{n} (-1)^{n-k} \binom{n}{k} \left(\frac{n!}{d}\right)^k,$$

where each term is divisible by $n!$, since $d^2 | n!$. It follows that $n! \nmid \left(\frac{n!}{d} - 1\right)^n + 1$; hence n cannot be composite.

It remains to show that if n is an odd prime and $n! | a^n + 1$, then $n! | a + 1$; therefore $a = n! - 1$. Choose a prime number $p \leq n$. If $p | \frac{a^n+1}{a+1}$, then $p|(-a)^n - 1$ and by Fermat's theorem, $p|(-a)^{p-1} - 1$; therefore $p|(-a)^{\gcd(n,p-1)} - 1 = -a - 1$, that is, $a \equiv -1 \pmod{p}$. But then

$$\frac{a^n + 1}{a + 1} = a^{n-1} - a^{n-2} + \cdots - a + 1 \equiv n \pmod{p},$$

which implies $p = n$. It follows that

$$\gcd\left(\frac{a^n + 1}{a + 1}, (n-1)!\right) = 1,$$

consequently, $(n-1)! | a + 1$. Moreover, $n | a + 1$, so that $n! | a + 1$.

176. For $p | \varphi(n)$, we need that n has a prime divisor $\equiv 1 \pmod{p}$ or $p^2 | n$. Let c(n) be the highest order any a coprime to n has mod n. Then $c(p^s) = \varphi(p^s)$, $c(2) = 1$, $c(4) = 2$, $c(2^{k+2}) = 2^k$, $k \geq 1$, and $c(ab) = $ lcm $(c(a), c(b))$ for all a, b.

We want $pc(n) | \varphi(n)$. Let $n = \prod p_i^{a_i}$ be the prime decomposition of n. Then

$$p \operatorname{lcm}(\varphi(p_i^{a_i})) | \prod \varphi(p_i^{a_i})$$

if and only if $p | \varphi(p_s^{a_s}), \varphi(p_t^{a_t})$ for some $s \neq t$, that is, there are two different primes $\equiv 1 \pmod{p}$ or $p^2 | n$.

177. (a) $A_n \neq \emptyset$ if and only if n is odd or n is even with $v_2(n) = 1$ and has only prime factors $p \equiv 1 \pmod 4$.

For n odd, $n - 1 \in A_n$. For n even, $a^n + 1$ is a sum of two squares, one of which is odd. It follows that the conditions $v_2(n) = 1$ and $p|n$ implies $p \equiv 1 \pmod 4$ are both needed. Conversely, we can choose n to be a square root modulo each prime power dividing n.

(b) $|A_n| \neq 0$ is even for all numbers in (a), except $n = 2$.

If $n \neq 2$ is even, then $a \in A_n$ implies $-a \in A_n$.

If n is odd, then $a^n \equiv -1 \pmod{p^e}$ has an odd number of solutions; $a \equiv -1 \pmod n$ is a solution and all other solutions are in pairs $\{a, \frac{1}{a}\}$.

(c) Suppose that such n exists. Then by (b) it is even. If n has at least two prime factors, then $4||A_n|$, contradiction.

178. Answer: $(x, y) = (2, 0), (0, -2)$.

From $(x - y)(x^2 + xy + y^2) = 2(xy + 4)$, it follows that $x - y$ is even; hence $x^2 + xy + y^2 | xy + 4$. If $xy + 4 = 0$, then $x = y$; therefore $x^2 = -4$, contradiction.

We have $4 \geq x^2 + y^2$, so we need to consider only cases $-2 \leq x, y \leq 2$.

179. Since $1003 = 17 \cdot 59$, observe that $17|n, 59|n$; otherwise the denominator of the sum is not divisible by 17 or 59. As n is even, $2006|n$. For $n = 2006$, we have

$$\frac{1}{1} + \frac{1}{2} + \frac{1}{17} + \frac{1}{59} + \frac{1}{2 \cdot 17} + \frac{1}{2 \cdot 59} + \frac{1}{17 \cdot 59} + \frac{1}{2 \cdot 17 \cdot 59} = \frac{1620}{1003}.$$

There are no other solutions, since multiplication of 2006 by an integer greater than 1 makes the LHS larger, by apparition of new divisors, which implies that the solution is unique.

180. The given sequence is the Lucas sequence. We shall prove a general property: a_p divides $a_{\frac{p+1}{2}}^p - 1$ for all primes $p > 3$.

For $\alpha = \frac{1+\sqrt{5}}{2}, \beta = \frac{1-\sqrt{5}}{2}$, we have $a_n = \alpha^n + \beta^n$. This equality remains true for negative n if we extend the definition of the sequence. The following properties hold:

i. $a_m a_n = a_{m+n} + (-1)^m a_{n-m}$.
ii. $a_{n+2p} \equiv a_n \pmod{a_p}$.

If $k = \frac{p+1}{2}$, then

$$a_k^p = (\alpha^k + \beta^k)^p \sum_{m=0}^{p} \binom{p}{m} \alpha^{(p-m)k} \beta^{mk} = \sum_{m=0}^{k-1} (-1)^{mk} \binom{p}{m} \alpha^{(p-2m)k} +$$

$$+ \sum_{m=k}^{p} (-1)^{(p-m)k} \binom{p}{m} \beta^{(2m-p)k} = \sum_{m=0}^{k-1} (-1)^{mk} \binom{p}{m} a_{(p-2m)k} =$$

$$= \sum_{m=0}^{k-1} (-1)^{mk} \binom{p}{m} a_{(k-m)p-m}. (*)$$

If $k - m$ is odd, then $(-1)^{mk} a_{(k-m)p-m} = a_{(k-m)p-m} \equiv a_{p-m} \pmod{a_p}$.
If $k - m$ is even, then $(-1)^{mk} a_{(k-m)p-m} = (-1)^m a_m \equiv a_m \pmod{a_p}$.
Substituting in (*), we get

$$2a_p^k \equiv 2 \sum_{0 \le m \le p,\, m-k \text{ even}} \binom{p}{m} a_m \equiv 2 \sum_{0 \le m \le p,\, m-k \text{ even}} (\alpha^m + \beta^m) \equiv$$
$$\equiv a_{2p} + (-1)^{k+p} a_p \equiv a_0 \equiv 2 \pmod{a_p}.$$

Since a_p is odd (a_n is even if and only if $3|n$), this implies $a_k^p \equiv 1 \pmod{a_p}$. For $p = 59$, we obtain the required result.

181. (a) Notice that the sequence may be extended to the left with $x_0 = 0$. Writing two consecutive squared recurrence relations yields

$$x_n^2 = x_{n+1} x_{n-1} + 1,\ x_{n+1}^2 = x_{n+2} x_n + 1, \text{ for } n \ge 1,$$

so, by subtracting,

$$x_{n+1}(x_{n+1} + x_{n-1}) = x_n(x_{n+2} + x_n),$$

that is,

$$\frac{x_{n+2} + x_n}{x_{n+1}} = \frac{x_{n+1} + x_{n-1}}{x_n} = \cdots = \frac{x_2 + x_0}{x_1} = 4,$$

whence $x_{n+1} = 4x_n - x_{n-1}$ (clearly $x_n \ne 0$ for $n \ge 1$). Since the first two terms are integers, it follows inductively that all terms are integers.

(b) We have

$$0 = x_{n+1}(x_{n+1} - 4x_n + x_{n-1}) = x_{n+1}^2 - 4x_{n+1}x_n + x_{n+1}x_{n-1} =$$
$$= x_{n+1}^2 - 4x_{n+1}x_{n-1} + x_n^2 - 1 = (x_{n+1} - x_n)^2 - (2x_n x_{n+1} + 1)$$

for $n \ge 1$; hence $2x_n x_{n+1} + 1$ is a square.

182. If $\sigma(n)$ is odd, then n has an odd number of odd divisors; hence $n = 2^k m^2$, with k nonnegative integer and m an odd integer.

We will prove that k is even. If not, then $\sigma(2^k) | \sigma(n)$ implies $3 | \sigma(n)$, since $3 | \sigma(2^k)$ for k odd. Then $n \equiv 1 \pmod 3$. Since k odd implies $2^k \equiv 2 \pmod 3$, it follows that $m^2 \equiv 2 \pmod 3$, contradiction.

Hence $k = 2l$, with $l \ge 0$. If $l > 0$, then $\sigma(2^{2l}) \equiv 7 \pmod 8$; therefore there is a prime p such that $p | \sigma(2^{2l})$ and $p \not\equiv 1, 3 \pmod 8$. But $p | 2n + 1 = 2x^2 + 1$; hence

$$x^2 \equiv \frac{p+1}{2} \pmod p,$$

contradiction, since $p \not\equiv 1, 3 \pmod 8$.

183. If $d_5 \geq \sqrt{n}$, then $d_5^2 + d_6^2 - 1 \geq n + (\sqrt{n}+1)^2 - 1 > 2n$; hence $d_5 < \sqrt{n}$.

If $d_6 \leq \sqrt{n}$, then $d_5^2 + d_6^2 - 1 < 2n - 1 < 2n$; hence $d_6 > \sqrt{n}$. This implies that $d_5 d_6 = n$ and that n has ten divisors.

From $(d_6 - d_5)^2 = d_5^2 + d_6^2 - 2d_5d_6 = 2n + 1 - 2n = 1$, it follows that $d_6 - d_5 = 1$ and $\gcd(d_5, d_6) = 1$. Since n has ten divisors, it has at least two prime factors (one of which is 2) and is of the form $n = pq^4$, p, q primes. Moreover, $\{p, q^4\} = \{d_5, d_6\}$; hence $q = 2, p = 17, n = 272$.

184. Answer: $k = 1$.

Suppose that $k > 1$. If k is even, then $\nu_2(3^k + 5^k) = 1$; therefore $3^k + 5^k$ cannot be a perfect power. If k is odd, then $\nu_2(3^k + 5^k) = 3$; hence $3^k + 5^k$ is a cube. Taking the residues mod 9, we conclude that $k = 3l$. But $(3^l)^3 + (5^l)^3 = t^3$ has no integer solutions.

185. Answer: $x_1 = 472$.

For $n \geq 2$, we have $x_{n+1} = x_n + x_n^2$; hence x_n is even for all $n \geq 3$. It suffices to find the smallest x_1 for which $p | x_{2006}$, where $p \in \{17, 59\}$. If $p | x_{n+1}$, then $p | x_n$ or $p | x_n + 1$.

If $p | x_n + 1$, then $p | (2x_{n-1} + 1)^2 + 3$, contradiction, since -3 is not a quadratic residue modulo 17 or 59.

If $p | x_n$, then finally $p | x_3 = x_1^2(x_1^2 + 1)$.

Case 1. $17 | x_1^2, 59 | x_1^2 + 1$. This case is impossible, since -1 is not a quadratic residue modulo 59.

Case 2. $59 | x_1^2, 17 | x_1^2 + 1$. Then $x_1 = 59k$. From $59^2 \equiv 13 \pmod{17}$, we deduce that $13k^2 + 1$ is divisible by 17; hence $k^2 \equiv 13 \pmod{17}$. Then $k \equiv 8, 9 \pmod{17}$.

The smallest value is $x_1 = 8 \cdot 59 = 472$.

186. For $B \equiv A \pmod 8$, we have $B = 4$ if n is even and $B = 2$ if n is odd. Therefore $k < 4$. If $k = 2$, then $A = x(x + 1)$ is impossible for $n > 0$. The solution is $n = 0, A = 12$.

If $k = 3$, then $A = (x - 1)x(x + 1)$ and x is even. From $A = x(x^2 - 1) \equiv 0, \pm 1 \pmod 5$, we conclude that there are no solutions in this case.

187. Squaring and replacing a_n^2 in function of A_n, we get

$$A_{n+1} = \frac{4A_n(A_n + 3)}{3}, n \geq 0. (*)$$

We have also

$$A_{n+1} = \frac{A_n(4A_{n-1} + 6)^2}{9}, n \geq 1. (**)$$

Indeed,

$$(4A_{n-1} + 6)^2 = 16A_{n-1}(A_{n-1} + 3) + 36 = 12A_n + 36 = 12 \cdot \frac{3A_{n+1}}{4A_n}.$$

From (∗) and (∗∗), it follows inductively that A_n is an iteger and a perfect square, since the base case is $A_1 = 3^2, A_2 = 12^2$.

2.35 2007

188. Answer: $\{1, 2, 3, 7, 13, 14\}$.
For each of these values $d = \gcd(x, y, z)$, we have the solutions (a, b, c, d):

$$(1, 2, 3, 1), (2, 4, 6, 2), (3, 6, 9, 3), (1, 4, 9, 7), (1, 3, 9, 13), (2, 8, 4, 14).$$

We shall prove that there are no other solutions. Let p be a prime dividing d. Then $p | x + y + z = 11(a + b + c)$ and $p \neq 11$, since all multtiples of 11 have equal digits. It follows that $p | a + b + c$; hence $p \leq 7 + 8 + 9 = 24$. We prove that $p \notin \{17, 19, 23\}$, writing in each case all multiples:

$$17, 34, 51, 68, 85; 19, 38, 57, 76, 95; 23, 46, 69, 92.$$

We cannot have $p = 5$, since in this case all digits are equal. If $d = 4$, then all digits are even, and $x = 10a + b$, for example, is divisible by 4, so that b (consequently all digits) is divisible by 4, impossible. If $d = 6$, then all digits are even, and $a + b, b + c, c + a$ are divisible by 3, impossible, since 46 and 64 are not divisible by 3. If $d = 9$, then $a + b = b + c = c + a = 9$; hence $a = b = c$, contradiction.

189. It is known that if a, b, c are positive integers such that $a^2 = b^2 + c^2$; then abc is divisible by 60 and one of the a, b, c is divisible by 4.
Let $D = \{d_1, \ldots, d_r\}$ be the set of divisors of n. Since $d_7^2 + d_{15}^2 = d_{16}^2$, it follows that n is divisible by 60; hence $d_k = k$, $1 \leq k \leq 6$ and $10 \in D$. We deduce that $7 \leq d_7 \leq 10$.

Case 1. $d_7 = 7$. Then $d_{16} - d_{15} = 1, d_{16} + d_{15} = 49$, so that $d_{15} = 24$, $d_{16} = 25$. If n is divisible by 9, 11, or 13, then $d_{15} < 24$, contradiction. As a result, the positive divisors of n are $d_k = k$ for $1 \leq k \leq 8$, $d_9 = 10, d_{10} = 12$, $d_{11} = 14, d_{12} = 15, d_{13} = 20, d_{14} = 21, d_{15} = 24$, and $d_{16} = 25$, and we deduce $d_{17} = 28$.

Case 2. $d_7 = 8$. Since $7 \notin D$, we deduce $d_{15} > 15, d_{16} > 16$; hence $d_{15} + d_{16} \geq 33$. The $(d_{16} - d_{15})(d_{16} + d_{15}) = 64$ has no solution, because the two factors have the same parity.

Case 3. $d_7 = 9$. Then $(d_{16} - d_{15})(d_{16} + d_{15}) = 81$. Since $d_{16} + d_{15} \geq 16 + 15 = 31$, it follows that $d_{16} - d_{15} = 1, d_{16} + d_{15} = 81$; whence $d_{15} = 40$, contradiction, since $8 \notin D$.

Case 4. $d_7 = 10$. Then we have $d_{16}^2 - d_{15}^2 = (d_{16} - d_{15})(d_{16} + d_{15}) = 100$. Since $d_{16} - d_{15}$ and $d_{16} + d_{15}$ have the same parity, we must have $d_{16} - d_{15} = 2$, $d_{16} + d_{15} = 50$; whence $d_{15} = 24$, $d_{16} = 26$. But then $8 \in D$, which contradicts $d_7 = 10$.

190. Answer: no.

Take $n = 10^{4k} + 2 \cdot 10^{3k} - 10^{2k} + 2 \cdot 10^k + 1$ for some large positive integer k. Then it is easy to see that we may make $s(n)$ as large as we wish. But

$$n^2 = 10^{8k} + 4 \cdot 10^{7k} + 2 \cdot 10^{6k} + 11 \cdot 10^{4k} + 2 \cdot 10^{2k} + 4 \cdot 10^k + 1$$

and $s(n^2) = 16$, when k is large. This implies the conclusion.

191. If $y \neq 0$, then

$$\left(y^2\right)^2 < (x+1)^2 = y^4 + y^2 + 1 < \left(y^2 + 1\right)^2.$$

No solutions. If $y = 0$, then $x = -2$ and $x = 0$ are solutions.

192. The following assertions hold:

a. If $p = 4k + 3$ is prime and $\gcd(x, p) = \gcd(y, p) = 1$, then $x^2 + y^2 \not\equiv 0 \pmod{p}$.

b. If $p = 4k + 3$ is prime, then there are positive integers x_0, y_0 such that

$$x_0^2 + y_0^2 + 1 \equiv 0 \pmod{p}, \gcd(x_0, p) = \gcd(y_0, p) = 1.$$

Now, if $(x^2 + y^2)^2 \equiv r \pmod{p}$ for $0 \le r < p$ and $\gcd(x, p) = \gcd(y, p) = 1$, then (i) implies that r is a quadratic residue, which shows that r has at most $\frac{p-1}{2}$ values.

From (ii) we deduce $(x_0^2 + y_0^2)^2 \equiv 1 \pmod{p}$, so that $\left((p_i x_0)^2 + (p_i y_0)^2\right)^2 \equiv q_i^2 \pmod{p}$ for $i = 1, 2, \ldots, \frac{p-1}{2}$ and $q_1, q_2, \ldots, q_{\frac{p-1}{2}}$ are all quadratic residues mod p,

$$q_i \equiv p_i^2 \pmod{p}, i = 1, 2, \ldots, \frac{p-1}{2}.$$

We deduce that there are at least $\frac{p-1}{2}$ values of r, since $q_1^2, q_2^2, \ldots, q_{\frac{p-1}{2}}^2$ are distinct modulo p.

193. For $a \in (\sqrt{n}, \sqrt{2n})$, we have $\text{lcm}(a, a+1) = a(a+1)$, so

$$\left| A \cap (\sqrt{n}, \sqrt{2n}) \right| \le \frac{(\sqrt{2}-1)\sqrt{n}}{2} + 1.$$

For $a \in (\sqrt{2n}, \sqrt{3n})$, we have $\text{lcm}(a+1, a+2) = (a+1)(a+2) > n$, $\text{lcm}(a, a+2) = a(a+2) \ge \frac{a(a+2)}{2} > n$, so

$$\left|A \cap \left(\sqrt{2n}, \sqrt{3n}\right)\right| \leq \frac{(\sqrt{3}-\sqrt{2})\sqrt{n}}{3} + 1.$$

For the same reason,

$$\left|A \cap \left(\sqrt{3n}, 2\sqrt{n}\right)\right| \leq \frac{(2-\sqrt{3})\sqrt{n}}{4} + 1.$$

So,

$$|A \cap (1, 2\sqrt{n})| \leq \sqrt{n} + \frac{(\sqrt{2}-1)\sqrt{n}}{2} + \frac{(\sqrt{3}-\sqrt{2})\sqrt{n}}{3} + \frac{(2-\sqrt{3})\sqrt{n}}{4} + 3 =$$

$$= \left(1 + \frac{\sqrt{6}}{2} + \frac{\sqrt{3}}{12}\right)\sqrt{n} + 3.$$

For positive integer k, assume $a, b \in \left(\frac{n}{k+1}, \frac{n}{k}\right), a > b$ and assume lcm $(a, b) = as = bt$; then

$$\frac{a}{\gcd(a,b)} \cdot s = \frac{b}{\gcd(a,b)} \cdot t.$$

Because $\frac{a}{\gcd(a,b)}, \frac{b}{\gcd(a,b)}$ are coprime, so s is a multiple of $\frac{b}{\gcd(a,b)}$; therefore

$$\text{lcm}(a, b) = as \geq \frac{ab}{\gcd(a,b)} \geq \frac{ab}{a-b} = b + \frac{b^2}{a-b} > \frac{n}{k+1} + \frac{\left(\frac{n}{k+1}\right)^2}{\frac{n}{k} - \frac{n}{k+1}} = n.$$

So, $\left|A \cap \left(\frac{n}{k+1}, \frac{n}{k}\right)\right| \leq 1$. Take positive integer T such that $\frac{n}{T+1} \leq 2\sqrt{n} < \frac{n}{T}$; then

$$|A \cap (2\sqrt{n}, n)| \leq \sum_{k=1}^{T} \left|A \cap \left(\frac{n}{k+1}, \frac{n}{k}\right)\right| \leq T < \frac{\sqrt{n}}{2}.$$

Finally, we get

$$|A| \leq \left(\frac{3}{2} + \frac{\sqrt{2}}{6} + \frac{\sqrt{3}}{12}\right)\sqrt{n} + 3 < 1,9\sqrt{n} + 5,$$

as desired.

194. We shall prove that all integer solutions (m, n) of the given equation are $(\pm 1, 0)$, $(\pm 4, 3)$, $(\pm 4, 5)$. Clearly, we must have $n \geq 0$. As the *RHS* $\neq 0$; then the *LHS* $\neq 0$; hence $|m| \geq n + 1$ or $|m| \leq n - 1$. In both cases, $(m^2 - n^2)^2 \geq (2n - 1)^2$; therefore $(2n - 1)^2 \leq 16n + 1$; hence $n \leq 5$.

Trying all such values of n, we find the above solutions.

195. It suffices to prove the assertion for the sequence (c_n), where $c_n = \lfloor n\sqrt{3} \rfloor - \lfloor n\sqrt{2} \rfloor$. Assume that the sequence contains only finite number of odd or even terms. Since each two consecutive terms differ by at most 1, the sequence must become constant after some point. But the sequence is unbounded, contradiction.

196. It is obvious that $b \geq 2$. If $b^n - 1 = (b - 1)p^l$, with p prime, then $n \geq 2$ and $l \geq 1$. Suppose that $n = xy$, with x, y greather than 1, and consider the representation

$$\frac{b^{xy} - 1}{b - 1} = \frac{b^{xy} - 1}{b^y - 1} \cdot \frac{b^y - 1}{b - 1} = \left(1 + b^y + \cdots + b^{y(x-1)}\right) \cdot \frac{b^y - 1}{b - 1}.$$

Both factors must be powers of p, since the product is a power of p. Both factors are multiples of p, since $x, y > 1$. Then $b^y - 1$ is a multiple of p, so that all the terms in the first factor are congruent to 1 modulo p; therefore the first factor is congruent to x modulo p. We deduce that x is divisible by p. But x is an arbitrary divisor of n, which shows that $n = p^m$ for a positive integer m. Consider the representation

$$\frac{b^{p^m} - 1}{b - 1} = \frac{b^{p^m} - 1}{b^{p^{m-1}} - 1} \cdot \ldots \cdot \frac{b^{p^2} - 1}{b^p - 1} \cdot \frac{b^p - 1}{b - 1}.$$

Each factor is a power of p greater than 1. Since $\frac{b^p - 1}{b - 1}$ is a positive integral power of p, it follows that the numerator is divisible by p, that is, $b^p \equiv 1 \pmod{p}$. By the Fermat's theorem, $b^p \equiv b \pmod{p}$; whence $b \equiv 1 \pmod{p}$, i.e., $b - 1$ is divisible by p. But then $b^p - 1$ is divisible by p^2. For $m \geq 2$ the above representation cantains the factor

$$\frac{b^{p^2} - 1}{b^p - 1} = 1 + b^p + \cdots + b^{p(p-1)}.$$

This number is congruent to p modulo p^2, since all terms are congruent to 1 modulo p^2. On the other hand, this factor is a power of p, contradiction. This shows that $m = 1$, which ends the proof.

197. Answer: no.

We will prove first that one of $a_1, \ldots, a_6$ has an even number of digits. Suppose the contrary and let a_1 has k digits, with k odd. If a_n has k digits, then a_{n+1} has at most $k + 1$ digits. We deduce that each $a_1, a_2, \ldots, a_6$ has exactly k digits. Let p and q be the first and the last digit of a_1, so a_2 starts and ends with

$p + q$. Similarly, a_3, a_4, a_5, and a_6 start and end with $2(p + q)$, $4(p + q)$, $8(p + q)$, and $16(p + q)$, respectively. But $16(p + q) > 10$, so that a_6 has more than k digits, contradiction.

Let a_i, $i \leq 6$, be the integer with an even number of digits. The digits in odd positions of a_i become digits in even positions of a'_i and conversely; therefore the sums of digits in odd and even positions of $a_i + a'_i$ are equal, thus $11 | a_{i+1}$. Note that if $11 | a_k$, then $11 | a'_k$; hence $11 | a_{k+1}$, which implies $11 | a_n$ for all $n \geq k$. Since $11 | a_{i+1}$, $i \leq 6$, then $11 | a_7$.

If a_7 is prime, then $a_7 = 11$. If a_6 has only one digit, then $a_6 = a'_6$ and $11 = a_7 = 2a_6$, contradiction. If a_6 has two digits, then $a_6 = 10$, $a_5 = 5$, which leaves no possible value for a_4.

198. The equation has no integer solutions.

We shall prove that each positive divisor d of $\frac{x^7 - 1}{x - 1}$ satisfies $d \equiv 0, 1 \pmod{7}$. Indeed, let p be a prime divisor of $\frac{x^7 - 1}{x - 1}$. Then $x^7 - 1$ and $x^{p-1} - 1$ are divisible by p, by hypothesis and by the Fermat's theorem. Suppose that 7 does not divide $p - 1$. Then $\gcd(7, p - 1) = 1$, so there exist integers k, m such that $7k + (p - 1)m = 1$. We have $x = x^{7k + (p-1)m} = (x^7)^k \cdot (x^{p-1})^m \equiv 1 \pmod{p}$, so that

$$\frac{x^7 - 1}{x - 1} = 1 + x + \ldots + x^7 \equiv 7 \pmod{p}.$$

It follows that p divides 7; hence $p = 7$. In conclusion, $p \not\equiv 1 \pmod{7}$ implies $p = 7$.

If (x, y) is an integer solution of the given equation, then $\frac{x^7 - 1}{x - 1} > 0$ for all $x \neq 1$; hence $y - 1 > 0$. Since $y - 1$ divides $\frac{x^7 - 1}{x - 1} = y^5 - 1$, it follows that $y \equiv 1, 2 \pmod{7}$.

If $y \equiv 1 \pmod{7}$, then $1 + y + y^2 + y^3 + y^4 \equiv 5 \pmod{7}$, contradiction.
If $y \equiv 2 \pmod{7}$, then $1 + y + y^2 + y^3 + y^4 \equiv 3 \pmod{7}$, contradiction.

199. Note that $A = \{\lfloor \frac{n}{3} \rfloor + 1, \ldots, n\}$ satisfies the given condition; hence $|A| \geq \frac{2n}{3}$.

Next, assume that $|A| \geq \frac{3n}{4} + 1$ and set $B = 2A = \{2a | a \in A\}$, $C = A \cap B$. Since $|A|, |B| \geq \frac{3n}{4} + 1$, we have $|C| \geq \frac{n}{2} + 2$. Then one can choose $c_1, c_2 \in C$ such that $c_1 | c_2$. But then there is $a \in A$ such that $c_1 = 2^l a$, with $l \geq 1$. Choose the smallest such a (i.e., the greatest l); then $a \notin B$; otherwise the minimality is contradicted. Then $a | c_1, c_2$, as desired.

200. We see that $x = 0$ or $y = 0$ are not solutions. From

$$y^x(14^x + y^y) = 2007 = 3^2 \cdot 223,$$

it follows that $y \in \{1, 3, 3^2, 223, 3 \cdot 223, 3^2 \cdot 223\}$.
The only solution is $(x, y) = (2, 3)$.

201. The equalities $7 \cdot 351^2 - 11 \cdot 280^2 = 7, 7 \cdot 4^2 - 11 \cdot 3^2 = 13$ show that the values 7 and 13 hold. We shall prove that there are no other values less than 13.

Consider the eq. $7m^2 - 11n^2 = c$, for a positive integer c. Modulo 7 this equation becomes $c \equiv -11n^2 \equiv 3n^2 \pmod 7$. Since $n^2 \equiv 0, 1, 2, 4 \pmod 7$, it follows that $c \equiv 0, 3, 5, 6 \pmod 7$. Analagously, modulo 11 the equation becomes $c \equiv 7m^2$ and from $m^2 \equiv 0, 1, 3, 4, 5, 9 \pmod{11}$; it follows that $c \equiv 0, 2, 6, 7, 8, 10 \pmod{11}$. We have also $c \equiv 3(m^2 - n^2) \equiv 0, 1, 3 \pmod 4$. These conditions together imply that the smallest two values are $c = 7, 13$.

202. More generally, let k be given, and let $a_1, a_2, \ldots, a_k$ be integers not divisible by the prime $p > 2k + 1$. Then for all integers n, there are integers $x_1, x_2, \ldots, x_k$ such that

$$a_1 x_1^{2k-1} + a_2 x_2^{2k-1} + \cdots + a_k x_k^{2k-1} \equiv n \pmod p.$$

203. Choose $n = (2k^2)^4$ for $k \equiv 1 \pmod{91}$. Then

$$n^2 + n + 1 = \left(4k^4 + 2k^2 + 1\right)\left(4k^4 - 2k^2 + 1\right)\left(4k^4 - 4k^3 + 2k^2 - 2k + 1\right) \times$$
$$\times \left(4k^4 + 4k^3 + 2k^2 + 2k + 1\right),$$

$$4k^4 + 2k^2 + 1 \equiv 0 \pmod 7, 4k^4 + 4k^3 + 2k^2 + 2k + 1 \equiv 0 \pmod{13}.$$

This implies that all prime divisors of $n^2 + n + 1$ are less than $\sqrt{n} = 4k^4$.

204. If $m = n$, then $m = n \in \{1, 2\}$.

If $m > n$, denote $\frac{m+1}{n} + \frac{n+1}{m} = k$. Then the equation

$$x^2 - (kn - 1)x + n^2 + n = 0$$

has a root $x = m$, so the other root is

$$x = kn - 1 - m = \frac{n^2 + n}{m}.$$

Note that $0 < \frac{n^2+n}{m} < n$; thus we can repeat this procedure until we reach $m = n$ or $k = 2, 3$. Now we will characterize the sequences satisfying

$$m_i = km_{i-1} - m_{i-2} = \frac{m_{i-1}^2 + m_{i-1}}{m_{i-2}}.$$

In particular, we will characterize $d_i = \gcd(m_i, m_{i+1})$. First, we show that $d_i d_{i-1} = m_i$. This follows from

$$d_i = \gcd(m_i, m_{i+1}) = \gcd\left(m_i, \frac{m_i(m_i+1)}{m_{i-1}}\right) = \gcd\left(m_i, \frac{m_i}{m_{i-1}}\right) =$$
$$= \frac{m_i}{\gcd(m_i, m_{i-1})} = \frac{m_i}{d_{i-1}}.$$

Next, we show that $d_i = \frac{m_i}{m_{i-1}} d_{i-2}$. This is also clear, as

$$\frac{d_i}{d_{i-2}} = \frac{d_i d_{i-1}}{d_{i-1} d_{i-2}} = \frac{m_i}{m_{i-1}}.$$

Finally, we show that $d_i = kd_{i-2} - d_{i-4}$, which follows from

$$d_i - kd_{i-2} + d_{i-4} = d_{i-4}\left(\frac{m_i m_{i-2}}{m_{i-1} m_{i-3}} - \frac{km_{i-2}}{m_{i-3}} + 1\right) =$$
$$= d_{i-4} \cdot \frac{m_i m_{i-2} - km_{i-1} m_{i-2} + m_{i-1} m_{i-3}}{m_{i-1} m_{i-3}},$$
$$0 = m_{i-1} m_{i-3} - m_{i-2}^2 - m_{i-2} = m_i m_{i-2} - km_{i-1} m_{i-2} + m_{i-1} m_{i-3}.$$

Now, the main problem. We can compute

$$f_i = \frac{m_i + m_{i+1}}{d_i^2} = \frac{d_{i-1} + d_{i+1}}{d_i} = \frac{5d_{i-1} - d_{i-3}}{4d_{i-2} - d_{i-4}}.$$

If $f_{i-2} = f_{i-4} = f$, then

$$\frac{4d_{i-1} + 4d_{i-3}}{4d_{i-2}} = \frac{d_{i-3} + d_{i-5}}{d_{i-4}} = f,$$

implying

$$f = \frac{4d_{i-1} + 4d_{i-3} - d_{i-3} - d_{i-5}}{4d_{i-2} - d_{i-4}} = \frac{5d_{i-1} - d_{i-3}}{4d_{i-2} - d_{i-4}}.$$

For $k = 3$, we get $m_1 = 2, m_2 = 2, m_3 = 3, m_4 = 6, m_5 = 14$; thus $f_1 = f_3 = 1$, $f_2 = = f_4 = 5$; hence $f_{2i+1} = 1, f_{2i} = = 5$.
For $k = 4$, we get $m_1 = 1, m_2 = 1, m_3 = 2, m_4 = 6, m_5 = 21$; hence $f_1 = f_3 = 2, f_2 = = f_4 = 3$; hence $f_{2i+1} = 2, f_{2i} = 3$.
This shows that $\frac{m+n}{(\gcd(m,n))^2} \in \{1, 2, 3, 5\}$.

205. Answer: no.

Since $\varphi(n) < n$ for all $n > 1$, consider $k \geq 0$ such that $\varphi^k(a) = 1$, which must exist. Let $n = {}^k b$ denote the tetration (superexponentiation), so $b^n = {}^{k+1}b$. By Euler totient,

$$^{k+1}b \equiv {}^k b \pmod{a} \Longleftarrow {}^k b \equiv {}^{k-1}b \pmod{\varphi(a)} \Longleftarrow \ldots \Longleftarrow {}^1 b = b$$
$$\equiv 1 = {}^0 b \pmod{\varphi^k(a)},$$

which is true, as $1 | b - 1$.

2.36 2008

206. Answer: (1) $p = 2$, $n = 8, 12$; (2) $p = 3$, $n = 9, 18, 24$; (3) $p > 3$, $n = 8p, 12p$.
If $n = p^r p_1^{r_1} \ldots p_k^{r_k}$, $p_1 < p_2 < \cdots < p_k$, then $n = p\tau(n)$ implies

$$p^{r-1} p_1^{r_1} \ldots p_k^{r_k} = (r+1)(r_1 + 1) \ldots (r_k + 1).$$

If $k = 0$, then $p^{r-1} = r + 1$; hence $(p, r) = (2, 3)$ or $(3, 2)$ and $n = 8$ or $n = 9$.

If $k = 1$, $p_1 \geq 3$, then $p_i^{r_i} > r_i + 1$ for all i and $p^{r-1} < r + 1$, which implies $r = 1$ or $(p, r) = (2, 2)$. But $r = 1$ implies $p_1 = 2$, contradiction; hence $p = r = 2$. Now $p_1 = 3, 2 \cdot 3^{r_1} \leq 3(r_1 + 1)$ hold if and only if $k = r = 1$, which implies $n = 12$.

If $p_1 = 2, r = 1$, then $2^{r_1 - 1} \leq r_1 + 1$; therefore $r_1 = 1, 2$ or 3; $r_1 = 1$ leads to a contradiction, if $r_1 = 2$, then $p_2 = 3$. The inequality $2 \cdot 3^{r_2 - 1} \leq r_2 + 1$ implies $r_2 = 1, k = 2$; hence $n = 12p$. If $r_1 = 3$, then $k = 1$; therefore $n = 8p$.

If $p_1 = 2, r > 1$, then $p^{r-1} \geq r + 1$ and $p_i^{r_i} > r_i + 1$ for all i; therefore $2^{r_1} \leq r_1 + 1$. The only solution is $r_1 = 1$, so that $k = 1, p^{r-1} = r + 1$ (only when $p = 3, r = 2$). Therefore, $n = 18$.

207. Answer: $(1, 2), (2, 1), (3, 2), (2, 3), (2, 2), (3, 3)$.

If $m = n$, then

$$m^2 - m | m^2 + m \Longrightarrow m - 1 | m + 1 \Longrightarrow m - 1 | 2 \Longrightarrow m \in \{2, 3\}.$$

If $m = n + k, k \geq 1$, then

$$m^2 - n | m + n^2 \Longrightarrow m^2 - n \leq m + n^2 \Longrightarrow (k - 1)(2n + k + 1) \leq 0 \Longrightarrow k = 1.$$

Now we have

$$m^2 - m - 1 | m^2 + 3m + 1 \Longrightarrow m^2 - m - 1 | 4m + 2 \Longrightarrow m \leq 5.$$

By checking all possibilities, we find the above solutions.

208. It is easy to see that

$$x_n = \frac{(3+2\sqrt{2})^n - (3-2\sqrt{2})^n}{2\sqrt{2}}, n \geq 1.$$

Let a_n, b_n be positive integers such that $a_n + b_n\sqrt{2} = (3+2\sqrt{2})^n$; then $a_n - b_n\sqrt{2} = (3-2\sqrt{2})^n$; whence $x_n = b_n$ and $a_n^2 - 2b_n^2 = 1, n \geq 1$.

Suppose that $q \neq 2$. Since $q|x_p$, we deduce that there are terms in the sequence $\{b_n\}$ which are divisible by q. Let d be the least number such that $q|b_d$. We shall prove that $q|b_n$ if and only if $d|n$. Indeed, for $a, b, c, d \in \mathbb{Z}$, denote $a + b\sqrt{2} \equiv c + d\sqrt{2} \pmod{q}$ when $a \equiv c \pmod{q}$ and $b \equiv d \pmod{q}$. If $d|n$, then $n = du$ and

$$a_n + b_n\sqrt{2} = (3+2\sqrt{2})^{du} \equiv a_d^u \pmod{q};$$

whence $b_n \equiv 0 \pmod{q}$. Conversely, if $q|b_n$, suppose that $n = du + r$, $1 \leq r \leq d-1$. From

$$a_n + b_n\sqrt{2} = (3+2\sqrt{2})^{du}(3+2\sqrt{2})^r \equiv a_d^u(a_r + b_r\sqrt{2}) \pmod{q},$$

we have $b_n \equiv a_d^u b_r \equiv 0 \pmod{q}$. But $a_d^2 - 2b_d^2 = 1$ and $q|b_d$; therefore $\gcd(q, a_d^u) = 1$, since q is a prime. It follows that $q|b_r$, which contradicts the definition of d.

Now, $q | \binom{n}{i}, 1 \leq i \leq q-1$, since q is a prime. By the Fermat's theorem,

$$3^q \equiv 3 \pmod{q}, 2^q \equiv 2 \pmod{q}.$$

As $q \neq 2$, we have $2^{\frac{q-1}{2}} \equiv \pm 1 \pmod{q}$; therefore

$$(3+2\sqrt{2})^q = \sum_{i=0}^{q} \binom{q}{i} \cdot 3^{q-i}(2\sqrt{2})^i \equiv 3^q + (2\sqrt{2})^q \equiv 3^q + 2^q \cdot 2^{\frac{q-1}{2}}\sqrt{2}$$

$$\equiv 3 \pm 2\sqrt{2} \pmod{q}.$$

By the same argument, we have

$$(3+2\sqrt{2})^{q^2} \equiv (3 \pm 2\sqrt{2})^q \equiv 3 + 2\sqrt{2} \pmod{q}.$$

so that

$$\left(a_{q^2-1} + b_{q^2-1}\sqrt{2}\right)\left(3 + 2\sqrt{2}\right) \equiv 3 + 2\sqrt{2} \pmod{q};$$

whence

$$3a_{q^2-1}^2 + 4b_{q^2-1}^2 \equiv 3 \pmod{q}, 2a_{q^2-1}^2 + 3b_{q^2-1}^2 \equiv 2 \pmod{q}.$$

We know that $q \mid b_{q^2-1}$. Since $q \mid b_p$, we have $d \mid p$. We deduce $d \in \{1, p\}$. If $d = 1$, then q divides $b_1 = 2$, contradiction. It follows that $d = p$; hence $p \mid b_{q^2-1}$, so that $p \mid q^2 - 1$. Thus, p divides $q - 1$ or $q + 1$, which are even, whence $q \geq 2p - 1$.

209. Let $\{p_1, p_2, \ldots, p_k\}$ be the set of prime divisors of n. The given condition implies $p_i \mid n - 1 - \frac{n}{p_i}$ for all i, so that for all i, p_i cannot divide $\frac{n}{p_i}$, which implies that n is square-free. It follows that $p_i \mid 1 + \frac{n}{p_i}$ for all i. Multiplying, we get the conclusion.

210. For $a_{-1} = 0$, the relation $a_n = a_{n-1} + a_{n-2} + 1$ holds also when $n = 1$. Consider the pairs (a_n, a_{n+1}) of consecutive terms of the sequence. There are m^2 pairs modulo m; hence there are pairs (a_k, a_{k+1}) and (a_l, a_{l+1}) with $k < l$ which are componentwise congruent modulo m. Since $a_{n-2} = a_n - a_{n-1} - 1$, two consecutive terms in the sequence determine the previous term uniquely. The same is true modulo m. It follows that the pairs (a_{k-1}, a_k) and (a_{l-1}, a_l) are componentwise congruent modulo m. Continuing, we see that (a_{-1}, a_0) and (a_{l-k-1}, a_{l-k}) are componentwise congruent modulo m. Since $a_{-1} = a_0 = 0$, the terms a_{l-k-1} and a_{l-k} are divisible by m, as required.

211. For $k = 3x + 1$, we get $3^{2005} \mid P(x) = x^3 - 11x^2 - 2x - 9$. If $3^n \mid P(x)$ and $3 \mid x$, then there is $y = x + 3^n a$ such that $3^{n+1} \mid P(y)$. For $n = 1$ we can take $x = 3$.

212. Answer: $(2, 4)$.

Let (a, b) be a pair which satisfies the given condition. Since $7^a - 3^b$ is even, $a^4 + b^2$ is also even; hence a, b have the same parity. If a, b are odd, then $a^4 + b^2 \equiv 2 \pmod{4}$, while $7^a - 3^b \equiv 7 - 3 \equiv 0 \pmod{4}$. We deduce that a, b are even.

Write $a = 2c, b = 2d$. Then $7^a - 3^b = \frac{7^c - 3^d}{2} \cdot 2(7^c + 3^d)$ and both factors are integers. So $2(7^c + 3^d) \mid 7^a - 3^b$ and $7^a - 3^b \mid a^4 + b^2 = 2(8c^4 + 2d^2)$; hence

$$7^c + 3^d \leq 8c^4 + 2d^4. \quad (*)$$

We prove by induction that $8c^4 < 7^c$ for $c \geq 4$, $2d^4 < 3^d$ for $d \geq 1$ and $2d^2 + 9 \leq \leq 3^d$ for $d \geq 3$. In the initial cases, $c = 4, d = 1, d = 2,$ and $d = 3$, we have

$$8 \cdot 4^4 = 2048 < 7^4 = 2401, 2 < 3, 2 \cdot 2^2 = 8 < 3^2 = 9, 2 \cdot 3^2 + 9 = 3^3 = 27.$$

If $8c^4 < 7^c (c \geq 4)$ and $2d^4 + 9 \leq 3^d (d \geq 3)$, then

$$8(c+1)^4 = 8c^4\left(\frac{c+1}{c}\right)^4 < 7^c \cdot \left(\frac{5}{4}\right)^4 < 7^{c+1},$$

$$2(d+1)^2 + 9 < (2d^2+9)\left(\frac{d+1}{d}\right)^2 \le 3^d \cdot \left(\frac{4}{3}\right)^2 < 3^{d+1},$$

as desired.

For $c \ge 4$ we obtain $7^c + 3^d > 8c^4 + 2d^2$ and the inequality (∗) cannot hold; hence $c \le 3$.

If $c = 1$, then $a = 2$ and $8 + 2d^2 \ge 7 + 3^d$; thus $2d^2 + 1 \ge 3^d$. This is possible only for $d \le 2$. If $d = 1$, then $b = 2$ and $\frac{a^4+b^2}{7^a-3^b} = \frac{2^4+2^2}{7^2-3^2} = \frac{1}{2}$ is not an integer. If $d = 2$, then $b = 4$ and $\frac{a^4+b^2}{7^a-3^b} = \frac{2^4+4^2}{7^2-3^4} = -1$, so $(a,b) = (2,4)$ is a solution.

If $c = 2$, then $a = 4$ and $a^4 + b^2 = 256 + 4d^2 \ge |7^4 - 3^b| = |49 - 3^d| \cdot (49 + 3^d)$. The smallest value of the first factor is 22, attained at $d = 3$, so $128 + 2d^2 \ge 11(49 + 3^d)$, which is impossible, since $3^d > 2d^2$.

If $c = 3$, then $a = 6$ and $a^4 + b^2 = 1296 + 4d^2 \ge |7^6 - 3^b| = |343 - 3^d|(343 + 3^d)$.

We have $|343 - 3^d| \ge 100$ and $324 + d^2 \ge 25(343 + 3^d)$, which is impossible.

We find that the solution $(a,b) = (2,4)$ is unique.

213. The difference of the two binomial coefficients can be written as

$$D = \binom{2^{k+1}}{2^k} - \binom{2^k}{2^{k-1}} = \frac{(2^{k+1})!}{(2^k)!\cdot(2^k)!} - \frac{1}{(2^k)!}\cdot\left(\frac{(2^k)!}{(2^{k-1})!}\right)^2 =$$

$$= \frac{2^{2^k}}{(2^k)!}\cdot(2^{k+1}-1)!! - \frac{2^{2^{k-1}\cdot 2}}{(2^k)!}\cdot((2^k-1)!!)^2 = \frac{2^{2^k}\cdot(2^k-1)!!}{(2^k)!}\cdot P(2^k),$$

for $P(x) = (x+1)(x+3)\ldots(x+2^k-1) - (x-1)(x-3)\ldots(x-2^k+1)$. The exponent of 2 in $(2^k)!$ is equal to $2^k - 1$; hence the exponent of 2 in D is bigger by 1 than the exponent of 2 in $P(2^k)$. Since $P(-x) = -P(x)$, we get $P(x) = \sum_{i=1}^{2^{k-1}} c_i x^{2i-1}$. Then c_1 satisfies

$$c_1 = 2\cdot(2^k-1)!!\cdot\sum_{i=1}^{2^{k-1}}\frac{1}{2i-1} = (2^k-1)!!\cdot\sum_{i=1}^{2^{k-1}}\left(\frac{1}{2i-1}+\frac{1}{2^k-2i+1}\right) =$$

$$= 2^k\cdot\sum_{i=1}^{2^{k-1}}\frac{(2^k-1)!!}{(2i-1)(2^k-2i+1)}.$$

Let a_i be the solution of $a_i \cdot (2i-1) \equiv 1 \pmod{2^k}$. Then

$$\sum_{i=1}^{2^{k-1}} \frac{(2^k-1)!!}{(2i-1)(2^k-2i+1)} \equiv -\sum_{i=1}^{2^{k-1}} (2^k-1)!! \cdot a_i^2 = -(2^k-1)!! \cdot \sum_{i=1}^{2^{k-1}} (2i-1)^2 =$$

$$= -(2^k-1)!! \cdot \frac{2^{k-1}(2^k+1)(2^k-1)}{3} \equiv 2^{k-1} \pmod{2^k}.$$

The exponent of 2 in c_1 is $2k - 1$. Now $P(2^k) = c_1 \cdot 2^k + 2^{3k}Q(2^k)$ for some polynomial Q. Clearly, $2^{3k-1} | P(2^k)$, but $2^{3k} \nmid P(2^k)$. Finally, the exponent of 2 in D is equal to $3k$.

214. (i) Suppose that for $a \neq 4$ the equation has a solution, and let (u, v) be the solution with the smallest $u + v$. Then $\left(\frac{v^2+2}{u}, v\right)$ is also a solution; hence $\frac{v^2+2}{u} \geq u$.

If $u \geq v + 1$, then $v^2 + 2 \geq v^2 + 2v + 1$, impossible. We deduce that $u = v$; whence $u = v = 1$ and $a = 4$, contradiction.

(ii) Define the sequence (a_n) by $a_1 = 1$, $a_2 = 3$, $a_{n+1} = 4a_n - a_{n-1}$ for $n \geq 2$. Then $(x, y) = (a_n, a_{n+1})$ is a solution of the equation for all positive integers n.

215. Let $M = \begin{pmatrix} a & b \\ c & d \end{pmatrix}$. Then $M^2 \equiv M \pmod{37}$ and $M^{-1} \pmod{37}$ exist, since $\det M \not\equiv 0 \pmod{37}$. Multiplying by M^{-1}, we get $M \equiv I_2 \pmod{37}$; whence

$$a \equiv d \equiv 1 \pmod{37}, b \equiv c \equiv 0 \pmod{37}.$$

216. If $p \equiv 5 \pmod 6$, then $\{x^3 | x = 0, 1, \ldots, p-1\}$ is a complete set of residues modulo p, since $x^3 \equiv 1 \pmod p$ implies $x \equiv 1 \pmod p$. From $\gcd(p, 3) = 1$, we deduce that the set $\{3x^3 | x = 0, 1, \ldots, p-1\}$ is also a complete set of residues modulo p. Therefore, for each pair (y, z), there is only an x modulo p such that p divides $3x^3 + 4y^3 + 5z^3 - y^4z$. There are exactly p^2 pairs (y, z) with y, z in the set $\{0, 1, \ldots, p-1\}$, so that the given equation has exactly p^2 solutions modulo p.

217. The following decomposition holds:

$$x^3 + y^3 - 3xy + 1 = (x + y + 1)(x^2 + y^2 + 1 - x - y - xy).$$

Since $x = y = 1$ is not a solution, we have $x + y \geq 1$, which implies

$$x^2 + y^2 + 1 - x - y - xy = 1.$$

If $x \geq 2$, $y \geq 2$, then $x^2 + y^2 \geq 2(x + y)$ and $x^2 + y^2 \geq 2xy$. The equality $x = y = 2$ case gives $p = 5$.

If $x = 1$, then only $y = 0$ is a solution, with $p = 2$.

In conclusion, the given equation has solutions only for $p = 2$ and $p = 5$.

218. Because $\gcd(n,p) = 1$, it follows that

$$\sum_{i \geq 1} \left\lfloor \frac{n}{p^i} \right\rfloor = \sum_{i \geq 1} \left\lfloor \frac{n-1}{p^i} \right\rfloor.$$

From

$$\nu_p\left(\binom{np^k}{p^k}\right) = \nu_p((np^k)!) - \nu_p((p^k)!) - \nu_p((np^k - p^k)!) =$$
$$= \frac{n(p^k - 1)}{p - 1} - \frac{p^k - 1}{p - 1} - \frac{(n-1)(p^k - 1)}{p - 1} + \sum_{i \geq 1}\left\lfloor \frac{n}{p^i} \right\rfloor - \sum_{i \geq 1}\left\lfloor \frac{n-1}{p^i} \right\rfloor = 0,$$

we deduce

$$\gcd\left(p, \binom{np^k}{p^k}\right) = 1.$$

219. From $a + b - c + d \mid ab + cd$, we deduce

$$a + b - c + d \mid ab + (a + b + d)d = (a + d)(b + d).$$

Since $a + b - c + d > b + d$, then there exists a divisor $k > 1$ of $a + b - c + d$ that divides $a + d$; hence k divides $b - c$.

Because n is odd, it follows that $a^n \equiv -d^n \pmod{k}$. From $b^m \equiv c^m \pmod{k}$, we deduce $a^n b^m + c^n d^m \equiv 0 \pmod{k}$; hence this number is composite.

220. If $n = 2k + 1$ and $a_n = x^2$, then $x^2 \equiv 2^n \pmod{p}$, since $p \mid a$; therefore $x^2 \equiv 2 \cdot (2^k)^2 \pmod{p}$. From $p \neq 2$ we deduce that 2^k has an inverse mod p, so that

$$\left(2^{-k}x\right)^2 \equiv 2 \pmod{p};$$

whence it follows $p \equiv \pm 1 \pmod{8}$, contradiction.

If $n = 2k$, then $2^{2k} + a = x^2$ implies $a = (x - 2^k)(x + 2^k)$. But a has finitely many divisors; whence it follows that there are only finitely many k for which a_{2k} is a square.

221. We shall prove the following properties:

If m is happy, then $k^2 m$ is happy, for all $k \in \mathbb{N}^*$.

Proof. There are integers $a > b > 0$ such that $m = \frac{a^2 b}{a - b}$; hence

$$k^2 m = \frac{(ka)^2 (kb)}{ka - kb}$$

is happy.

1. Every happy number m has a prime divisor p such that $2 | \nu_p(m)$.

   Proof. If $m = \frac{a^2 b}{a-b}$, let $\gcd(a,b) = d$, $a = da_1$, $b = db_1$, $\gcd(a_1, b_1) = 1$. Then

   $$m = \frac{d^2 a_1^2 b_1^2}{a_1 - b_1}.$$

   We see that $a_1 > 1$ (otherwise, if $a_1 = 1$, then $a \leq b$, contradiction), so there exists a prime $p | a_1$. Since $p \nmid b_1(a_1 - b_1)$, it follows that $2 \lfloor \nu_p(m) \rfloor$.

   Now, let m be an evil and happy number. If there exist odd divisors of m, i.e.,

   $$m = 2^k(2c+1), c \geq 1,$$

   then we shall find a happy number n such that $\tau(n) = m$. Indeed, we start with

   $$l = \frac{3^2 \cdot 2}{3 - 2} = 3^2 \cdot 2,$$

   and using property 1 we get a happy number $n = 3^{2c} \cdot 2^{2^k - 1}$ with $\tau(n) = m$, contradiction. We conclude that $p = 2$ is the only prime divisor of m. Applying the property 2, we conclude that $m = 4^k$.

   It remains to verify that 4^k is happy and evil. From

   $$4^k = \frac{(2^k)^2 \cdot 2^{k-1}}{2^k - 2^{k-1}}$$

   it follows that 4^k is happy. The property 2 says that for every happy number n, there is an odd prime divisor of $\tau(n)$; therefore 4^k is evil.

222. If $(n,p) = 1$, then $n^{p-1} \equiv 1 \pmod{p}$; hence

$$\sum_{n=1}^{103} n^{p-1} \equiv 103 - \left\lfloor \frac{103}{p} \right\rfloor \pmod{p}.$$

For $103 = pq + r$, $0 \leq r \leq p - 1$, it follows that $r \equiv q \pmod{p}$.

Case 1. $q < p$. Then $q = r$; whence it follows $103 = (p+1)r$. But 103 is a prime, so that $r = 1$ and $p = 103 - 1 = 102$, contradiction.

Case 2. $q \geq p$. Then $pq \geq p^2$, so that $103 \geq p^2$; whence $p \leq \lfloor \sqrt{103} \rfloor = 10$. Checking $p \in \{3, 5, 7\}$, we conclude that only $p = 3$ is a solution.

223. Answer: $(2, 3, 5)$, $(1, 1, n)$, $n \geq 1$ and all permutations.

Clearly, a, b, c are pairwise coprime; otherwise a common divisor $d > 1$ of a, b, for example, must divide 1, since $a|bc - 1$, contradiction.

From $a|bc - 1$, it follows that $a|ab + bc + ca - 1$ and similarly for b, c; hence

$$abc| \, ab + bc + ca - 1.$$

From this, dividing by abc, we deduce

$$\frac{1}{a} + \frac{1}{b} + \frac{1}{c} - \frac{1}{abc} \geq 1.$$

Suppose wlog that $a \leq b \leq c$.
Case 1. $a \geq 3$. Then LHS <1, contradiction.
Case 2. $a = 2$. If $b \geq 4$, then LHS < 1, which implies $b = 2, 3$. But $\gcd(a, b) = 1$; hence $b = 3$ and $c = 5$.
Case 3. $a = 1$. Then $c|b - 1$ and $c \geq b$; therefore $b = 1$ and c is arbitrary.

224. Let a be an even number and $b = a^{a+2} - a^a + a$. Then $a^a + b = a^{a+2} + a$ and $b^b + a \equiv (-a^a)^b + a \equiv a^{ab} + a \pmod{a^a + b}$, since b is even. We have to prove that $a^{a+1} + 1 | a^{ab-1} + 1$. Since $a + 1$ is odd, it suffices to show that $a + 1 | ab - 1$, which is a consequence of the equalities:

$$ab - 1 = a^{a+3} - a^{a+1} + a^2 - 1 = (a-1)(a+1)(a^{a+1} + 1).$$

225. From $3|(2m + 3)^n + 1$, it follows that $m \equiv 1 \pmod 3$ and n is odd. We know that $m|3^n + 1$, but every odd prime divisor of $3^n + 1$ has the form $3k + 1$. If m is even, then $4|m$; hence $8|(2m + 3)^n + 1$; therefore $8|3^n + 1$, which contradicts the fact that n is odd. Hence m is odd, but $2m|(2m + 3)^n + 1$ implies $4m|3^n + 1$.

2.37 2009

226. If (n, m) is a solution, then n is odd and $m \geq 2$. We have $(n - 1)(n + 1) = 5 \cdot 2^m$.
From $(n - 1, n + 1) = 2$, it follows the following cases:
Case 1. $n - 1 = 2$. No solution.
Case 2. $n - 1 = 2 \cdot 5$. No solution.
Case 3. $n - 1 = 2^{m-1}$, which implies $(n, m) = (9, 4)$.
Case 4. $n - 1 = 2^{m-1} \cdot 5$. No solution.

227. Answer: $n = 6k \pm 1$.

Observe that $309 = 3 \cdot 103$. Taking mod 3, we have

$$f(n) = 20^n - 13^n - 7^n \equiv (-1)^n - 2 \pmod 3;$$

whence it follows that $3|f(n)$ if and only if $n \equiv 1 \pmod 2$.

Case 1. $n = 6k + 1$. Then $f(n) \equiv 20 \cdot 23^k - 13 \cdot 23^k - 7 \cdot 23^k \equiv 0 \pmod{103}$.
Case 2. $n = 6k + 3$. Then $f(n) \equiv 69 \cdot 23^k - 38 \cdot 23^k - 38 \cdot 23^k \equiv 23^k \pmod{103}$. No solution.
Case 3. $n = 6k + 5$. Then $f(n) \equiv 99 \cdot 23^k - 81 \cdot 23^k - 18 \cdot 23^k \equiv 0 \pmod{103}$.

228. Answer: $(a, b) = (6, 10)$.

The given condition is $a^2(b - a) = p^2(a + b)$.

Case 1. $p | a$. Then $a = pc$ and $c^2(b - pc) = b + pc$; whence $b = pc + \frac{2pc}{c^2 - 1}$.
Since $c^2 - 1 = 1, 2$ are impossible, we must have $c^2 - 1 \in \{p, 2p\}$. If $c^2 - 1 = p$, then $c - 1 = 1$, since p is a prime, so that $p = 3$ and $(a, b) = (6, 10)$.

Case 2. $p \nmid a$. Then $p | b - a$, so that $b = a + p^2c$ and $a^2c = 2a + p^2c$. It follows that $a | p^2c$, so that $| c, c = ka$, and we get $ka^2 = 2 + p^2k$, implying $k | 2$.
If $k = 1$, then $a^2 = 2 + p^2$, impossible.
If $k = 2$, then $a^2 = 1 + p^2$, impossible.

229. We can write

$$n = \prod_{k=1}^{g} q_k \prod_{i=1}^{h} p_i^{a_i},$$

with p_i, q_k primes, $a_i > 1$ and $q_1 < q_2 < \cdots < q_g$. We take $d = \prod_{i=1}^{h} p_i^{b_i}$, where $b_i = \lfloor \frac{a_i}{2} \rfloor$. If $g \geq 2$, we take $d' = \prod_{k=1}^{\lfloor \frac{g}{2} \rfloor} q_k$ and $m = dd' \leq \sqrt{n}$. Then

$$\tau^3(m) = 2^{3\lfloor \frac{g}{2} \rfloor} \prod_{i=1}^{h} (b_i + 1)^3 \geq 2^g \prod_{i=1}^{h} (a_i + 1) = \tau(n).$$

230. Let P be the set of all prime divisors of b. For $p \in P$, let s_p be the least integer such that $p | a^{s_p} - 1$. Then $s_p | n$, $s_p | p - 1$ and $n \leq \nu_p(a^{s_p} - 1) + \nu_p(n/s_p)$. Observe that

$$b \geq \prod_{p \in P} p > \prod_{p \in P} s_p = S$$

and $a^{s_p} - 1 | a^S - 1$ for all $p \in P$. Thus,

$$a^b > a^S - 1 \geq \prod_{p \in P} p^n \prod_{p \in P} p^{-\nu_p(n)} \geq \frac{3^n}{n}.$$

231. The equation $n^2 + 1 = 5m^2$ is a negative Pell equation and has the solution $(2, 1)$. It follows that the equation has infinitely many solutions. If (m, n) is a solution, then $n^2 + 1 = m \cdot 5m$, and since $m < n$, $5m < n$, it follows that $n^2 + 1 | n!$.

On the other hand, if $p \equiv 1 \pmod 4$, then -1 is a quadratic residue mod p, so that there is a positive integer n such that $p | n^2 + 1$. We can choose $n < p$ and then $n^2 + 1 \nmid n!$. There are infinitely many primes $p \equiv 1 \pmod 4$; therefore n can be arbitrarily large.

232. Answer: $(a, 2^a \pm 1, 1)$, $(1, 1, c)$, $(3, 3, 2)$.

If $c = 1$, then $|2^a - b| = 1$, which yields the solutions $(a, 2^a \pm 1, 1)$.

Assume $c > 1$. If $a = 1$, then $b = 1$, which gives the solutions $(1, 1, c)$. If $a > 1$, then we have two consecutive perfect powers, which must be 8 and 9, by the Mihăilescu's theorem, so $(a, b, c) = (3, 3, 2)$.

233. (a) Let $A = \{k \cdot n! + 1 | 1 \leq k \leq n\}$. For any two elements $k \cdot n!$, $l \cdot n!$, $1 \leq k \leq n$,

$$\gcd(k \cdot n! + 1, l \cdot n! + 1) = \gcd(k \cdot n! + 1, (l - k) \cdot n!).$$

The number $k \cdot n! + 1$ is not divisible by none of the prime numbers $2 \leq p \leq n$, whereas all prime divisors of $(l - k) \cdot n!$ are the primes $2 \leq p \leq n$. It follows that every two elements of A are coprime, that is, A is independent.

For arbitrary m elements $k_i \cdot n! + 1$, $1 \leq i \leq m$, of A where $1 \leq m \leq n$, we get

$$\sum_{i=1}^{m} (k_i \cdot n! + 1) = m \left(\frac{n!}{m} \sum_{i=1}^{m} k_i + 1 \right).$$

Since $\frac{n!}{m}$ is an integer, the sum of the m elements is divisible by m; hence A is nice.

(b) Answer: no. The difference of any two elements of an n-element nice set is divisible by every positive integer less than n. Indeed, let X be an n-element nice set. Fix an integer i, $2 \leq i \leq n - 1$ and arbitrary $a, b \in X$. If s is the sum of every $i - 1$ elements of X different from a, b, then both $a + s$ and $b + s$ are divisible by i, so that $a - b$ is also divisible by i.

Suppose that there is an infinite set B of positive integers such that for any positive integer n, there is an n-element-independent subset and that every independent subset is nice. We show that every finite independent subset A can be extended to larger independent subsets by adding new numbers. Indeed, there is only a finite number, say k, of primes which divide some element of A. By Pigeonhole principle, a $k + 1$-element-independent subset of B has an element that is coprime with all elements of A, and we can add this element to A to form a larger independent set.

Let a, b be coprime elements of B (they exist, since $\{a, b\}$ is a 2-element independent subset of B) and consider an infinite subset of B containing a, b and having pairwise coprime elements (it can be constructed by infinitely repeating the steps described above starting from $\{a, b\}$). Since any of its finite subsets containing a, b is nice, $a - b$ is divisible by any positive integer; therefore $a = b$, contradiction.

234. Take
$$y = c(n) \cdot \left(\frac{n}{c(n)} + 1\right).$$

Then
$$n = c(n) \cdot \frac{n}{c(n)} < y < (c(n)+1)\left(\frac{n}{c(n)}+1\right),$$

while
$$y \mid c(n) \cdot \frac{n}{c(n)} \cdot (c(n)+1) \cdot \left(\frac{n}{c(n)}+1\right) = n \cdot (c(n)+1) \cdot \left(\frac{n}{c(n)}+1\right).$$

This implies
$$s(n) \le (c(n)+1)\left(\frac{n}{c(n)}+1\right).$$

For minimality, choose x such that
$$n < x < (c(n)+1)\left(\frac{n}{c(n)}+1\right).$$

It has to be proven that no integer strictly between n and x divides nx. Let $n < y < x$, $d = \gcd(y, n)$, $y = dy'$, $n = dn'$. As $c(n)$ and $\frac{n}{c(n)}$ are central divisors of n, we have
$$c(n) + \frac{n}{c(n)} \le d + \frac{n}{d}. (*)$$

Indeed, the function $f(z) = z + \frac{1}{z}$ is decreasing on $(0, 1)$. Then
$$c(n) + \frac{n}{c(n)} = \sqrt{n} \cdot f\left(\frac{c(n)}{\sqrt{n}}\right), \quad d + \frac{n}{d} = \sqrt{n} \cdot f\left(\frac{\min(d, \frac{n}{d})}{\sqrt{n}}\right),$$

as well as $\min(d, \frac{n}{d}) \le c(n) \le \sqrt{n}$.
Adding $n + 1$ to both sides of $(*)$ and factorising, we get
$$(c(n)+1)\left(\frac{n}{c(n)}+1\right) \le (d+1)\left(\frac{n}{d}+1\right). (**)$$

But $d|y$ implies $d|y - n$ and furthermore $d \leq y - n$. Thus $n + d \leq y$, which implies

$$\frac{n}{d} + 1 \leq \frac{y}{d} = y'.$$

Then

$$dy' = y < x < (c(n) + 1)\left(\frac{n}{c(n)} + 1\right) \leq (d+1)\left(\frac{n}{d} + 1\right) \leq (d+1)y'.$$

Hence x is strictly between two consecutive multiples of y'; therefore $y' \nmid x$. If $y|nx$, then $y'|n'x$. As $\gcd(y', b') = 1$, it follows that $y'|x$, contradiction.

235. Define the sequence $b_n = a_n^2 + 1$. Then $a_{n+1} = b_n^2 - a_n(b_n - 1)$; therefore $a_{n+1} \equiv a_n \pmod{b_n}$ and $a_{n+1}^2 + 1 \equiv a_n^2 + 1 \equiv 0 \pmod{b_n}$. It follows that $b_n | b_{n+1}$ for all n. Then $b_{n+1} = b_n c_n$ and $\gcd(b_n, c_n) = 1$ for all n. Indeed, $a_{n+1}^2 + 1 \equiv a_n^2(b_n - 1)^2 \equiv b_n(1 - 2a_n^2) \pmod{b_n^2}$. We show that

$$\gcd(1 - 2a_n^2, b_n) = 1 \text{ for all } n.$$

Let p be a prime such that $p| 1 - 2a_n^2, p| b_n$. Then $p| 1 - 2a_n^2 + 2(a_n^2 + 1) = 3$;. whence $p = 3$, so that $a_n^2 \equiv -1 \pmod 3$, contradiction, since -1 is not a square mod 3. Therefore, $\gcd(1 - 2a_n^2, b_n) = 1$ for all n. Since $b_n | b_{n+1}$ and $b_n^2 \nmid b_{n+1}$, we conclude that $b_{n+1} = b_n c_n$, with $\gcd(b_n, c_n) = 1$ for all n.

Choose primes $p_k | c_k$, and then $p_k \nmid b_k$; moreover, $p_k \nmid b_1\ldots b_k$, since $b_n | b_{n+1}$. Finally, $p_k \nmid c_1 c_2 \ldots c_{k-1}$, so that the primes p_k are distinct.

It remains to show that any prime of the sequence satisfies the required condition. Let p an element of the sequence, $p|b_n$ for some $n \geq 1$ and $p|a_k$ for some $k \geq 1$. By induction we prove that $a_{n+j} \equiv a_j \pmod{a_n}$. The base case is obvious. Suppose that it is true for some j, then $a_{n+j+1} \equiv a_{j+1} \equiv a_j^4 - a_j^3 + 2a_j^2 + 1 \pmod{a_n}$.

Choose now j such that $kj \geq n$. It follows that $a_{kj} \equiv 0 \pmod{a_k}$. But $b_{kj} = a_{kj}^2 + +1 \not\equiv 0 \pmod p$, contradiction, since $p|b_n|b_{kj}$.

In conclusion, there are infinitely many primes that not divide any of the numbers $a_1, a_2, a_3, \ldots$.

236. Assume the contrary. If two of the numbers are the same, then all three are equal. If the numbers are different, then the given condition implies that

$$\frac{a^n - b^n}{a - b} \cdot \frac{b^n - c^n}{b - c} \cdot \frac{c^n - a^n}{c - a} = -p^3,$$

which implies that some of a, b, c are negative. Also, n is even; otherwise each fraction is positive.

If p is odd, then $2 \nmid a - b$, since $2 \nmid \frac{a^n - b^n}{a-b} = a^{n-1} + a^{n-2}b + \ldots + b^{n-1}$ and similarly $2 \nmid b - c$, $2 \nmid c - a$, contradiction.

If $p = 2$, for $n = 2m$ we have

$$(a^m + b^m)(b^m + c^m)(c^m + a^m) \cdot \frac{a^m - b^m}{a - b} \cdot \frac{b^m - c^m}{b - c} \cdot \frac{c^m - a^m}{c - a} = -8;$$

whence it follows $a^m + b^m = \pm 2$, $a^m - b^m = \pm(a - b)$ and analogous equalities for the other two pairs of numbers. If m is even, then $|a| = |b| = |c| = 1$, which means that at least two of a, b, c are equal. If m is odd, then $a + b \mid a^m + b^m = \pm 2$. Since $a^m + b^m \equiv a + b \pmod{2}$, we conclude that $a + b = \pm 2$ and similarly $b + c = \pm 2$, $c + a = \pm 2$. Then at least two of a, b, c are equal, contradiction.

237. If $a = 2k$, then $A = (2k)^n + (2k)^{n-1} + \cdots + (2k) + 1 = (2m + 1)^2$; whence

$$(2k)^n + (2k)^{n-1} + \cdots + (2k) = 4m(m + 1).$$

We deduce $4 \mid k$, that is, $k = 4s$, since RHS is a multiple of 8, so that $a = 2k = 8s$.

238. For $a = 1$, we know that the Fermat numbers are pairwise coprime. Assume $a > 1$, and suppose that the set of primes that divide at least one term of the sequence is finite. If $\{p_1, \ldots, p_k\}$ is this set, then write each term as $p_1^{e_1} \ldots p_k^{e_k}$, and choose the index i, $1 \leq i \leq k$, such that $p_i^{e_i}$ is maximal. Since we have assigned to each term one of k indices, the Pigeonhole principle implies that $x_n = 2^{2^n} + a$ and $x_m = 2^{2^m} + a$ have assigned the same index i for some $n + k > m > n > N$, where $N > 0$ is a sufficiently large constant.

Observe that

$$p_i^{a_i} = p_i^{\nu_{p_i}(\gcd(x_m, x_n))} \geq \min(\sqrt[k]{x_n}, \sqrt[k]{x_m}), \; p_i^{a_i} \mid 2^{2^n}\left(\left(2^{2^n}\right)^{2^{m-n}-1} - 1\right),$$

which implies

$$1 \equiv \left(2^{2^n}\right)^{2^{m-n}-1} \equiv (-a)^{2^{m-n}-1} \pmod{p_i^{a_i}};$$

therefore $p_i^{a_i} \leq \left|(-a)^{2^{m-n}-1} - 1\right|$. This is a contradiction, since $m - n \leq k$, so that the number from the right remains bounded.

239. We will prove that $f(x) = x^n$ and $f(x) = -x^n$ are the only solutions.

Let n be the least positive integer for which there is a solution

$$f(x) = a_0 x^n + a_1 x^{n-1} + \cdots + a_{n-1} x + a_n, \; a_0 \neq 0.$$

Then $a_n \neq 0$; otherwise we can reduce the degree. From the given condition, we have that $p | g(x)$, where

$$g(x) = (a_0 x^n + \cdots + a_n)(a_n x^n + \cdots + a_0) - x^n, \forall x \in \mathbb{Z}.$$

Take $x = 1, 2, \ldots, 2n + 1$, and choose p greater than the largest of $|g(1)|$, $|g(2)|, \ldots, |g(2n + 1)|$. This implies $g = 0$ for all $x \in \mathbb{Z}$. Then the leading coefficient of g, $a_0 a_n = 0$, contradiction.

240. Note that $\nu_2(3^{2^n} - 1) = \nu_2(5^{2^n} - 1) = n + 2$, so both exponents are odd positive integers. Since $ord_{2^{n+4}}(5)$ must be some power of 2, say it is 2^l. Then $2^{n+4} | 5^{2^l} - 1$ implies $l \geq n + 2$; hence $l = n + 2$.

The 2^{n+2} residues modulo 2^{n+4} of the numbers 5^i, where $0 \leq i < 2^{n+2}$ are distinct and $\equiv 1 \pmod 4$. Since $-3 \equiv 1 \pmod 4$, there exists a positive integer t such that $5^t \equiv -3 \pmod{2^{n+4}}$. Then

$$(-3)^{\frac{5^{2^n}-1}{2^{n+2}}} \equiv 5^{\frac{3^{2^n}-1}{2^{n+2}}} \pmod{2^{n+4}}$$

if and only if

$$5^{t \cdot \frac{5^{2^n}-1}{2^{n+2}}} \equiv 5^{\frac{3^{2^n}-1}{2^{n+2}}} \pmod{2^{n+4}}.$$

Since $ord_{2^{n+4}}(5) = 2^{n+2}$, we have to show that $t(5^{2^n} - 1) \equiv 3^{2^n} - 1 \pmod{2^{2n+4}}$.

Since $\nu_2(5^{2^n} - 1) = n + 2$, then

$$1 + t(5^{2^n} - 1) \equiv (5^{2^n})^t \equiv (5^t)^{2^n} \pmod{2^{2n+4}}.$$

Now we need to show that $(5^t)^{2^n} \equiv 3^{2^n} \pmod{2^{2n+4}}$. Note that $2^{n+4} | 5^t - (-3)$; hence

$$\nu_2\left((5^t)^{2^n} - (-3)^{2^n}\right) = \nu_2(5^t - (-3)) + \nu_2(2^n) \geq 2n + 4.$$

We deduce $(5^t)^{2^n} \equiv 3^{2^n} \pmod{2^{2n+4}}$.

241. Answer: $(x, y) = (0, -1), (0, 0), (1, 2)$.

The equation $y^3 = 8x^6 + 2x^3 y - y^2$ can be written as

$$(64x^3 + 8y)^2 = (8y)^2(8y + 9),$$

so that $8y + 9 = z^2$, with $z > 1$. Then

$$64x^3 + z^2 - 9 = \pm z(z^2 - 9)$$

and a simple framing between consecutive cubes leads to checking for z only values $z \in \{1, 3, 5, 7, 9\}$. Finally, we find the above solutions.

242. Take d as the product of all sums with p terms from the set $\{1, 2, \ldots, n\}$, and choose $A = \{d, 2d, \ldots, nd\}$.

243. Choose a prime $p > 7$, and let q a prime that divides $2^{p-1} - 1$ and not divides $2^a - 1$ for all $a < p - 1$ (such a prime exists, by the Zsigmondy's theorem). The order of 2 mod q is $p - 1$, which means that $p - 1 | q - 1$. From this we deduce that p divides $2^{q-1} - 1$. Since the set of primes is infinite, this finishes the proof.

244. Answer: $N = 911$, $(b_1, \ldots, b_6) = (25, \ldots, 30)$,

$$(a_1, \ldots, a_6) = (286, 235, 182, 127, 70, 11),$$

where $b_i^2 = N - a_i$, $1 \leq i \leq 6$.

245. For $x^{41} \equiv y^{49} \equiv a \pmod{p}$, we want to find z with $z^{2009} \equiv a \pmod{p}$.

If $p | a$, then $z \equiv 0 \pmod{p}$.

If $p \nmid a$, then let b, c be integers such that $49b + 41c = 1$ (which exist, since $\gcd(49, 41) = 1$). Choose $z \equiv x^b y^c$ and then

$$z^{2009} \equiv x^{2009b} y^{2009c} \equiv a^{49b} a^{41c} \equiv a^{49b+41c} \equiv a \pmod{p}.$$

Now we need $z^{2009} \geq a$. Assume that $z \equiv d \pmod{p}$ and pick a number $\equiv d \pmod{p}$ and $\geq a$. Then $z^{2009} \geq a^{2009} \geq a$.

246. Since $m \neq n$, it follows that $\gcd(m, n) = \gcd(m, m^2 + 8) = \gcd(n, n^2 + 8) = 1$. If (m, n) satisfies the required condition, then $mn | (m^2 + 8)(n^2 + 8)$, so that

$$m^2 + n^2 + 8 = kmn,$$

with k positive integer. Fix k, suppose, due to symmetry, that $m > n$, and write the above equality as a quadratic equation in m:

$$m^2 - kn \cdot m + n^2 + 8 = 0.$$

This equation has the solutions m, m' and the Viète's relations are

$$m + m' = kn, \quad mm' = n^2 + 8.$$

It follows that m' is a positive integer. For $n \geq 4$, both m, m' cannot be greater than n; otherwise $n^2 + 8 = mm' \geq (n+1)^2 \geq n^2 + 9$, contradiction. Then, for every solution (m, n), we find another smaller solution $(n, kn - m)$. Finally, we arrive at $n = 1$ if and only if the discriminant $\Delta = k^2 - 36$ is a square, that is, for $k \in \{6, 10\}$.

For $k = 10$, we find the initial solution $(m, n) = (1, 1)$ and all solutions

$$(1, 1), (9, 1), (89, 9), (881, 89), (8721, 881), \ldots$$

that is, solutions of the form (a_{k+1}, a_k) given by the recurrent sequence

$$a_0 = a_1 = 1, a_i = 10a_{i-1} - a_{i-2}, i \geq 2.$$

For $k = 6$, we find the initial solution $(m, n) = (3, 1)$ and all solutions

$$(3, 1), (17, 3), (99, 17), (567, 99), (3303, 567), \ldots$$

that is, solutions of the form (a_{k+1}, a_k) given by the recurrent sequence

$$a_0 = 1, a_1 = 3, a_i = 6a_{i-1} - a_{i-2}, i \geq 2.$$

2.38 2010

247. If $n = \prod_i p_i^{\alpha_i}$ is the canonical decomposition, then

$$\sigma(n) = \prod_i \sigma(p_i^{\alpha_i}) = \prod_i (1 + p_i + \cdots + p_i^{\alpha_i}).$$

Moreover, if each factor is a Mersenne prime, then $p_i = 2^{\beta_i} - 1$. Then $\alpha_i = 1$ for all i implies $\sigma(n) = \prod_i 2^{\beta_i} = 2^k$.

Conversely, each prime factor p of n must satisfy $1 + p + \cdots + p^\alpha = 2^s$, for some positive integer s. This implies that $\alpha = 2q + 1$ for some integer q and furthermore $2^s = (1 + p)(1 + p^2 + p^4 + \cdots + p^{2q})$. In particular, $1 + p$ is a power of 2; thus p is a Mersenne prime, $p = 2^u - 1$. Suppose that $q > 0$; then q must be odd, $q = 2r + 1$ and

$$2^t = \left(1 + p^2\right)\left(1 + p^4 + p^8 + \cdots + p^{4r}\right)$$

for some positive integer t. This means that $1 + p^2$ is a power of 2, but

$$1 + p^2 = 1 + (2^u - 1)^2 = 2 - 2^{u+1} + 2^{2u}$$

is not divisible by 4, contradiction. We conclude that $\alpha = 1$, which ends the proof.

248. Answer: $(p,q) = (2,7)$.

Case 1. p, q odd. The sequence mod 2 is

$$1, 1, 0, 1, 1, 0, 1, 1, 0, \ldots;$$

therefore x_{3k} is even for every k and cannot be -3.

Case 2. $p = q = 2$. The sequence mod 2 is $1, 0, 0, 0, 0, 0, 0, \ldots$ and as above, $x_{3k} \neq -3$.

Case 3. $p = 2$, q arbitrary. The sequence mod $x_3 = p^2 - q$ is

$$x_{3k} = 0, k \geq 1, x_{3k+1} = (-1)^k p^{3k}, k \geq 0, x_{3k+2} = (-1)^k p^{3k+1}, k \geq 0.$$

Since x_3 divides x_{3k} for every $k \geq 1$, we must have $x_3 = \pm 3$, so that $q = 2^2 \pm 3$, with the unique acceptable solution $q = 7$.

Case 4. $q = 2$, p arbitrary. As above, $p^2 = 2 \pm 3$, with no solutions.

249. (a) We deduce that $p + q^2$ divides $p(p + q^2) - (p^2 + q) = q(pq - 1)$. But p, q are distinct primes, so that $\gcd(q, p + q^2) = 1$; whence $p + q^2 | pq - 1$.

(b) Clearly, $p = 11$ is a solution. If $p \neq 11$ is a solution, then $p + 11^2 | 11p - 1$. From $11p - 1 = 11(p + 11^2) - (11^3 + 1)$, it follows that $p + 11^2$ divides $11^3 + 1 = 1332 = 2^2 \cdot 3^2 \cdot 37$; whence

$$p + 11^2 \in \{1, 2, 3, 4, 6, 9, 12, 18, 36, 37, 74, 111, 148, 222, 333, 444, 666, 1332\},$$

with solution $p = 222 - 11^2 = 101$.

250. We shall prove firstly that n must be even. Indeed, let $\nu_2(n + 1) = r$; then $m^2 + 1$ divides mp^a, and since $\gcd(m^2 + 1, m) = 1$, it follows that $m^2 + 1 | p^a$, contradiction.

This forces $m + 1 = p$. Expanding the given equation, we get

$$q + m^2 + m^3 + \cdots + m^{2^n - 2} = m^n + p^a.$$

If $n \geq 3$, then $LHS \equiv q + 4 \pmod 8$ and $RHS \equiv q \pmod 8$, contradiction. Hence $n = 2$ and the equation reduces to $m + 1 = p^a$, so that $a = 1$ and $(m, n) = (p - 1, 2)$.

251. The result follows from *On a Problem in the Elementary Theory of Numbers* by Paul Erdös and Paul Turán.

252. **Lemma.** Let $\{p_1, \ldots, p_t\}$ be a finite set of primes. Then for any integer $k \geq 0$, there exists an integer $n > 1$ such that

$$\nu_{p_i}\bigl(f^j(n)\bigr) = 1 \text{ for } 1 \leq i \leq t, 0 \leq j \leq k.$$

Proof. Induction on k: the base case $k = 0$ is obvious, assume the $k - 1$ case, and prove it for k. By the Dirichlet theorem, we can choose primes $q_1, \ldots, q_t$ such that

$$q_i \equiv -1 \pmod{p_i^2}.$$

The induction hypothesis show that there exists an integer $n > 1$ such that

$$\nu_{p_i}(f^j(n)) = \nu_{q_i}(f^j(n)) = 1, \ 1 \leq i \leq t, 0 \leq j \leq k-1.$$

This implies $p_i^2 \mid \sigma(f^{k-1}(n))$; hence

$$\nu_{p_i}(f^k(n)) = \nu_{p_i}(f^{k-1}(n)) = 1$$

implying the conclusion.

Back to the problem, apply the lemma for the set $\{2, 3\}$; then there exists integer $n > 1$ such that

$$\nu_2(f^i(n)) = \nu_3(f^i(n)) = 1 \text{ for } 0 \leq i \leq k.$$

Now it suffices to prove that $f^i(n)$ is abundant, for $1 \leq i \leq k$. Let $r = f^i(n)$, $r = 6s$ with $\gcd(s, 6) = 1$. Hence $\sigma(r) = \sigma(6)\sigma(s) > 12s = 2r$, which finishes the proof.

253. **Lemma**. Let p be a prime number. Then

$$\nu_p\left(\binom{n}{k}\right) \leq n.$$

Proof. Denote $s = \lfloor \log_p n \rfloor$. Since

$$\nu_p(n!) = \left\lfloor \frac{n}{p} \right\rfloor + \left\lfloor \frac{n}{p^2} \right\rfloor + \cdots + \left\lfloor \frac{n}{p^s} \right\rfloor,$$

we get

$$\nu_p\left(\binom{n}{k}\right) = \sum_{i=1}^{s}\left(\left\lfloor \frac{n}{p^i} \right\rfloor - \left\lfloor \frac{k}{p^i} \right\rfloor - \left\lfloor \frac{n-k}{p^i} \right\rfloor\right)$$

$$= \sum_{i=1}^{s}\left(\left\{\frac{k}{p^i}\right\} + \left\{\frac{n-k}{p^i}\right\} - \left\{\frac{n}{p^i}\right\}\right) \leq s.$$

Back to the problem, assume the contrary. If

$$\frac{n(n-1)\ldots(n-k+1)}{k!} = \binom{n}{k} = p_1^{\alpha_1}\ldots p_j^{\alpha_j},$$

then the lemma implies

$$\binom{n}{k} \leq n^j \leq n^{k-1},$$

contradiction.

254. Let N be one large integer which satisfies the property, and suppose that $N! \equiv -1 \pmod{p}$, with p minimal. Then by Wilson's theorem, $N < p$ and

$$N!(p-N-1)! \equiv (-1)^N \pmod{p}.$$

We have $p - N \geq 3$. Take $a_n = \max\{m | m! \leq n+1\}$; then $p \geq a_N + N$, which finishes the proof.

255. Note that $da \ominus db = a \ominus b$ for all positive integers a, b, d. Indeed,

$$da \ominus db = \frac{da - db}{\gcd(da, db)} = \frac{d(a-b)}{d \cdot \gcd(a,b)} = \frac{a-b}{\gcd(a,b)} = a \ominus b.$$

If n is a prime power and $m < n$, then $n \ominus m$ is relatively prime to n. Indeed, let $n = p^k$ with p prime and let $m = p^i s$, where $\gcd(p, s) = 1$. Then $m < n$ implies $i < k$. Now

$$n \ominus m = p^{k-i} \ominus s = \frac{p^{k-i} - s}{\gcd(p^{k-i}, s)} = p^{k-i} - s,$$

because $\gcd(p^{k-i}, s) = 1$ by the choice of s. Also, $\gcd(p, p^{k-i} - s) = 1$; hence

$$\gcd(n, n \ominus m) = \gcd(p^k, p^{k-i} - s) = 1.$$

It remains to show that if n is not a prime power, then there exists a positive integer m such that $m < n$, and the integers $n \ominus m$ and n share a common prime factor. Since n is not a prime power, it has at least two different prime factors. Let p and q be some prime factors of n, whereby $p < q$. Let $n = p^k t$, where $\gcd(p, t) = 1$. Take $m = n - p^{k+1}$. As n is divisible by both p^k and q which are relatively prime, it is also divisible by their product $p^k q$. Consequently, $p^{k+1} < p^k q \leq n$, i.e., $0 < m < n$. Now

$$n \ominus m = n \ominus (n - p^{k+1}) = t \ominus (t-p) = \frac{t-(t-p)}{\gcd(t, t-p)} = \frac{p}{\gcd(t,p)} = p,$$

since $\gcd(t, p) = 1$. We see that $n \ominus m$ and n have a common prime factor p.

256. (a) To each nonnegative integer n, we associate a sequence $(s_1, \ldots, s_{50})$ of numbers from $\{0, 1\}$ such that

$$s_i = \begin{cases} 0, & \text{if } n+i \text{ is good,} \\ 1, & \text{if } n+i \text{ is not good.} \end{cases}$$

Since there are at most 2^{50} such sequences, we see that there are nonnegative integers a, b that correspond to the same sequence. For such a, b, the number $P(i)$ is good for each $1 \leq i \leq 50$.

(b) Assume that $a < b$ and that $P(n)$ is good for each nonnegative integer n. If $k > a$ and $x = k(b-a) - a$, then $P(x) = (b-a)^2 k(k+1)$ is good, which means that $k, k+1$ are both good or both not good. It follows that the numbers greater than a are all good or not good, which is clearly false, since squares are good, but primes are not.

257. The side lengths of such triangle satisfy

$$a : b : c = \sin \frac{\pi}{7} : \sin \frac{2\pi}{7} : \sin \frac{4\pi}{7}.$$

From

$$\frac{\sin \frac{2\pi}{7}}{\sin \frac{\pi}{7}} = 2 \cos \frac{\pi}{7} \notin \mathbb{Q},$$

we conclude that such triangle does not exist. Indeed, if $\cos \frac{\pi}{7} = x \in \mathbb{Q}$, then $\cos \frac{2\pi}{7} = 2x^2 - 1 \in \mathbb{Q}$ and also $\cos \frac{4\pi}{7} \in \mathbb{Q}$. Choose coprime positive integers a, b such that $\cos \frac{2\pi}{7} = \frac{a}{b}$. Then

$$\cos^2 \frac{\pi}{7} = \frac{a+b}{2b}, \quad \cos \frac{4\pi}{7} = \frac{2a^2 - b^2}{b^2}.$$

The equality $\cos \frac{\pi}{7} \cos \frac{2\pi}{7} \cos \frac{4\mu}{7} = -\frac{1}{8}$ implies, after squaring and eliminating denominators,

$$2^7 (a+b) a^2 (2a^2 - b^2)^2 = b^7.$$

This equality shows that a is odd and b is even. We have $\nu_2(LHS) = 9$ and $7 | \nu_2(RHS)$, contradiction.

258. Answer: $n = 11$.

Denote
$$f(a,m) = \frac{a^m + 3^m}{a^2 - 3a + 1}, (a,m) \in A \times \mathbb{Z}^+.$$

If $a \equiv 5 \pmod{11}$, then $11 \mid a^2 - 3a + 1$, but $a^m + 3^m \equiv 5^m + 3^m \equiv 3^m(9^m + 1) \not\equiv 0$ (mod 11) for all $m \in \mathbb{Z}^+$. Therefore $f(a,m)$ is not an integer for any $a \equiv 5$ (mod 11) and $m \in \mathbb{Z}^+$.

On the other hand, $f(a,m)$ is an integer for (a,m) in the set

$\{(-5, 20), (-4, 14), (-3, 1), (-1, 2), (0, 1), (1, 1), (2, 1), (3, 1), (4, 2), (6, 9), (7, 2),$
$(8, 20), (12, 9)\}.$

259. Answer: no.

If there exists such a k, then
$$\binom{3k}{0} \equiv \binom{3k}{k} \equiv \binom{3k}{2k} \equiv \binom{3k}{3k} \equiv 1 \pmod{p}.$$

Let g be a primitive root mod p and let $\omega = g^6$; then
$$\frac{(1 + \omega^0 x)^{3k} + (1 + \omega^1 x)^{3k} + \cdots + (1 + \omega^{k-1} x)^{3k}}{k} \equiv 1 + x^k + x^{2k} + x^{3k} \pmod{p}.$$

Set $x = 1$ and let $a_i = (1 + \omega^i)^{3k} = (1 + \omega^i)^{\frac{p-1}{2}}$. We have
$$a_0 + a_1 + \cdots + a_{k-1} \equiv 4k \pmod{p}.$$

Each a_i is congruent to $b_i \in \{-1, 0, 1\}$ mod p for all i. Since
$$|b_0 + b_1 + \ldots + b_{k-1}| < p, 4k < p,$$

then either $b_0 + b_1 + \ldots + b_{k-1} = 4k$ or 0. The last case is impossible and the first leads to a contradiction:
$$|b_0 + \cdots + b_{k-1}| = |-(b_0 + \cdots + b_{k-1})| = |6k + 1 - 4k| = 2k + 1,$$

but $|b_0 + \cdots + b_{k-1}| \leq |b_0| + \cdots + |b_{k-1}| \leq k$.

Note. There are integers a, b such that $p = a^2 + 3b^2$, $a \equiv 1 \pmod 3$ and
$$\binom{3k}{k} \equiv 2a \not\equiv 1 \pmod{\mathsf{p}}.$$

2.39 2011

260. Clearly, $p = 2$ is not a solution, but $p = 3$ is a solution, which is unique, since for $p > 3$, $2^p + p^2 \equiv -1 + 1 \equiv 0 \pmod{3}$.

261. Denote $a_n = n(n+2)(n+4)$ and let b_n be the number of positive divisors of a_n. The values of b_1 to b_{10} are 4, 10, 8, 14, 12, 24, 12, 28, 12, and 40. Suppose that $n \geq 11$.

    Case 1. n even. Then $n = 2k$, $a_n = 2^3 k(k+1)(k+2)$. At least one of the numbers $k, k+1, k+2$ is divisible by 2, and exactly one of them is divisible by 3. Since $k \geq 6$, the numbers $k, k+1$, and $k+2$ cannot be all powers of 2 or 3; therefore their product has a prime divisor p different from 2 and 3. Hence, $2^4 \cdot 3p | a_n$, so that

    $$b_n \geq 5 \cdot 2 \cdot 2 = 20.$$

    Case 2. n odd. Then the numbers n, $n+2$ and $n+4$, are pairwise coprime. One of these three numbers is divisible by 3. This number has at least one other prime divisor p or is a power of 3. In the latter case, it is divisible by 3^3, since $n \geq 11$. Let q, r be prime divisors of the other two numbers. In the first case, a_n is divisible by $3pqr$; therefore $b_n \geq 2 \cdot 2 \cdot 2 \cdot 2 = 16$, since 3, p, q, and r are distinct. In the second case, a_n is divisible by $3^3 qr$; therefore $b_n \geq 4 \cdot 2 \cdot 2 = 16$.

    In conclusion, a_n has at most 15 positive divisors only for $n = 1, 2, 3, 4, 5, 7, 9$.

262. Let $n = p_1^{\alpha_1} \ldots p_k^{\alpha_k}$ be the canonical factorization of n. Since $p_1^{\alpha_1}, \ldots, p_k^{\alpha_k}$ are pairwise coprime, the Chinese remainder theorem implies that the system of congruences $x \equiv 1 \pmod{p_i^{\alpha_i}}$, $x \equiv 0 \pmod{p_j^{\alpha_j}}$, $j \neq i$, has a solution x_i, for $1 \leq i \leq k$.

    For any solution of the congruence $x^2 \equiv x \pmod{n}$ and for each $i = 1, 2, \ldots, k$, we have either $x \equiv 1 \pmod{p_i^{\alpha_i}}$ or $x \equiv 0 \pmod{p_i^{\alpha_i}}$. Let A be a subset of $\{x_1, x_2, \ldots, x_k\}$ and let $S(A)$ be the sum of elements of A. In particular, $S(\emptyset) = 0$. Then.

    $$S(A)(S(A) - 1) \equiv 0 \pmod{n},$$

    since each x_i is either 0 or 1 modulo $p_i^{\alpha_i}$. Moreover, $A \neq A'$ implies $S(A) \not\equiv S(A') \pmod{n}$. We deduce that the solutions of the congruence $x(x-1) \equiv 0 \pmod{n}$ are all sums $S(A)$. Let $S_0 = n$, S_r be the remainder of $x_1 + x_2 + \cdots + x_r$ modulo n for $1 \leq r \leq k$. Thus, $S_k = 1$ and $S_r \neq 0$ for $1 \leq r \leq k - 1$. Since $k + 1$ numbers $S_0, S_1, \ldots, S_k$ are in the interval $[1, n]$, the Pigeonhole principle implies that there are $0 \leq l < m \leq k$ such that S_l, S_m belong to the same interval $\left(\frac{jn}{k}, \frac{(j+1)n}{k}\right]$, $0 \leq j \leq k - 1$, where $l = 0$ and $m = k$ do not hold simultaneously.

Thus, $|S_l - S_m| < \frac{n}{k}$. Denote $y_1 = S_1, y_r = S_r - S_{r-1}, 2 \le r \le k$. Then any sum of y_r, $1 \le r \le k$ meets the requirement.

If $S_m - S_l > 1$, then $a = y_{l+1} + \cdots + y_m = S_m - S_l \in \left(1, \frac{n}{k}\right)$ is the solution of $x^2 \equiv x \pmod{n}$.

If $S_m - S_l = 1$, then n divides $(y_1 + \cdots + y_l) + (y_{m+1} + \cdots + y_k)$, that is, n divides $(x_1 + \cdots + x_l) + (x_{m+1} + \cdots + x_k)$. Notice that $m > l$, which contradicts the definition of x_i.

If $S_m - S_l = 0$, then $n | y_{l+1} + \cdots + y_m$, that is, $n | x_{l+1} + \cdots + x_m$ which contradicts the definition of x_i.

If $S_m - S_l < 0$, then

$$a = (y_1 + \cdots + y_l) + (y_{m+1} + \cdots + y_k) = S_k - (S_m - S_l) = 1 - (S_m - S_l)$$

is the solution of the equation $x^2 \equiv x \pmod{n}$ and $1 < a < 1 + \frac{n}{k}$.

Summing up, there exists a satisfying the required condition.

263. Suppose that $n \ge 2k$ (if $n < 2k$, then interchange the roles of k and $n - k$). Let p be an arbitrary prime. Consider the numbers that remain into the numerator of

$$\binom{n}{k} = \frac{n(n-1)\ldots(n-k+1)}{k(k-1)\cdot\ldots\cdot 1}$$

after reducing all factors by the highest power of p by which they are divisible. Suppose that some of the k factors resulting after this operation are equal. Then the corresponding initial factors are of the form $p^i s$ and $p^j s$ with $i > j$. But then

$$n \ge p^i s \ge ps \cdot p^j > p(n-k) \ge 2(n-k),$$

which contradicts the choice $n \ge 2k$. It follows that the k new factors are distinct. As $1 < k < n - 1$, the numerator contains initially two consecutive factors, at least one of which is not divisible by p. This number does not change in the operation described above. From $n \ge 2k$, it follows that this number is greater than k; whence it follows that the product remaining in the numerator after elimination of powers of p is greater than the denominator. This means that the powers of p in the initial numerator cannot be completely cancelled out with the denominator. It follows that the canonical representation of $\binom{n}{k}$ cannot consist only of a power of p.

264. Clearly, $q > p$.
  Case 1. $p = 2$. Then $q(q + 1) = 30$; whence $q = 5$.
  Case 2. $p > 2$, $q \mid p + 1$. Then $p < q \leq p + 1$, impossible, since q is a prime.
  Case 3. $p > 2$, $q \mid p^2 + 1$. Then q divides $\frac{p^2+1}{2}$, so that $q \leq \frac{p^2+1}{2}$. We have

$$p(p+1)(p^2+1) = q(q+1) \leq \frac{p^2+1}{2} \cdot \frac{p^2+3}{2};$$

whence $4p(p+1) \leq p^2 + 3$, contradiction.
The only solution is $(p, q) = (2, 5)$.

265. Answer: $n = 1$.
From (a) we have $n^2 = 3a^2 + 3a + 1$, which implies that

$$3(2n+1)(2n-1) = 12n^2 - 9 = (6a+3)^2$$

is a square. Since $\gcd(2n+1, 2n-1) = 1$, there are two cases:
Case 1. $2n + 1$ is a square. Then $2n + 1 = u^2$, $2n + 119 = v^2$, so that

$$v^2 - u^2 = 119 - 1 = 118 \equiv 2 \pmod{4},$$

which is impossible, since the squares mod 4 are only 0 and 1.
Case 2. $2n - 1$ is a square. Analogously, $2n - 1 = u^2$, $2n + 119 = v^2$, with u, v odd. Then $v - u$, $v + u$ are even and

$$v^2 - u^2 = 119 + 1 = 120 = 2 \cdot 60 = 4 \cdot 30 = 6 \cdot 20 = 10 \cdot 12;$$

whence we find $n \in \{1, 25, 85, 421\}$. Only for $n = 1$, we have $3(2n+1) = 3^2$ square.

266. Working modulo 4, n must be odd; otherwise we have a perfect square $\equiv 3$ (mod 4), contradiction. Suppose that $n^7 + 7 = x^2$. Then

$$x^2 + 11^2 = n^7 + 2^7 = (n+2)(n^6 - 2n^5 + 4n^4 - 8n^3 + 16n^2 - 32n + 64).$$

The second factor is always $\equiv 3$ (mod 4), since $n^6 \equiv 1$ (mod 4), $2n^5 \equiv 2$ (mod 4) and the rest vanishes mod 4. It follows that this factor has a prime divisor $p \equiv 3$ (mod 4). Then $x^2 \equiv -11^2 \pmod{p}$; whence

$$\left(\frac{-121}{p}\right) = \left(\frac{-1}{p}\right) = 1,$$

contradiction, since $p \equiv 3$ (mod 4).

267. Answer: $(6, 3)$, $(9, 3)$, $(9, 5)$, $(54, 5)$.

For fixed values of n, the given equation is a quadratic equation in m. For $n \leq 5$ there are only the above solutions. We shall prove that there are no other solutions.

Suppose that (m, n) satisfies the given equation and $n \geq 6$. Since

$$m \mid 2 \cdot 3^n = m(2^{n+1} - 1) - m^2,$$

we have $m = 3^p$, $0 \leq p \leq n$ or $m = 2 \cdot 3^q$, $0 \leq q \leq n$.
In the first case, let $q = n - p$; then

$$2^{n+1} - 1 = m + \frac{2 \cdot 3^n}{m} = 3^p + 2 \cdot 3^q.$$

In the second case, let $p = n - q$. Then

$$2^{n+1} - 1 = m + \frac{2 \cdot 3^n}{m} = 2 \cdot 3^q + 3^p.$$

Hence, in both cases we need to find the nonnegative integer solutions of

$$3^p + 2 \cdot 3^q = 2^{n+1} - 1, p + q = n. \quad (*)$$

Next we prove bounds for p, q. From $(*)$ we get

$$3^p < 2^{n+1} = 8^{\frac{n+1}{3}} < 9^{\frac{n+1}{3}} = 3^{\frac{2(n+1)}{3}}$$

and

$$2 \cdot 3^q < 2^{n+1} = 2 \cdot 8^{\frac{n}{3}} < 2 \cdot 9^{\frac{n}{3}} = 2 \cdot 3^{\frac{2n}{3}} < 2 \cdot 3^{\frac{2(n+1)}{3}};$$

therefore $p, q < \frac{2(n+1)}{3}$. Combining these inequalities with $p + q = n$, we obtain

$$\frac{n-2}{3} < p, q < \frac{2(n+1)}{3}.$$

If $h = \min(p, q)$, then $h > \frac{n-2}{3} > 1$. On the LHS of $(*)$, both terms are divisible by 3^h; therefore $9 \mid 3^h \mid 2^{n+1} - 1$. It is easy check that $\text{ord}_9(2) = 6$, so $9 \mid 2^{n+1} - 1$ if and only if $6 \mid n + 1$. Therefore $n + 1 = 6r$ for some positive integer r, and we can write

$$2^{n+1} - 1 = 4^{3r} - 1 = (4^{2r} + 4^r - 1)(2^r - 1)(2^r + 1).$$

The factor $4^{2r} + 4^r + 1 = (4^r - 1)^2 + 3 \cdot 4^r$ is divisible by 3 and is not divisible by 9. The other two factors are coprime; both are odd and their difference is 2. Since the whole product is divisible by 3^h, we have either $3^{h-1} | 2^r - 1$ or $3^{h-1} | 2^r + 1$. In any case we have $3^{h-1} \leq 2^r + 1$. Then

$$3^{h-1} \leq 2^r + 1 \leq 3^r = 3^{\frac{n+1}{6}}, \frac{n-2}{3} - 1 < h - 1 \leq \frac{n+1}{6}, n < 11,$$

which is impossible, since $n \geq 6$ and $6 | n + 1$.

268. Let $P(a,b)$ be the given property.

Lemma 1. For every nonnegative integer a, $af(a)$ is a square.

Proof. $P(a,p)$ implies $af(a) + pf(p) + 2ap \equiv af(a) \pmod{p}$ is a quadratic residue for every prime p. By a well-known result, $af(a)$ is a square.

Lemma 2. For every sufficiently large prime p, $f(p) = p$.

Proof. It is easy to see that $p | f(p)$. Let $pf(p) = (pk)^2$, $k \in \mathbb{N}$. For primes p sufficiently large, $P(p, 1)$ implies

$$pf(p) + 2p + f(1) = (pk)^2 + 2p + f(1) < (pk + 2)^2$$

and then $(pk)^2 + 2p + f(1) = (pk + 1)^2$; hence $k = 1$; therefore $f(p) = p$.

Back to the problem, let $xf(x) = m^2$. Then for large enough prime p, $P(x, p)$ implies

$$(p + x - 1)^2 < p^2 + 2px + m^2 < (p + x + 1)^2,$$

so that $p^2 + 2px + m^2 = (p + x)^2$; hence $m^2 = x^2$ and $f(x) = x$.

269. It is well known that

$$\nu_2(n!) = \sum_{k \geq 1} \left\lfloor \frac{n}{2^k} \right\rfloor.$$

Assume that the representation of n in base 2 is $n = \overline{d_l d_{l-1} \ldots d_0}_{(2)}$, with $d_k \in \{0, 1\}$, $0 \leq k \leq l - 1$ and $d_l = 1$. Then

$$\nu_2(n!) = \sum_{k=1}^{l} \left\lfloor \frac{\overline{d_l d_{l-1} \ldots d_0}_{(2)}}{2^k} \right\rfloor = \sum_{k=1}^{l} \overline{d_l d_{l-1} \ldots d_k}_{(2)} = \sum_{k=1}^{l} \sum_{i=k}^{l} 2^{i-k} d_i =$$

$$= \sum_{i=1}^{l} d_i \left(\sum_{k=1}^{i} 2^{i-k} \right) = \sum_{i=1}^{l} (2^i - 1) d_i.$$

Now we will construct an infinite increasing sequence $(i_k)_{k \geq 1}$ of positive integers such that $2^{i_1} - 1 \equiv 2^{i_2} - 1 \equiv \ldots \equiv 2^{i_n} - 1 \equiv \ldots \equiv r \pmod{m}$, where r is a positive integer coprime with m. Assume that $m = 2^t u$ with u odd and choose $i \geq t$ such that $\varphi(u) | i - 1$. Then Euler's theorem implies $u | 2^{\varphi(u)} - 1 | 2^{i-1} - 1$, so

$$2^i - 1 = 2(2^{i-1} - 1) + 1 \equiv 1 \pmod{u}, 2^i - 1 \equiv -1 \pmod{2^t}, i \geq t.$$

From this we deduce $\gcd(2^i - 1, m) = 1$. We can choose $i_k = (t+k)\varphi(n) + 1$ for every positive integer k and t a sufficiently large positive integer. Since gcd $(r, m) = 1$, then there exists a positive integer u such that $ur \equiv a \pmod{m}$.
Consider $n_1 = 2^{i_1} + 2^{i_2} + \cdots + 2^{i_u}$. Then

$$\nu_2(n_1!) = \sum_{k=1}^{u} (2^{i_k} - 1) \equiv r + r + \cdots + r \equiv ur \equiv a \pmod{m}.$$

Similarly, we can construct $n_k = 2^{i_k} + 2^{i_{k+1}} + \cdots + 2^{i_{k+u}}$, and we see that (n_k) is increasing and $\nu_2(n_k!) \equiv a \pmod{m}$.

270. (a) We show that the pair $(1, -51^2)$ is good, but not very good. If $P(x) = x^3 - 51^2 x$, then $P(51) = P(0)$, so that this pair is n-good if and only if $n|51$. We conclude that it is not very good.

If $P(m) \equiv P(k) \pmod{51}$, then $m^3 \equiv k^3 \pmod{51}$. The Fermat's theorem implies

$$m \equiv m^3 \equiv k^3 \equiv k \pmod{3}, m \equiv m^{33} \equiv n^{33} \equiv n \pmod{17}.$$

Then we have $m \equiv k \pmod{51}$, so that the pair $(1, -51^2)$ is 51-good.
(b) We will show that if a pair (a, b) is 2010-good, and then it is 67^i-good for all positive integers i. The following properties are true:

1. If (a, b) is 2010-good, then (a, b) is 67-good. Indeed, suppose that $P(m) \equiv P(k) \pmod{67}$. Since $\gcd(30, 67) = 1$, there exist integers m', k' such that $k' \equiv k \pmod{67}, k' \equiv 0 \pmod{30}, m' \equiv m \pmod{67}$, and $m' \equiv 0 \pmod{30}$. Then

$$P(m') \equiv P(0) \equiv P(k') \pmod{30}, P(m') \equiv P(m) \equiv P(k) \equiv P(k') \pmod{67};$$

hence $P(m') \equiv P(k') \pmod{2010}$, which implies $m' \equiv k' \pmod{2010}$, as (a, b) is 2010-good. It follows that $m \equiv m' \equiv k' \equiv k \pmod{67}$; therefore (a, b) is 67-good.

2. If (a, b) is 67-good, then $67|a$. Indeed, suppose that $67 \nmid a$, and consider the sets of residues mod 67 $\{at^2 | 0 \leq t \leq 33\}$ and $\{-3as^2 - b | 0 \leq s \leq 33\}$. Since $a \not\equiv 0 \pmod{67}$, each of these sets has 34 elements; therefore they have at least one common element. If $at^2 \equiv -3as^2 - b \pmod{67}$, then for $m = t \pm s, k = \mp 2s$, we have

$$P(m) - P(k) = a(m^3 - k^3) + b(m-k) = (m-k)(a(m^2 + mk + k^2) + b) =$$
$$= (t \pm 3s)(at^2 + 3as^2 + b) \equiv 0 \pmod{67}.$$

Since (a, b) is 67-good, we must have $m \equiv k \pmod{67}$ in both cases, that is, $t \equiv 3s \pmod{67}$ and $t \equiv -3s \pmod{67}$, which implies $t \equiv s \equiv 0 \pmod{67}$ and

$$b \equiv -3as^2 - at^2 \equiv 0 \pmod{67}.$$

But then $67 | P(7) - P(2) = 67 \cdot 5a + 5b$ and $67 \nmid 7 - 2$, contradiction.

3. If (a, b) is 2010-good, then (a, b) is 67^i-good for all $i \geq 1$. Indeed, the property 2 implies $67 | a$. If $67 | b$, then $P(x) \equiv P(0) \pmod{67}$ for all x, contradicting that (a, b) is 67-good. Hence, $67 \nmid b$.

Suppose that $67^i | P(m) - P(k) = (m - k)(a(m^2 + mk + k^2) + b)$. Since $67 | a, 67 \nmid b$, it follows that $\gcd(67, a(m^2 + bk + k^2) + b) = 1$; hence $67^i | m - k$, so that (a, b) is 67^i-good.

271. If (d, n) is a solution, that is, $d(d + n) = n^2 + 1$, then $(d + n, d + 2n)$ is a solution:

$$(d + n)(2d + 3n) = (d + 2n)^2 + 1.$$

Since $(2, 3)$ is a solution, it follows that there are infinitely many solutions.

272. Answer: $(p, q) = (3, 7)$.

Assume that $p^3 + 1 = qx^2$, with x positive integer. From

$$p^2 - p + 1 - (p - 2)(p + 1) = 3 \text{ and } 3 \nmid p + 1,$$

it follows that $\gcd(p + 1, p^2 - p + 1) = 1$. Since $p^3 + 1 = (p + 1)(p^2 - p + 1)$, we have

Case 1. $p + 1 = u^2, p^2 - p + 1 = qv^2$. Then $p = (u - 1)(u + 1)$, so that p is a prime number if and only if $u = 2$; hence $p = 3, q = 7$.

Case 2. $p + 1 = qu^2, p^2 - p + 1 = v^2$. Then $(2p - 1)^2 + 3 = (2v)^2$; hence $p = 1$, which is not a prime number.

273. (a) For $b_i = 1 + 2^i + 2^{2i} + \cdots + 2^{(k-1)i}$, we have $t(b_i) = k$. Let

$$a_1 = b_1, a_i = \frac{b_{(k+1)^i}}{b_{(k+1)^{i-1}}}, i > 1.$$

Then a_i are integers and the sequence $(a_i)_{i \geq 1}$ satisfies the required condition.

(b) The following lemma solves the problem.

Lemma. If a positive integer n is divisible by $2^k - 1, k \in \mathbb{N}^*$, then $t(n) \geq k$.

Proof. Assume for contradiction that n_0 is the smallest positive integer such that $2^k - 1 | n_0$ and $t(n_0) < k$. Let $n_0 = 2^{\alpha_1} + 2^{\alpha_2} + \cdots + 2^{\alpha_l}$, where $0 \leq \alpha_1 \leq \alpha_2 < \cdots < \alpha_l$ and $l < k$. If $\alpha_l \geq k$, then $n_1 = 2^{\alpha_1} + 2^{\alpha_2} + \cdots + 2^{\alpha_{l-1}} + 2^{\alpha_l - k} < n_0$ is divisible by $2^k - 1$ and $t(n_1) \leq t(n_0) < k$, contradiction. It follows that $\alpha_l \leq k - 1$; hence

$$n_0 \leq 2^{k-1} + 2^{k-2} + \cdots + 2^{k-l} \leq 2^{k-1} + \cdots + 2^l < 2^k - 1.$$

But $2^k - 1 | n_0$ implies $n_0 \geq 2^k - 1$, contradiction.

274. **Answer:** $(x, y) = (a, a + 1), (a + 1, a)$, with a positive integer.

Denote $s = x + y, p = xy$. Then $s + 1 | 2p, s - 1 | s^2 - 2p - 1$; hence $s - 1 | 2p$. We have $s + 1 - (s - 1) = 2$, so that $\gcd(s + 1, s - 1) \in \{1, 2\}$. If $\gcd(s + 1, s - 1) = 1$, then $s^2 - 1 | p$; hence $s^2 - 1 \leq p$, which is impossible. If $\gcd(s + 1, s - 1) = 2$, then $\frac{s+1}{2}, \frac{s-1}{2}$ are coprime; hence $s^2 - 1 | 4p$, so that $s^2 \leq 4p + 1$.

Case 1. $s^2 = 4p$. Then $x = y$, but this case is impossible, since s is odd.

Case 2. $s^2 = 4p + 1$. Then $(x - y)^2 = 1$; hence $x - y = \pm 1$. Every such pair satisfies the required condition.

275. Consider the sequence (b_n) defined by $b_0 = 1, b_1 = 3, b_{n+2} = 4b_{n+1} - 2b_n$, $n \geq 0$. By induction, we shall show that

(a) $b_{n+2}b_n - b_{n+1}^2 = 2^n$. Using the recurrence relation, we have

$$b_{n+2}b_n - b_{n+1}^2 = (4b_{n+1} - 2b_n)b_n - b_{n+1}^2 = (4b_n - b_{n+1})b_{n+1} - 2b_n^2 =$$
$$= 2(b_{n+1}b_{n-1} - b_n^2) = \cdots = 2^{n-k+1}(b_{k+1}b_{k-1} - b_k^2)$$
$$= 2^n(b_2 b_0 - b_1^2) = 2^n.$$

(b) $b_{n+1} > 2b_n$. We have $b_1 = 3 > 2b_0 = 2$. If $b_{n+1} > 2b_n$; then

$$b_{n+2} - 2b_{n+1} = 2(b_{n+1} - b_n) > 2b_n > 0,$$

which ends the proof by induction.

It follows that $b_n > 2^n$ for all n; therefore

$$\left\lfloor b_{n+2} - \frac{b_{n+1}^2}{b_n} \right\rfloor = 0,$$

and since the sequence contains only positive integers, it follows that

$$b_{n+2} = 1 + \left\lfloor \frac{b_{n+1}^2}{b_n} \right\rfloor;$$

hence $b_n = a_n$ for all n, which ends the proof.

276. If $n \leq 5$, then $n = 3, 5$.

If $n > 5$, then $(2^{n+1} - 1)^2 - 8 \cdot 3^n = m^2$ implies

$$(2^{n+1} + m - 1)(2^{n+1} - m - 1) = 8 \cdot 3^n.$$

Write $m = 2t + 1$ and deduce upon division by 4: $(2^n + t)(2^n - t - 1) = 2 \cdot 3^n$. Then

$$\{2^n + t, 2^n - t - 1\} = \{2 \cdot 3^k, 3^{n-k}\}$$

for some k. Adding, it follows that $2^{n+1} - 1 = 2 \cdot 3^k + 3^{n-k}$. Assume $n > 10$. By size considerations, $k \notin \{0, n\}$. Let $1 \leq l := \min(k, n - k)$. Then $3^l | 2^{n+1} - 1$. By lifting the exponent lemma,

$$\nu_3(2^{n+1} - 1) = 1 + \nu_3(n + 1) \leq 1 + \log_3(n + 1);$$

hence $l \leq 1 + \log_3(n + 1)$, so

$$M := \max(k, n - k) = n - l \geq n - 1 - \log_3(n + 1).$$

Note that

$$2^{n+1} - 1 = 2 \cdot 3^k + 3^{n-k} \geq 3^M \geq 3^{n-1} \cdot \frac{1}{n+1};$$

whence $4(n + 1) \geq (1,5)^{n-1}$, contradiction with $n > 10$.

Analyzing the small cases, we conclude that there are no solutions for $5 < n \leq 10$.

2.40 2012

277. If $d = \gcd(a, b)$, then $a = dx$, $b = dy$, $\gcd(x, y) = 1$ and $\text{lcm}(a, b) = dxy$. From

$$7dxy = 1 + 9(xy + x + y) \leq 28xy,$$

it follows that $d \leq 4$. For $d = 4$ we get the solution $(a, b) = (4, 4)$.

For $d \in \{2, 3\}$, we find another solution $(a, b) = (38, 4)$.

278. There are integers x, y such that $x^2 + y^2 + 1 \equiv 0 \pmod{p}$.

Indeed, let Q be the set of squares modulo p. Then $|Q| = \frac{p+1}{2}$, $|-1 - Q| = \frac{p+1}{2}$, so that $Q \cap (-1 - Q) \neq \emptyset$, since $|\mathbb{Z}_p| = p < 2 \cdot \frac{p+1}{2}$. We conclude that there exist integers x, y such that $x^2 \equiv -1 - y^2 \pmod{p}$. We can choose $0 \leq x, y < \frac{p}{2}$, since $x^2 \equiv (p - x)^2 \pmod{p}$.

Now, we have $x^2 + y^2 + 1 = mp < \frac{p^2}{2} + 1$, which implies $m < \frac{p}{2}$.

279. Case 1. $p = 2$. Then $f(1) = 3$, so that $f(n) \equiv 2 \pmod 3$ for all $n > 1$.
    Case 2. $p = 4k + 1$. Then $f(1) \equiv 2 \pmod 4$, $f(n) \equiv 3 \pmod 4$ for all $n > 1$.
    Case 3. $p = 4k + 3$. Then $f(1) \equiv 0 \pmod 4$, $f(n) \equiv 3 \pmod 4$ for all $n > 1$.
    In all cases, excepting $n = 1$, the values are quadratic nonresidues.
    For $n = 1$, if $x^2 = p + 1$, then $p = 3$.
280. Choose

$$A = \{1, 5, 252, k+1, 2k, 8k^2 + 8k + 2\}, B = \{251, k, 2k+2, 8k^2 + 8k + 2\}, k > 252.$$

Another possible construction is

$$A = \{5, 252, 2k-1, 3k, 6k-1\}, B = \{1, 251, 2k, 3k-1, 6k-2\}, k > 1000.$$

281. Let G be the set of all good numbers.
    Lemma 1. If $p_1 < p_2 < \cdots$ are all prime numbers, α, r are positive integers and $a \in G$ satisfies $p_r^\alpha \mid a$, then $\left(\prod_{i=1}^{r} p_i\right)^\alpha \mid a$.
    Proof. If not, there exists $1 \le i < r$ which satisfies $\nu_{p_i}(a) = \beta < \alpha$; then

$$b = \frac{a}{p_i^\beta p_r^\alpha} p_i^\alpha p_r^\beta < a$$

and $\tau(b) = \tau(a)$, contradiction.
    Lemma 2. For a prime number p_t, there exists a positive integer M such that if $a \in G$, $a > M$; then $p_t \mid a$.
    Proof. Take a positive integer α such that $2^\alpha > p_t$ and let $M = (p_1 \cdots p_{t-1})^{3\alpha}$. For any $a \in G$, if $a > M$ and $\gcd(a, p_t) = 1$, by Lemma 1 all prime divisors of a belong to $\{p_1, p_2, \ldots, p_{t-1}\}$, so there exists $1 \le i \le t-1$ such that $\nu_{p_i}(a) = \beta > 3\alpha$. Since $p_i^\alpha \ge 2^\alpha > p_t$, we have $b = \frac{a}{p_i^\alpha} p_t < a$. As $2(\beta - \alpha + 1) > \beta + 1$, it follows $\tau(b) > \tau(a)$, contradiction.
    Lemma 3. For a prime number p_t and a positive integer α, there exists a positive integer M such that if $a \in G$, $a > M$; then $p_t^\alpha \mid a$.
    Proof. Take a prime $p_r > p_t^\alpha$. By Lemma 2, there exists a positive integer M such that if $a \in G$, $a > M$, then $\nu_{p_r}(a) = \gamma \ge 1$. If $\nu_{p_t}(a) = \beta < \alpha$; then $b = \frac{a}{p_r} p_t^\alpha < a$. As

$$\frac{(\alpha + \beta +)\gamma}{(\beta+)(\gamma + 1)} \ge \frac{2\gamma}{\gamma + 1} \ge 1,$$

we have $\tau(b) \ge \tau(a)$, contradiction.
    Back to the problem, let p_t be the largest prime divisor of k, and let α be a positive integer such that $2^\alpha > k$. Then for any prime $p \mid k$, we have $\nu_p(k) < \alpha$; hence k divides $n = \left(\prod_{i=1}^{t} p_i\right)^\alpha$. By Lemma 3, there is an integer M such that if $a \in G$, $a > M$; then $n \mid a$, so $k \mid a$.

282. If k is divisible by a square $t^2 > 1$, then we can take $m = n = k - \frac{k}{t} + 1$ and we see that k has the required properties.

If k is square-free and it has two prime factors p_1, p_2 such that $(p_1 - 2)(p_2 - 2) \geq 4$, then at least one of the numbers $\frac{(p_1-1)k}{p_1} + 1$ and $\frac{(p_1-2)k}{p_1} + 1$ is coprime with p_1; otherwise p_1 divides their difference $\frac{k}{p_1}$, contradiction. Taking this number as m, we have $1 < m < k$ and $\gcd(m, k) = 1$. Similarly, we can take a number n from $\frac{(p_2-1)k}{p_2} + 1$ or $\frac{(p_2-2)k}{p_2} + 1$ such that $1 < n < k$ and $\gcd(n, k) = 1$. Therefore m and n satisfy the conditions, since

$$m + n \geq \frac{(p_1 - 2)k}{p_1} + 1 + \frac{(p_2 - 2)k}{p_2} + 1 = k + \frac{((p_1 - 2)(p_2 - 2) - 4)k}{p_1 p_2} > k.$$

If k is square-free and there are no prime divisors p_1, p_2 of k such that $(p_1 - 2) \times (p_2 - 2) \geq 4$, then it is easy to verify that k can be only $15, 30, p$ or $2p$, where p is an odd prime. For $k = 30, p, 2p$, there are no m, n satisfying the condition. If $k = 15$, then $m = 11, n = 13$ satisfy the conditions.

In conclusion, an integer k satisfies the conditions if an only if k is not $30, p$ or $2p$, where p is an odd prime.

283. Answer: for all primes.

Let n be a prime. By condition (1) the sequence $a_0, a_1, \ldots$ contains only finitely many different numbers. If a_m is maximal of them, then by condition (3) a_{m+a_m} must also be maximal. If a_m is maximal of the numbers, then a_{m+ka_m} is also maximal, for every $k \geq 0$. Indeed, this assertion is valid for $k = 0$. If it holds for k, then $a_{m+(k+1)a_m} = a_{m+ka_m+a_m} = a_{m+ka_m+a_{m+ka_m}} = a_{m+ka_m} = a_m$; therefore it holds also for $k+1$. By condition (2) a_m is not divisible by n. Since n is prime, the numbers a_m and n are relatively prime. Hence among the numbers $m + ka_m$ with $0 \leq k < n$, there is one in each congruence class modulo n. Hence all terms of the sequence are maximal, i.e., they are equal.

If n is composite, let m be a proper divisor of n. For any $k < m$, choose $a_k = m + kn$ and continue the sequence with period m. Condition (1) holds, since n is a multiple of m. Condition (2) holds, since all terms of the sequence are congruent to m modulo n. For the condition (3), notice that all terms of the sequence are divisible by m; hence i and $i + a_i$ are always congruent modulo m; therefore $a_i = a_{i+a_i}$. At the same time, not all the numbers are equal.

284. Answer: $\left(2^{\frac{c-1}{2}}, 2^{\frac{c-1}{2}}, c, 2\right)$, c odd, $(1, 1, 1, 2)$, $(2, 1, 2, 3)$, $(1, 2, 2, 3)$.

If $a = b = 1$, then $(a, b, c, p) = (1, 1, 1, 2)$.

If $a = b > 1$, then $p = 2$, c must be odd and $(a, b, c, p) = \left(2^{\frac{c-1}{2}}, 2^{\frac{c-1}{2}}, c, 2\right)$.

If $a > b$ and $(a, b, c, p) \neq (2, 1, 2, 3)$, then the Zsigmondy's theorem implies that $a^p + b^p$ has at least two prime divisors, contradiction.

285. Answer: $(m, n, p) = (3, 0, 2), (2, 1, 3), (4, 2, 2)$.
Factorizing, we get $(n + 2)(n^2 - 2n + 4) = p^m$. Clearly, $m \geq 2$; hence

$$n + 2 = p^a, n^2 - 2n + 4 = p^b.$$

The inequality $n + 2 \geq n^2 - 2n + 4$ is true only for $n = 1, 2$. Analysing cases $n = 0, 1, 2$, we find the above solutions. Suppose that $n \geq 3$. Then $a < b$ and

$$p^a(p^a - 6) + 12 = p^b.$$

We deduce that $p | 12$; hence $p = 2, 3$.
Case 1. $p = 2$. Then $a = 0, 1, 2$, which implies $n = 0, 2$.
Case 2. $p = 3$. Then $a = 0, 1$, which implies $n = 1$.
In both cases, we find the same solutions as above.

286. Clearly, $p \geq 5$.
(a)$\Longrightarrow$(b). It follows that $x^3 \equiv y^3 \equiv z^3 \pmod{p}$. Subtracting and decomposing, we get

$$(x - y)(x^2 + xy + y^2) \equiv 0 \pmod{p}, (x - z)(x^2 + zx + z^2) \equiv 0 \pmod{p}.$$

Since $x - y \not\equiv 0 \pmod{p}$ and $x - z \not\equiv 0 \pmod{p}$, we deduce

$$x^2 + xy + y^2 \equiv x^2 + xz + z^2 \equiv 0 \pmod{p}.$$

Subtracting and decomposing again, we obtain

$$(y - z)(x + y + z) \equiv 0 \pmod{p},$$

so that $x + y + z \equiv 0 \pmod{p}$, since $y - z \not\equiv 0 \pmod{p}$. The identity

$$2(x^3 + y^3 + z^3 - 3xyz) = (x + y + z)\left((x - y)^2 + (y - z)^2 + (z - x)^2\right)$$

shows that $x^3 + y^3 + z^3 \equiv 3xyz \pmod{p}$, since $p \neq 2$. Using (a), we have $x^3 \equiv y^3 \equiv z^3 \equiv xyz \pmod{p}$, since $p \neq 3$. Simplifying, we get (b)
(b)$\Longrightarrow$(a). This implies $(yz)^2 \equiv x^2 yz \pmod{p}$, so that $x^2 \equiv yz \pmod{p}$ and

$$x^3 \equiv y^3 \equiv z^3 \equiv xyz \pmod{p}.$$

287. It is easy to see that $a + b | b^2 - 1, a - b | b^2 - 1$; therefore $a^2 - b^2 | 2(b^2 - 1)$, as $\gcd(a, b) = 1$. We can take $3b^2 = a^2 + 2$ and the condition is satisfied. Let

$$a_n = x_n + 3y_n, b_n = x_n + y_n, \text{ where } x_n^2 - 3y_n^2 = 1.$$

As the Pell equation has infinitely many solutions, then there are infinitely many pairs (a_n, b_n).

288. The equality $(1 + x)^n = 1 + x^n$ holds in $\mathbb{Z}_2[x]$ if and only if n is a power of 2.
Indeed, the direct implication is obvious. For the converse, write $n = 2^m l$ with l odd. Then $(1 + x)^{2^m l} = (1 + x^{2^m})^l$, which upon expansion contains the term x^{2^m}. Then $1 + x^{2^m l}$ also contains the term x^{2^m}; hence $l = 1$.

The given condition requires that $\binom{n}{i}$ and i have opposite parity, for $1 \leq i \leq n$, which forces that n is even. Equivalently, we wish to find all even n for which

$$(1 + x)^n = 1 + x^2 + x^4 + \cdots + x^n$$

in $\mathbb{Z}_2[x]$. We can multiply both sides by $(1 + x)^2 = 1 + x^2$ to find

$$(1 + x)^{n+2} = 1 + n^{n+2};$$

whence it follows that $n = 2^m - 2$. All steps are bidirectional; therefore $n = 2^m - 2$ is a solution for all m.

289. The number $\sigma(m)$ is odd if and only if the power of every odd prime dividing m is even, that is, $m = x^2$ or $2x^2$; hence

$$f(n) = \lfloor \sqrt{n} \rfloor + \left\lfloor \sqrt{\frac{n}{2}} \right\rfloor.$$

If (x, y) is a solution of the Pell equation $y^2 - 2x^2 = 1$ (there are infinitely many) and $y^2 \leq n \leq y^2 + 2y$, then for at least one n,

$$f(n) = y + x \mid n.$$

290. Observe that

$$\frac{S_r(n)}{n^{r+1}} = \frac{1}{n}\left(\left(\frac{1}{n}\right)^r + \left(\frac{2}{n}\right)^r + \cdots + \left(\frac{n}{n}\right)^r\right)$$

is a Riemann sum of $f(x) = x^r$ on $[0, 1]$, so that

$$\lim_{n \to \infty} \frac{S_r(n)}{n^{r+1}} = \int_0^1 f(x)dx = \frac{1}{r+1}.$$

If $S_a(n) = (S_b(n))^c$ for infinitely many values of n, then

$$\frac{S_a(n)}{n^{a+1}} = \frac{(S_b(n))^c}{n^{c(b+1)}} \cdot n^{c(b+1)-(a+1)}.$$

The LHS converges to $\frac{1}{r+1}$, but the RHS has a nonzero limit only when $a + 1 = c(b + 1)$. In this case the RHS tends to $\frac{1}{(b+1)^c}$, so that $a + 1 = (b + 1)^c$. This implies $(b + 1)^{c-1} = c$. For $c > 1$, $(b + 1)^{c-1}$ is an integer if and only if b is an integer. Then $b \geq 1$ and the Bernoulli's inequality implies

$$c = (b+1)^{c-1} \geq (1+1)^{c-1} \geq 1 + (c-1) = c.$$

The inequalities are strict for $c \geq 3$; therefore $c = 1$ or $c = 2$.

If $c = 1$, we must have $a = b$, and then $S_a(n) = (S_b(n))^c$ for all n.

If $c = 2$, we must have $b + 1 = 2$; therefore $b = 1$. Then $a + 1 = c(b+1) = 4$, so that $a = 3$. In this case

$$S_3(n) = \left(\frac{n(n+1)}{2}\right)^2 = (S_1(n))^2.$$

The solutions are $(a, a, 1)$, $a \in \mathbb{Q}_+$ and $(3, 1, 2)$.

291. Answer: yes.

A solution is $(a, b, c) = (56, 36, 25)$. If (a, b, c) is a solution, then

$$(a + 2b + 2c, a + b + 2c, a + b + c)$$

is also a solution; therefore the given equation has solutions with arbitrary large components.

292. Answer: $(a, n) = (2^k - 2, 2), (1, n)$.

By the Zsigmondy's theorem, there exists a primitive prime divisor of $a^n - 1$ for all a, n, except for $(a, n) = (2, 6), (2^k - 1, 2)$. In the first case, we can list out all of the possible $a^m - 1$ with $m < n$ and find that none works: the complete list is 31, 15, 7, 3, 1. In the second case, any prime divisor must divide $(2^k - 1)^2 - 1 = 2^{k+1}(2^{k-1} - 1)$. The set of all prime divisors of $a^n - 1$ is the set of all prime divisors of $2^{k-1} - 1$ and 2. The only possible m is $m = 1$, which yields $(2^k - 1) - 1 = 2(2^{k-1} - 1)$; thus $a^m - 1$ is divisible by all prime divisors of $a^n - 1$.

293. Denote $A = \{bx\} \pmod{p}$, $B = \{cy\} \pmod{p}$, where $x, y \in \{-k, -k+1, \ldots, k+1\}$ and $k = \lfloor \frac{p}{t} \rfloor - 1$. Then we have $|A| = |B| = 2k + 1$. The Cauchy-Davenport theorem implies

$$|A + B| \geq \min\{4k + 1, p\}, \quad 4k + 1 = 4\left\lfloor \frac{p}{3} \right\rfloor - 3 \geq \frac{4(p-2)}{3} - 3 \geq p,$$

for $p \geq 17$.

2.41 2013

294. We shall prove that $a_{n+1}a_{n-1} = (a_n+2)^2$, for every positive integer n. Indeed,

$$a_{n+1}a_{n-1} - (a_n+2)^2 - \left(a_n a_{n-2} - (a_{n-1}+2)^2\right) =$$
$$= (14a_n - a_{n-1} - 4)a_{n-1} - a_n^2 - 4a_n - 4 - a_n a_{n-2} + a_{n-1}^2 + 4a_{n-1} + 4 =$$
$$= 14a_n a_{n-1} - a_n a_{n-2} - a_n^2 - 4a_n = (a_n+4)a_n - (a_n+4)a_n = 0.$$

It follows that $a_{n+1}a_{n-1} - (a_n+2)^2 = \cdots = a_2 a_0 - (a_1+2)^2 = 9 \cdot 1 - (1+2)^2 = 0$.

Now, we prove by induction that a_n is a square for every nonnegative n. From $a_0 = a_1 = 1$, we see that the property is true for $n = 0, 1$. Suppose that it is true for $n-1$ and n. Then

$$a_{n+1} = \frac{(a_n+2)^2}{a_{n-1}}$$

is a square, since it is an integer.

295. We need $p \neq q$. Then $pq | 30p + 30q - 1$; whence $30p + 30q - 1 = kpq$, with $\gcd(30, k) = 1$ and $k < 30$, which implies $k \in \{1, 7, 11, 13, 17, 19, 23, 29\}$. It is easy to see that $p, q \geq 7$.

If $k \geq 13$, then $k(p-1)(q-1) > (p+11)(q+11)$. Since $k > 9$, $3(p-1) \geq p+11$, $3(q-1) \geq q+11$, we deduce

$$k(p-1)(q-1) = (30-k)(p+q) + k - 1 > (p+11)(q+11),$$
$$17(p+q) + 28 > (p+11)(q+11),$$

so that $(p-6)(q-6) + 47 < 0$, contradiction. It remains $k \in \{1, 7, 11\}$.

Case 1. $k = 1$. Then $(p-30)(q-30) = 899$, with solutions

$$(p, q) = (59, 61), (61, 59), (31, 929), (929, 31).$$

Case 2. $k = 7$. Then $(7p-30)(7q-30) = 893$, with solutions

$$(p, q) = (7, 11), (11, 7).$$

Case 3. $k = 11$. Then $(11p-30)(11q-30) = 889$. No solution.

296. Since $\Omega(ab) = \Omega(a) + \Omega(b)$, it follows that $\Omega(p(x_1)p(x_2)\ldots p(x_n))$ is even. Clearly,

$$p(1)\prod_{i=1}^{n}p(a_i+2) = 2^n \prod_{i=1}^{n}(a_i+1)^2 \prod_{1\leq i<j\leq n}(a_i+a_j+1)^2,$$

which implies that n is even.

297. **Answer: 2003.**

Subtracting the second equation from the first one gives

$$a_1(p-1) + \cdots + a_m(p^m-1) = 2002.$$

It follows that $2002 = 2 \cdot 7 \cdot 11 \cdot 13$ is divisible by $p-1$; therefore

$$p-1 \in \{1,2,7,11,13,14,22,26,77,91,143,154,182,286,1001,2002\}.$$

But p is a prime number; whence we deduce $p \in \{2,3,23,2003\}$.

Case 1. $p=2$. Then $m=10$, since $2^{10} < 2013 < 2^{11}$. The second equation implies $a_i = 1$ for all i, but $1+2+2^2+\cdots+2^{10} = 2^{11} - 1 = 2047 \neq 2013$. No solution.

Case 2. $p=3$. Then $2013 = 2\cdot 3 + 3^2 + 2\cdot 3^3 + 2\cdot 3^5 + 2\cdot 3^6$ and $2+1+2+2+2 = 9 \neq 11$. No solution.

Case 3. $p=23$. Then $2013 = 12 + 18\cdot 23 + 3\cdot 23^2$ and $12+18+3 \neq 11$. No solution.

Case 4. $p=2003$. Then $2013 = 10 + 2003$ and $10+1 = 11$, so that 2003 is a solution.

298. **Answer:** $(2^0+1, 2^1+1, 2^2+1, 2^3+1, 2^4+1)$.

We see that $(2^0+1, 2^1+1, 2^2+1, 2^3+1, 2^4+1)$ is a perfect tuple with length 5. We shall prove that there are no other perfect tuples of length 5 or larger.

Let $(a^m+1, a^{m+1}+1, \ldots, a^n+1)$ be an arbitrary perfect tuple of length at least 5. There must exist at least two odd exponents among $m, m+1, \ldots, n$; let k and $k+2$ be the largest odd exponents. As a^k+1 and $a^{k+2}+1$ are prime powers having common divisor $a+1$, these two integers are powers of a prime q. Thus, $a^{k+2}+1$ is divisible by a^k+1, which implies that a^k+1 divides the difference

$$a^2(a^k+1) - (a^{k+2}+1) = a^2 - 1;$$

hence $a^k + 1 \leq a^2 - 1$. This implies $k < 2$, so that $k = 1$, being odd. We deduce that the only odd exponents that occur in the chosen tuple are 1 and 3 and the tuple is of the form $(a^0+1, a^1+1, a^2+1, a^3+1, a^4+1)$. Since $a+1$ and a^3+1 are powers of q, their quotient a^2-a+1 is also a power of q. From $a \geq 2$, it follows that

$$a^2 - a + 1 \geq a + 1;$$

hence $a^2 - a + 1$ is divisible by $a + 1$. Then the difference

$$(a^2 - a + 1) - (a + 1)(a - 2) = 3$$

is divisible by $a + 1$, which implies $a = 2$.

299. From

$$n(n+2) = 4m^4 + 4m^2 - 4m + 4,$$

$$(2m^2)^2 < (n+1)^2 = 4m^4 + 4m^2 - 4m + 5 < (2m^2 + 1)^2, m > 1,$$

it follows that the solution $(m, n) = (1, 2)$ is unique.

300. We have given that $p \mid 2013^{p^n} - 1$. Taking $\mathrm{ord}_p(2013) = k$, we have $k \mid p^n$; hence $k = p^r$, $0 \leq r \leq n$. From $k \mid \varphi(p) = p - 1$, it follows that $r = 0$, since $\gcd(p^r, p-1) = 1$. This means that $p \mid 2013 - 1 = 2 \cdot 503$. Using the lifting the exponent lemma, one gets the solutions $(p, n) = (2, 98), (503, 99)$.

301. (a) Every a of the form $4k - 3$, $k \geq 2$, is friendly. Indeed, the numbers $m = 2k - 1$ and $n = k - 1$ satisfy the given equation:

$$(m^2 + n)(n^2 + m) = \left((2k-1)^2 + (k-1)\right)\left((k-1)^2 + (2k-1)\right) = (4k-3)k^3 = $$
$$= a(m-n)^3.$$

Hence $5, 9, \ldots, 2009$ are friendly, and so $\{1, 2, \ldots, 2012\}$ contains at least 502 friendly numbers.

(b) We show that $a = 2$ is not friendly. Consider the equation with $a = 2$ and rewrite its left-hand side as a difference of squares:

$$\left(\frac{m^2 + n + n^2 + m}{2}\right)^2 - \left(\frac{m^2 + n - n^2 - m}{2}\right)^2 = 2(m-n)^3.$$

Since $m^2 + n - n^2 - m = (m-n)(m+n-1)$, we can reformulate the equation as

$$(m^2 + n + n^2 + m)^2 = (m-n)^2 \left(8(m-n) + (m+n-1)^2\right).$$

It follows that $8(m-n) + (m+n-1)^2$ is a square. Clearly, $m > n$; hence there is an integer $s \geq 1$ such that

$$(m+n-1+2s)^2 = 8(m-n) + (m+n-1)^2.$$

Subtracting the squares gives $s(m+n-1+s) = 2(m-n)$. Since $m+n-1+s > m-n$, we conclude that $s < 2$. Therefore the only possibility is $s = 1$ and $m = 3n$. Then the left-hand side of the given equation for $a = 2$ is greater than $m^3 = 27n^3$, whereas its right-hand side equals $16n^3$. The contradiction shows that $a = 2$ is not friendly.

302. We show that the number

$$\frac{a^2 + b^2 + c^2}{3(ab + bc + ca)}$$

is never an integer. Indeed, if $a^2 + b^2 + c^2 = 3n(ab + bc + ca)$, then

$$(a+b+c)^2 = (3n+2)(ab+bc+ca).$$

Choose a prime $p \equiv 2 \pmod{3}$ that divides $3n+2$. Then $p | a+b+c$; whence p divides $ab+bc+ca$ (choose an odd power of p which divides $3n+2$).

It follows that $c \equiv -a-b \pmod{p}$, so that $p | a^2 + ab + b^2$, which implies that p divides $(2a+b)^2 + 3b^2$. But $\left(\frac{-3}{p}\right) = 1$ is a contradiction with $p \equiv 2 \pmod{3}$.

303. The number of solutions with $a \neq b$ is even, since if (a,b,c,d,e) is a solution, then (b,a,c,d,e) is also a solution. For $a = b$ and $c = d$, the equation becomes

$$\frac{2}{a} + \frac{2}{c} + \frac{1}{e} = 1.$$

As above, we take $a = c$ and then $(a-4)(e-1) = 4$, which has the solutions

$$(a,e) = (5,5), (6,3), (8,2).$$

The total number of solutions is odd.

304. The point $P = (0, a)$ is an integral point on the elliptic curve

$$y^2 = x^3 + x + a^2.$$

Then $3P = P \oplus P \oplus P$ is another integral point:

$$3P = \left(8a^2(8a^4+1), 4a(8a^4+1)(16a^4+1) - a\right).$$

Note. The second point can be found also using elementary methods.

305. Answer: prime numbers.

If p is a prime, then every n such that $1 \leq n < \frac{p}{2}$ divides $\binom{n}{p-2n}$.

We require that $n | p$; whence $n \in \{1, p\}$. If $n = 1$, then $p + 1 = a^2$ so that $p = 3$. Analogously, if $n = p$, then $p = 3$. For $p = 2$ this is true, $n = 1$ being the only possibility. If p is an odd prime, then choose n, $1 \leq n < \frac{p}{2}$ and observe that

$$(p - 2n)\binom{n}{p-2n} = n \cdot \binom{n-1}{p-2n-1}.$$

Since $p \geq 2n$ and p is odd, all factors are nonzero. Denote $d = \gcd(p - 2n, n)$. Then $d | p$, so that $d = 1$. We deduce $n | \binom{n}{p-2n}$. We will prove that no composite number m satisfies the given property.

Case 1. m even. Then choose $n = \frac{m}{2}$ and observe that $\binom{n}{m-2n} = \binom{\frac{m}{2}}{0} = 1$ is not divisible by $\frac{m}{2} > 1$.

Case 2. m odd. Then there exist an odd prime p and an integer $k \geq 1$ such that $m = (2k+1)p$. Choose $n = pk$, for which $\frac{m}{3} \leq n \leq \frac{m}{2}$ and observe that

$$\frac{1}{n}\binom{n}{m-2n} = \frac{1}{pk}\binom{pk}{p}$$

is not an integer, since p divides the denominator, but not the numerator.

306. More generally, we will find pairs (n, p) for which $\sqrt{n + \frac{p}{n}}$ is rational. Let

$$\sqrt{n + \frac{p}{n}} = \frac{a}{b}, a, b \in \mathbb{N}^*, \gcd(a, b) = 1.$$

Then

$$\frac{n^2 + p}{n} = \frac{a^2}{b^2}.$$

Let $d = \gcd(n^2 + p, n)$; therefore $d | n$, $d | n^2 + p$, so that $n | p$; hence $d \in \{1, p\}$.

If $d = 1$, then $n = b^2$, $n^2 + p = a^2$; hence $p = (a - n)(a + n)$, implying $p = 2n + 1$. Therefore $(n, p) = (b^2, 2b^2 + 1)$.

If $d = p$, then $n^2 + p = pa^2$, yielding p is a square, contradiction, or $p = a^2 - 1$, hence $n = p = 3$.

307. Consider the polynomial $f = (X-1)^n - c_n X^2 - b_n X - a_n \in \mathbb{Z}[X]$. Since $X^3 - 2$ is irreducible in $\mathbb{Z}[X]$ and $f(\sqrt[3]{2}) = 0$, it follows that $X^3 - 2$ divides f in $\mathbb{Z}[X]$; therefore $g_n = a_n + b_n X + c_n X^2$ is the remainder of the division $(X-1)^n$ by $X^3 - 2$ in $\mathbb{Z}[X]$. If $n = 3q + r$, with $r \in \{0, 1, 2\}$, then $(X - \hat{1})^n = (X^3 - \hat{1})^q (X - \hat{1})^r = (X^3 - \hat{2})g + (X - \hat{1})^r$ in $\mathbb{Z}_3[X]$, so that $g_n = (X - \hat{1})^r$ in $\mathbb{Z}_3[X]$. Consequently, $c_n \equiv 0 \pmod 3$ if $r \in \{0, 1\}$ and $c_n \equiv 1 \pmod 3$ if $r = 2$.

Alternate solution. The binomial expansion of $(\sqrt[3]{2} - 1)^n$ yields

$$c_n = \sum_{k \equiv 2 \pmod 3} (-1)^{n-k} \cdot 2^{\frac{k-2}{3}} \cdot \binom{n}{k} \equiv (-1)^n \sum_{k \equiv 2 \pmod 3} \binom{n}{k} \pmod 3.$$

Since

$$\sum_{k \equiv 2 \pmod 3} \binom{n}{k} = \frac{(1+1)^n + \varepsilon(1+\varepsilon)^n + \varepsilon^2(1+\varepsilon^2)^n}{3} = \frac{2^n + 2\cos\frac{(n+2)\pi}{3}}{3},$$

where $\varepsilon^2 + \varepsilon + 1 = 0$ and the condition $c_n \equiv 1 \pmod 3$ is equivalent to

$$3c_n \equiv (-1)^n \left(2^n + 2\cos\frac{(n+2)\pi}{3}\right) \equiv 3 \pmod 9.$$

Consideration of n modulo 6 yields $3c_n \equiv 3 \pmod 9$ for $n \equiv 2, 5 \pmod 6$ and $3c_n \equiv 0 \pmod 9$ otherwise, hence the conclusion.

308. With $d = \gcd(y, z)$, $y = dr$, $z = ds$, r, s, positive integers, the equation becomes

$$s(dsx + 1)^2 = d(5s + 2r)(2s + r);$$

hence $d|s$, $s = dt$, t positive integer, so that

$$t(d^2 tx + 1)^2 = (5dt + 2r)(2dt + r).$$

Then $t | 2r^2$, yielding $t | 2$, since $\gcd(r, s) = \gcd(r, dt) = 1$ implies $\gcd(r, t) = 1$.

Assume that $t = 2$. Then

$$(2d^2 x + 1)^2 = (5d + r)(4d + r).$$

If p is a prime divisor of $5d+r$ and $4d+r$, then $p\mid(5d+r)-(4d+r)=d$ and $p\mid 5(4d+r)-4(5d+r)=r$, implying $p\mid\gcd(r,d)=1$, contradiction. Hence

$$\gcd(5d+r,4d+r)=1,$$

so that

$$5d+r=u^2, 4d+r=v^2, 2d^2x+1=uv, \text{ with } \gcd(u,v)=1.$$

We obtain $u^2-v^2=(5d+r)-(4d+r)=d$, which gives us

$$uv>2d^2x>d^2=(u^2-v^2)^2=(u-v)^2(u+v)^2\geq 4uv,$$

contradiction. Consequently, $t\neq 2$.

We conclude that $t=1$ and $(d^2x+1)^2=(5d+2r)(2d+r)$. Moreover, $z=sd=td^2=d^2$. If d is even, then LHS is odd and RHS is even, contradiction. Therefore, $z=d^2$ is an odd square.

309. **Answer: 37.**

Let a_n be the given number, where n is the number of digits. Then

$$9a_n=5(10^n-1)-9\cdot 18=5\cdot 10^n-167.$$

Case 1. $n\equiv 0 \pmod 3$. Then the sum of digits of a_n is $5(n-2)+3+7=5$, so that a_n is divisible by 3.

Case 2. $n\equiv 1 \pmod 3$. If $n=6k+1$, since $10^6\equiv 1 \pmod{13}$, we have

$$9a_n\equiv 5\cdot 10-167\equiv 0 \pmod{13},$$

so that a_n is divisible by 13. If $n=6k+4$, since $10^6\equiv 1 \pmod 7$, we have

$$9a_n\equiv 5\cdot 10^4-167\equiv 5\cdot 3^4+1\equiv 5\cdot 2^2+1\equiv 0 \pmod 7,$$

so that a_n is divisible by 7.

Case 3. $n\equiv 2 \pmod 3$. Since $10^3\equiv 1 \pmod{37}$, we have $9a_n\equiv 5(10^2-1)-167\equiv 0 \pmod{37}$, so that a_n is divisible by 37.

310. **Answer:** $(m,n)=(2,2),(2,4)$.

For $n=\prod_{i=1}^{k} p_i^{a_i}$, we have

$$n-\varphi(n)-1=n\left[1-\prod_{i=1}^{k}\left(1-\frac{1}{p_i}\right)\right]-1\geq \frac{n}{p_1}-1,$$

where p_1 is the smallest prime divisor of n. Denote $p_1 = p$. If $p \leq n - \varphi(n) - 1$, then $2^n \equiv 1 \pmod{p}$. By the Fermat's theorem, $2^{p-1} \equiv 1 \pmod{p}$. But then $p | 2^1 - 1 = 1$, contradiction. So,

$$p \geq n - \varphi(n) - 1 \geq \frac{n}{p} - 1,$$

which implies $n \leq p(p+1)$.

If $n = p(p+1)$, since $p \nmid p+1$, then $p+1$ must be a prime; otherwise it has a prime factor less than p. Therefore, $p = 2$, but $n = 6$ is not a solution.

From $p | n$, it follows $n \leq p^2$, so that $n = p$ or $n = p^2$, since n cannot have a prime divisor less than p. In the first case, $2^p = p^m$, implying $n = p = 2$, and $(m, n) = (2, 2)$. In the second case, $2^{p^2} + (p-1)! = p^{2m} + 1$. If p is odd, taking modulo 4 gives

$$(p-1)! \equiv 2 \pmod{4},$$

so $p = 3$, which is not a solution. So, $p = 2$ and $(m, n) = (2, 4)$.

311. (a) Consider the Pell equation $x^2 - (4025^2 - 1)y^2 = 1$. It has the solution $(4025, 1)$, so it has infinitely many solutions. For any solution (x, y), let $t = 4xy + 16100y^2$ and we have

$$2012t + 1 = (x + 4024y)^2, \quad 2013t + 1 = (x + 4026y)^2.$$

(b) There exist positive integers $x, y, x > y$, such that

$$4mn + 4 = (x - y)^2, \quad 4mn + 4n + 4 = (x + y)^2, \quad n = xy.$$

The equation $4mxy + 4 = (x - y)^2$ is equivalent to

$$x^2 - 2(2m + 1)xy + y^2 - 4 = 0.$$

Let $x, y, x > y$ be a solution of this equation with $x + y$ minimal. If $y > 2$, then

$$z = (2m + 1)y - x = \frac{y^2 - 4}{x} < y$$

is a positive integer, and (y, z) is a solution which contradicts the minimality, so $y \leq 2$.

If $y = 1$, then $x | y^2 - 4 = -3$ implies $x = 3$, but then $2m + 1 = 1$; hence $m = 0$, which is not possible.

If $y = 2$, then $x^2 - 4(2m + 1)x = 0$; hence $x = 4(2m + 1)$. All solutions of the equation can be obtained by the transformation $(x, y) \to (2(2m + 1)x - y, x)$ starting with the initial solution. We can see that $2 | x, 2 | y$ for every solution.

We need to prove that $8(2m+1) | n = xy$. For the initial solution, this is true, since $xy = 8(2m+1)$. If (x, y) is an arbitrary solution and $8(2m+1)|xy$, then

$$(2(2m+1)x - y)y = 2(2m+1)x^2 - xy$$

and $2|x$, so $8(2m+1)|2(2m+1)x^2$ and $8(2m+1)|xy$, so it is true for the solution

$$(2(2m+1)x - y, x);$$

hence it is true for all solutions.

2.42   2014

312. From $7^x \equiv -1 \pmod{5}$, it follows that $x \equiv 2 \pmod{4}$. If $x > 2$, then by Zsigmondy's theorem, $7^x + 1$ has a prime divisor p which not divides $7^2 + 1 = 50$; hence $p \neq 2, 5$. We deduce that the given equation has only the solutions

$$(x, y) = (0, 0), (2, 2).$$

313. Let $n = \prod_{i=1}^{m} p_i^{a_i}$ be the canonical factorization. Then $\prod_{i=1}^{m}(a_i + 1) = a$, so that there are only a finite number of tuples $(m, a_1, \ldots, a_m) \in \mathbb{N}^* \times \mathbb{N}^m$ satisfying the condition. Fixing such a tuple, it suffices to show that there are only a finite number of sets of primes $\{p_1, \ldots, p_m\}$ such that $n = \prod_{i=1}^{m} p_i^{a_i}$ and $n | \varphi(n) + \sigma(n)$.

Assume that there are infinitely many such sets, and let S be the collection of these sets. Pick an element from S. From

$$\varphi(n) = \prod_{i=1}^{m} p_i^{a_i - 1}(p_i - 1), \sigma(n) = \prod_{i=1}^{m} \frac{p_i^{a_i+1} - 1}{p_i - 1},$$

we have

$$\frac{\varphi(n) + \sigma(n)}{n} = \prod_{i=1}^{m} \frac{p_i - 1}{p_i} + \prod_{i=1}^{m}\left(1 + \frac{1}{p_i} + \ldots + \frac{1}{p_i^{a_i}}\right).$$

Then there exists some $N \in \mathbb{N}$ such that one of these primes is less than N; otherwise

$$1 < \frac{\varphi(n) + \sigma(n)}{n} < 2,$$

contradiction. By the infinite pigeonhole principle, there are an infinite number of sets in S that all contain the same prime less than N. Looking at these sets and repeating the argument, we get finally that every element of a set in S is bounded by some number, so S is finite, contradiction.

314. Denote $a = d_1$, $b = d_2$, a, b odd and $\gcd(a, b) = 1$. We have to find infinitely many pairs (a, b) such that $a + b - k = n$ and $ab | (a + b - k)^2 + 1$. If (a, b) is a solution with $a \leq b$, then (b, a') is also a solution with

$$b < a' = (k^2 - 4k + 3)b + 2k - a.$$

We can repeat this process starting with $a = b = 1$, and from some time we have $a + b > k$; hence $n = a + b - k$ is positive and we can find infinitely many n.

315. Answer: $n = 1, 3$.

For $n = 1$, $x = y = z = 3$ is a solution.
For $n = 3$, $x = y = z = 1$ is a solution.

Case 1. n even. If there exists a solution, then RHS is even, which is possible only if at least one of the numbers x, y, and z is even. Then RHS is divisible by 4. Since the remainders modulo 4 of the squares are 0 and 1, all numbers x, y, and z must be even. Take $x = 2a$, $y = 2b$, $z = 2c$. Then $a^2 + b^2 + c^2 = 2nabc$; hence (a, b, c) satisfies a similar equation with doubled n, so they must be even. Continuing this process we conclude that x, y, and z are divisible by arbitrary power of 2, which is impossible for positive integers. Consequently, there are no solutions for even n.

Case 2. $n > 3$ odd. Suppose that (x, y, z) is a solution. The given equation is equivalent to $z^2 - nxyz + x^2 + y^2 = 0$; let z' be the other root of this quadratic equation. Then $zz' = x^2 + y^2$; therefore $z' > 0$. we have also $z' = nxy - z$. We can suppose that $z = \max(x, y, z)$; then $x^2 \leq xz \leq xyz$ and $y^2 \leq yz \leq xyz$; whence $z^2 \geq (n - 2)xyz$ and $z \geq (n - 2)xy$. Therefore $z' \leq 2xy < (n - 2)xy$, which implies $x + y + z' < x + y + z$. Thus we can infinitely reduce the sum of the components of a solution, which is impossible.

Note. Using the transformation $(x, y, z) \to (y, z, nyz - x)$, it is easy to show that for $n = 1$ and $n = 3$ the given equation has infinitely many solutions.

316. Let p_n be the largest prime divisor of $n^4 + n^2 + 1$, and let q_n be the largest prime divisor of $n^2 + n + 1$. Then $p_n = q_{n^2}$, and from

$$n^4 + n^2 + 1 = \left((n-1)^2 + (n-1) + 1\right)(n^2 + n + 1),$$

it follows that $p_n = \max\{q_n, q_{n-1}\}$ for $n \geq 2$. Since $n^2 - n + 1$ is odd, we have

$$\gcd(n^2 + n + 1, n^2 - n + 1) = \gcd(2n, n^2 - n + 1)$$
$$= \gcd(n, n^2 - n + 1) = 1;$$

therefore $q_n \neq q_{n-1}$. Now it suffices to prove that the set S of positive integers $n \geq 2$ for which $q_n > q_{n-1}$ and $q_n > q_{n+1}$ is infinite, since for $n \in S$ one has

$$p_n = \max\{q_n, q_{n-1}\} = q_n = \max\{q_n, q_{n+1}\} = p_{n+1}.$$

Suppose that S is finite. Since $q_2 = 7 < 13 = q_3$ and $q_3 = 13 > 7 = q_4$, the set S is nonempty. Let m be the largest element of S. We cannot have

$$q_m > q_{m+1} > q_{m+2} > \cdots$$

because all these numbers are positive integers, so there exists $k \geq m$ such that $q_k < q_{k+1}$. Observe that it is impossible to have $q_k < q_{k+1} < q_{k+2} < \cdots$, because

$$q_{(k+1)^2} = p_{k+1} = \max\{q_k, q_{k+1}\} = q_{k+1}.$$

Take the smallest $l \geq k+1$ such that $q_l > q_{l+1}$. By the minimality of l, we have $q_{l-1} < q_l$; therefore $l \in S$. Since $l \geq k+1 > k \geq m$, this contradicts the maximality of m; hence S is infinite.

317. (a) From $2x_2 = 3x_1 + \sqrt{5x_1^2 - 4} = 4$, we deduce $x_2 = 2$. For all $n \geq 1$, we have $2x_{n+1} = 3x_n + \sqrt{5x_n^2 - 4}$; whence we see that the sequence has only positive terms, so that it is increasing. Then

$$(2x_{n+1} - 3x_n)^2 - 5x_n^2 = -4 \Leftrightarrow x_{n+1}^2 - 3x_{n+1}x_n + x_n^2 = -1.$$

Now,

$$x_{n+2}^2 - 3x_{n+2}x_{n+1} + x_{n+1}^2 = -1 = x_{n+1}^2 - 3x_{n+1}x_n + x_n^2 \Rightarrow$$
$$\Rightarrow (x_{n+2} - x_n)(x_{n+2} + x_n - 3x_{n+1}) = 0.$$

As $x_{n+2} > x_{n+1} > x_n$, we get $x_{n+2} = 3x_{n+1} - x_n$, $n \geq 1$. The first two terms of the sequence are positive integers; therefore by induction it follows that all terms are positive integers.

(b) If $x_1 = 1$, $x_2 = 2$, $x_n = 3x_{n-1} - x_{n-2}$, $n \geq 3$, then it is easy to prove by induction the equality $x_n^2 + 1 = x_{n-1}x_{n+1}, n \geq 2$. Suppose that $2011 | x_{n-1}$; then $2011 | x_n^2 + 1$, contradiction, since $2011 \equiv 3 \pmod 4$, but $x_n^2 + 1$ can have only prime factors $p \equiv 1 \pmod 4$.

318. Let S be the set of odd numbers r such that $pr + 1 | p^p - 1$, and for the sake of contradiction, suppose that $S \neq \emptyset$. Let $r_0 \in S$ be minimal and let d be the order of p modulo $pr_0 + 1$. Then $d | p$, so that $d = p$. Also, $p^{\varphi(pr_0+1)} \equiv 1 \pmod{(pr_0 + 1)}$; hence p divides $\varphi(pr_0 + 1)$. Therefore, if the prime factorization of $pr_0 + 1$ is

$$pr_0 + 1 = q_1^{a_1} \ldots q_k^{a_k}, a_1, \ldots, a_k > 0,$$

then there exists a q_i, say q_1 such that $p | q_1 - 1$; whence $q_1 = mp + 1$. Then $k \geq 2$ and $q_1^{a_1} = Mp + 1$, with M even. It follows that $q_2^{a_2} \ldots q_k^{a_k}$ is also of the form $r_1 p + 1$ and $r_1 M p + M + r_1 = r_0$. We deduce that r_1 is odd and $r_1 p + 1 | r_0 p + 1 | p^p - 1$, contradicting the minimality of r_0.

Thus, $S = \emptyset$.

319. (a) For $b = -1, c = 2$, it follows that $a^2 - 14a + 33 = 0$; whence $a \in \{3, 11\}$.
   (b) The given equation is equivalent to $(a + b + c)^2 = 16(ab + bc + ca)$, so that

$$ab + bc + ca = \left(\frac{a+b+c}{4}\right)^2$$

   is a square.
   (c) If $S = n^2$, then $(a, b, c) = (3n, -n, 2n)$ is a solution.

320. The numbers with less than three digits will be completed with 0's to become three digits numbers. The sum of the products of digits of all these numbers is

$$0 \cdot 0 \cdot 0 + \cdots + 9 \cdot 9 \cdot 9 = (0 + 1 + \cdots + 9)^3.$$

The sum of the products of nonzero digits can be found by replacing 0 by 1; hence

$$S = 46^3 - 1 = 3^3 \cdot 5 \cdot 7 \cdot 103,$$

so that the largest prime factor is 103.

321. We first show that

$$N_r \leq \frac{r}{2} \binom{2r}{r}$$

by proving that

$$K(n, r) = \frac{r}{2(n+r)} \binom{2n}{n} \binom{2r}{r}$$

is an integer for all $n \geq 0$ and $r \geq 1$. Notice that

$$K(0, r) = \binom{2r-1}{r}, K(n, 1) = \frac{1}{n+1} \binom{2n}{n},$$

the latter being a Catalan number, so $K(0, r), K(n, 1)$ are integers for all r, n. The recurrence relation $K(n, r + 1) - K(n + 1, r) = 2K(n, 1)K(0, r)$ shows by induction on r that $K(n, r)$ is an integer for all n, r.

To prove the recurrence relation, notice that

$$\binom{2m+2}{m+1} = 2\binom{2m+1}{m} = \frac{2(2m+1)}{m+1}\binom{2m}{m},$$

to get

$$K(n,r+1) - K(n+1,r) = \frac{r+1}{2(n+r+1)}\binom{2n}{n}\binom{2r+2}{r+1} - \frac{r}{2(n+r+1)}\binom{2n+2}{n+1}\binom{2r}{r} =$$

$$= \frac{1}{n+r+1}\binom{2n}{n}\binom{2r}{r}\left(2r+1 - r\cdot\frac{2n+1}{n+1}\right)$$

$$= \frac{1}{n+1}\binom{2n}{n}\binom{2r}{r} = 2K(n,1)K(0,r).$$

Suppose now that, for some r, $N_r < M_r = \frac{r}{2}\binom{2r}{r}$. The integer N_r is the least common multiple, over all $n \geq 0$, of the denominators of the numbers $\frac{1}{n+r}\binom{2n}{n}$ when written in lowest terms. By the argument above, M_r is a multiple of all these denominators; hence $N_r | M_r$. By the definition of N_r, any prime p that divides $\frac{N_r}{M_r}$ also divides $K(n,r)$ for each $n \geq 0$.

Since $K(n, s+1) = K(n+1, s) + 2K(n,1)K(0,s)$, induction on m shows that p divides $K(n,m)$ for all $n \geq 0$, $m \geq r$. Now choose k such that $p^k \geq r$. Since $p \mid \binom{p^k}{j}$, $j = 1, 2, \ldots, p^k - 1$, the identity $\binom{2n}{n} = \sum_{j=1}^{n}\binom{n}{j}^2$ yields $\binom{2p^k}{p^k} \equiv 2 \pmod{p}$. Therefore p does not divide $\frac{1}{2}\binom{2p^k}{p^k} = K(0,p^k)$, which contradicts the preceding paragraph.

It follows that $N_r = \frac{r}{2}\binom{2r}{r}$ for all r.

322. Let $p_1 = 2, p_2 = 3, p_3 = 5, \ldots$ be the sequence of primes, and let $q < r$ be the first two primes which do not divide n. A necessary and sufficient condition that n be of the required type is that $n < qr$. Each of the primes less than r and different from q divides n and so does their product. Therefore the product of all primes less than r does not exceed $nq < q^2 r$. If $r = p_m$, then $q \leq p_{m-1}$, so $p_1 p_2 \cdots p_{m-2} < p_{m-1} p_m$.

Notice that 6 is the first index k such that $p_1 p_2 \cdots p_{k-2} > p_{k-1} p_k$ (*). If (*) holds for some index $k \geq 6$, then, by the Bertrand's postulate,

$$p_1p_2 \ldots p_{k-1} > p_{k-1}^2 p_k > 2p_{k-1} \cdot 2p_k > p_k p_{k+1};$$

therefore the inequality (∗) holds for all indices $k \geq 6$.

Consequently, $m \leq 5$, $r = p_m \leq p_5 = 11$, $q \leq p_4 = 7$, $n < qr \leq p_4 p_5 = 77$.

Examination of the positive integers less than 77 quickly yields the required numbers: 2, 3, 4, 5, 6, 8, 9, 10, 12, 14, 18, 20, 24, 30, 42, and 60.

323. Let a and b be positive integers which satisfy the hypothesis and $d = \gcd(a,b)$. Then $a = dx$, $b = dy$ and the arithmetic mean shows that x, y are odd. From $\gcd(x,y) = 1$, the geometric mean shows that x, y are squares, $x = x_1^2, y = y_1^2$. The quadratic mean shows that there is a positive integer z such that $x_1^2 + y_1^2 = 2z^2$.

Since x_1, y_1 are odd, this equation is equivalent to

$$\left(\frac{x_1^4 - y_1^4}{2}\right)^2 = z^4 - (x_1 y_1)^2.$$

Now we have to solve in positive integers the equation

$$r^2 = p^4 - q^4.$$

Suppose that (p,q,r) is a solution with pqr minimal.

Case 1. q even. Then Pythagorean triples formula implies that there are positive integers m, n such that $\gcd(m,n) = 1$, $p^2 = m^2 + n^2$, $q^2 = 2mn$. Now there are positive integers m_1, n_1 such that $mn = 2m_1 n_1 (m_1^2 - n_1^2)$; hence

$$\left(\frac{q}{2}\right)^2 = m_1 n_1 (m_1^2 - n_1^2).$$

From $\gcd(m_1, n_1) = 1$, it follows that the three factors on the right are squares, $m_1 = p_1^2, m_2 = q_1^2, m_1^2 - n_1^2 = r_1^2$, so that

$$r_1^2 = p_1^4 - q_1^4.$$

Now we have a solution (p_1, q_1, r_1) with $p_1 q_1 r_1 = \frac{q}{2} < pqr$, contradiction.

Case 2. q odd. Then we have similarly $p^2 = m^2 + n^2$, $q^2 = m^2 - n^2$; whence

$$(pq)^2 = m^4 - n^4.$$

From $pqr > 2mnpq > mnpq$, we deduce that the solution (p,q,r) is not minimal, contradiction.

324. Answer: no solution.

From the table

m	$p(m)$	$p(m+1)$
1	$-68 = -2^2 \cdot 17$	$-64 = -2^6$
2	$-64 = -2^6$	$-58 = -2 \cdot 29$
3	$-58 = -2 \cdot 29$	$-50 = -2 \cdot 5^2$
4	$-50 = -2 \cdot 5^2$	$-40 = -2^3 \cdot 5$
5	$-40 = -2^3 \cdot 5$	$-28 = -2^2 \cdot 7$
6	$-28 = -2^2 \cdot 7$	$-14 = -2 \cdot 7$
7	$-14 = -2 \cdot 7$	2

we see that there are no eligible $n > m$ for $1 \leq m \leq 7$.

Assume $m \geq 8$, thus $p(m+1) > p(m) > 0$. We need $n|p(m) = m^2 + m - 70$ and $n+1|p(m+1) = m^2 + m - 70 + 2(m+1)$. Let $m^2 + m - 70 = kn$ for some positive integer k, so $n+1|kn + 2(m+1) = k(n+1) + 2(m+1) - k$; hence

$$n+1 \mid 2(m+1) - k.$$

We have

$$n \geq m+1 \Rightarrow k \leq \frac{m^2 + m - 70}{m+1} < m.$$

Take $n = m + l$, for some integer $l \geq 1$. Then

$$m + l + 1 \mid 2(m+1) - k = 2(m+l+1) - k - 2l,$$

so $m + l + 1 | k + 2l$. But $k + 2l < m + 2l < 2(m + l + 1)$; therefore $m + l + 1 = k + 2l$, that is, $k = m - l + 1$.

Finally, $m^2 + m - 70 = kn = (m - l + 1)(m + l) = m^2 - n^2 + m + l$, i.e.

$$l^2 - l - 70 = 0,$$

which has no integer roots.

325. Answer: $(1, 1)$, $(43, 13)$, $(49, 37)$.

Let $3m + 1 = kn$, k even. We get

$$\frac{kn - 1}{3} \mid n^2 + 3 \Rightarrow kn - 1 \mid 3kn^2 + 9k \Rightarrow kn - 1 \mid 9k + 3n.$$

It follows that

$$kn - 1 \leq 9k + 3n \iff (n-9)(k-3) \leq 28.$$

If $k = 2$, no solution.
If $k = 4$, then $4n - 1 | 3n + 36 \implies 4n - 1 | 147 \implies n \in \{1, 37\}$.
If $k \geq 6$, then

$$n - 9 \leq \frac{28}{k-3} \leq \frac{28}{3} \implies n \leq 18.$$

Since n is odd, we deduce $n \in \{1, 3, 5, 7, 9, 11, 13, 15, 17\}$. We find that $n = 13$ is a solution.

2.43 2015

326. By lifting the exponent lemma, we have $\nu_2(3^2 - 1) = 3$, $\nu_2(3^4 - 1) = 4$, $\nu_2(3^8 - 1) = 5, \ldots, \nu_2\left(3^{2^k} - 1\right) = k + 2$. We claim that all numbers of the form

$$n = 3^{2^k} - 2^{2^k} \equiv 1 \pmod{2^{k+2}}$$

for $k \geq 2$, which are clearly composite, satisfy the condition.
We have

$$3^{n-1} - 2^{n-1} = 3^{\left(2^{k+2}x+1\right)-1} - 2^{\left(2^{k+2}x+1\right)-1} = 3^{2^k \cdot 4x} - 2^{2^k \cdot 4x} \equiv 0 \pmod{n},$$

since $n = 3^{2^k} - 2^{2^k}$. Thus, $n | 3^{n-1} - 2^{n-1}$ for infinitely many composite integers n.

327. For every integer x and a positive integer m that is relatively prime with x, there exists a positive integer t such that $x^t \equiv 1 \pmod{m}$. If $p \neq 2$ and $p \neq 5$, then for every positive integer k, there exists some integer $t > 0$ satisfying $p^t \equiv 1 \pmod{10^k}$. As obviously $p^t > 10^k$, the number p^t ends with $k - 1$ zeros followed by one. Thus choosing $k > a$ and $n = t$ solves the problem. If $p = 2$ or $p = 5$, then denote $q = \frac{10}{p}$. For every positive integer k, there exists some integer $t > 0$ satisfying $p^t \equiv 1 \pmod{q^k}$. This implies $p^{k+t} \equiv p^k \pmod{10^k}$. As obviously $p^{k+t} > 10^k > p^k$, the ending digits of the number p^{k+t} are precisely all the digits of number p^k preceded by zeros until the kth digit from the end. Choose k in such a way that $q^k > 10^a$; then $p^k < 10^{k-a}$. This means that the number of consecutive zeros must be at least a and taking $n = k + t$ solves the problem.

328. The set $\{2^{102!}+1, 2^{102!}+2, \ldots, 2^{102!}+2014\}$ contains 102 naughty numbers

$$2^{1 \cdot 102!}+1, 2^{2 \cdot \frac{102!}{2}}+2, \ldots, 2^{102 \cdot \frac{102!}{102}}+102.$$

Let $f(n)$ be the number of naughty numbers among $n, n+1, \ldots, n+101$. Notice that $|f(n)-f(n-1)| \leq 1$. It is easy to see that there are fewer than 100 naughty numbers from 1 to 102. By the "discrete continuity" of f, we deduce that there exists an $n < 2^{102!}+1$ which satisfies the required condition.

329. Consider the given equation As a quadratic equation in x:

$$(y-3)x^2 + (y^2-10y+21)x - (2y^2-16y+60) = 0.$$

The discriminant Δ satisfies

$$\left((y-3)^2\right)^2 < \Delta = (y-3)^4 + 120(y-3) < \left((y-3)^2+1\right)^2;$$

for $y \geq 11$, therefore it suffices to check $1 \leq y \leq 10$. The solutions (x,y) are

$$(3,8), (4,6), (5,5), (7,4).$$

330. By symmetry, we can suppose that $a \leq b$. Notice that $b^2 - a \geq 0$, so that if $\frac{a^2+b}{b^2-a}$ is a positive integer, then $a^2 + b \geq b^2 - a$; whence $(a+b)(a-b+1) \geq 0$.

Case 1. $a = b$. Then we have

$$\frac{a^2+a}{a^2-a} = 1 + \frac{2}{a-1},$$

which is a positive integer only for $a = 2, 3$.

Case 2. $a = b - 1$. Then we have

$$\frac{b^2+a}{a^2-b} = \frac{(a+1)^2+1}{a^2-(a+1)} = 1 + \frac{4a+2}{a^2-a-1}.$$

Since the last fraction is a positive integer, we must have $4a+2 \geq a^2-a-1$, that is, $a^2 - 5a - 3 \leq 0$; whence we deduce $1 \leq a \leq 5$. Checking these values, we find that only $a = 1, 2$ satisfy the given conditions.

In conclusion, the solutions (a,b) are (including those obtained by symmetry):

$$(2,2), (3,3), (1,2), (2,3), (2,1), (3,2).$$

331. Take $n = \frac{5(2m^2-1)}{7}$, where $m \equiv 2 \pmod 7$. The squares modulo 3 are 0 and 1, so that n is not divisible by 3. Since $5|n$ and $2|n+1$, we have

$$d_n + d_{n+1} = \frac{n}{5} + \frac{n+1}{2} = \frac{2m^2-1}{7} + \frac{5m^2+1}{7} = m^2.$$

332. Answer: $2^n(n-2) + 1$.

We shall prove by induction on n that every integer greater than $2^n(n-2) + 1$ can be represented as such a sum.

For $n = 2$, we have $A_2 = \{2, 3\}$ and $m = 2 + 2 + \cdots + 2$ (m even) and $m = 3 + 2 + \cdots + 2$ ($m \geq 3$ odd).

If $n > 2$ and $m > 2^n(n-2) + 1$ is even, then

$$\frac{m}{2} \geq \frac{2^n(n-2)+2}{2} = 2^{n-1}(n-2) + 1 > 2^{n-1}(n-3) + 1.$$

By the induction hypothesis, there is a representation of the form

$$\frac{m}{2} = (2^{n-1} - 2^{k_1}) + (2^{n-1} - 2^{k_2}) + \cdots + (2^{n-1} - 2^{k_r}),$$

with $0 \leq k_i < n-1$. It follows that m is a sum of elements of A_n:

$$m = (2^n - 2^{k_1+1}) + (2^n - 2^{k_2+1}) + \cdots + (2^n - 2^{k_r+1}).$$

If m is odd, then consider

$$\frac{m - (2^n - 1)}{2} > \frac{2^n(n-2) + 1 - (2^n - 1)}{2} = 2^{n-1}(n-3) + 1.$$

By the induction hypothesis, there is a representation of the form

$$\frac{m - (2^n - 1)}{2} = (2^{n-1} - 2^{k_1}) + (2^{n-1} - 2^{k_2}) + \cdots + (2^{n-1} - 2^{k_r}),$$

with $0 \leq k_i < n-1$. It follows that m is a sum of elements of A_n:

$$m = (2^n - 2^{k_1+1}) + (2^n - 2^{k_2+1}) + \cdots + (2^n - 2^{k_r+1}) + (2^n - 1).$$

It remains to show that there is no representation for $2^n(n-2) + 1$. Let N be the smallest positive integer which satisfies $N \equiv 1 \pmod{2^n}$ and which can be represented as a sum of elements of A_n:

$$N = (2^n - 2^{k_1}) + (2^n - 2^{k_2}) + \cdots + (2^n - 2^{k_r}),$$

where $0 \leq k_1, k_2, \ldots, k_r < n$. Suppose first that two of the terms of the sum coincide, that is, $k_i = k_j$ for some $i \neq j$. If $k_i = k_j = n-1$, then we can remove these two terms to obtain a representation for $N - 2(2^n - 2^{n-1}) = N - 2^n$ as a sum of elements of A_n, which contradicts the choice of N. If $k_i = k_j = k < n-1$, then replace the two terms by $2^n - 2^{k+1}$, which is an element of A_n, to obtain a representation of

$$N - 2(2^n - 2^k) + (2^n - 2^{k+1}) = N - 2^n,$$

contradiction. It follows that all k_i are distinct; hence

$$2^{k_1} + 2^{k_2} + \cdots + 2^{k_r} \leq 2^0 + 2^1 + \cdots + 2^{n-1} = 2^n - 1.$$

On the other hand,

$$2^{k_1} + 2^{k_2} + \cdots + 2^{k_r} \equiv -N \equiv -1 \pmod{2^n},$$

which implies $2^{k_1} + 2^{k_2} + \cdots + 2^{k_r} = 2^n - 1$. This fact is possible only when

$$\{k_1, k_2, \ldots, k_r\} = \{0, 1, \ldots, n-1\},$$

so that $N = 2^n n - (2^n - 1) = 2^n(n-1) + 1$.

In particular, $2^n(n-2) + 1$ cannot be represented as a sum of elements of A_n.

333. Answer: $(p, x, y) \in \{(3, 2, 5), (3, 5, 2)\} \cup \{(2, n, 2^k - n) | 0 < n < 2^k\}$.

For $p = 2$, clearly all pairs (x, y) of positive integers whose sum is a power of 2 satisfy the condition.

Assume that $p > 2$, and let a, b be positive integers such that

$$x^{p-1} + y = p^a, \quad x + y^{p-1} = p^b.$$

Assume wlog that $x \leq y$, so that $p^a = x^{p-1} + y \leq x + y^{p-1} = p^b$, which implies $a \leq b$. Now we have

$$p^b = y^{p-1} + x = (p^a - x^{p-1})^{p-1} + x.$$

We take this equation modulo p^a and take into account that $p - 1$ is even, which gives us $0 \equiv x^{(p-1)^2} + x \pmod{p^a}$. If $p | x$, then $p^a | x$, since $x^{(p-1)^2 - 1} + 1$ is not divisible by p in this case. However, this is impossible, since $x \leq x^{p-1} < p^a$. Thus we know that $p \nmid x$, which means that

$$p^a \mid x^{(p-1)^2 - 1} + 1 = x^{p(p-2)} + 1.$$

By the Fermat's theorem, $x^{(p-1)^2} \equiv 1 \pmod{p}$; thus p divides $x+1$. Let p^r be the highest power of p that divides $x+1$. By the binomial formula, we have

$$x^{p(p-2)} = \sum_{k=0}^{p(p-2)} \binom{p(p-2)}{k}(-1)^{p(p-2)-k}(x+1)^k.$$

Except for the terms corresponding to $k = 0, 1, 2$, all terms in the sum are divisible by p^{3r} and thus by p^{r+2}. The remaining terms are

$$\frac{p(p-2)(p^2-2p-1)}{2}(x+1)^2,$$

which is divisible by p^{2r+1}, and thus by p^{r+2},

$$p(p-2)(x+1),$$

which is divisible by p^{r+1}, but not by p^{r+2} and the initial term -1 corresponding to $k=0$. It follows that the highest power of p that divides $x^{p(p-2)}+1$ is p^{r+1}.

We know that p^a divides $x^{p(p-2)}+1$, which means that $a \leq r+1$. Moreover,

$$p^r \leq x+1 \leq x^{p-1} + y = p^a,$$

hence $a = r$ or $a = r+1$. If $a = r$, then $x = y = 1$, which is impossible for $p > 2$. Thus, $a = r+1$. Now since $p^r \leq x+1$, we get

$$x = \frac{x^2+x}{x+1} \leq \frac{x^{p-1}+y}{x+1} = \frac{p^a}{x+1} \leq \frac{p^a}{p^r} = p,$$

so we must have $x = p-1$ for p to divide $x+1$.

It follows that $r = 1$, $a = 2$. If $p \geq 5$, then

$$p^a = x^{p-1} + y > (p-1)^4 = (p^2 - 2p + 1)^2 > (3p)^2 > p^2 = p^a,$$

contradiction. The only case that remains is $p = 3$, and indeed $x = 2$, $y = p^a - x^{p-1} = 5$ satisfy the conditions.

334. The difference of the numbers $5(y^2 + 202^2)$ and $5(y^2 + 201^2)$ is

$$5(202 - 201)(202 + 201) = 2015.$$

Both numbers are sums of two squares, since $5 = 1^2 + 2^2$, and the set of sums of two squares is closed under multiplication.

2.43 2015

Choose 2014 prime numbers such that $p_i \equiv 3 \pmod 4$, where p_i is much larger than $5 \cdot 202^2$ and

$$\left(\frac{(5 \cdot 201^2 + 1) \cdot 5^{-1}}{p_i}\right) = -1.$$

Select y as the solution of the congruences

$$y^2 \equiv \frac{p_i - 1}{5} - 201^2 \pmod{p_i^2}, 1 \leq i \leq 2014.$$

As $y^2 = -\frac{5 \cdot 201^2 + 1}{5}$ has a solution in $\mathbb{Z}_{p_i}$, by Hensel's Lemma it has one in $\mathbb{Z}_{p_i^2}$, so that y exists, by the Chinese Remainder Theorem.

No number in the interval $(5(y^2 + 201^2), 5(y^2 + 202^2))$ is sum of two squares. This follows from the construction, as if we have $a \equiv p_i \pmod{p_i^2}$ for some i, and for a in the interval, this is a contradiction, as it is well known that ν_{p_i} is either 0 or even.

Now just pick another 2014 primes $\equiv 3 \pmod 4$ to construct another y; hence there are infinitely many of them.

335. From $(b-2)^2 + b^2 + (b+2)^2 = 3b^2 + 8$, the given condition becomes

$$3b^2 + 8 = 1111k.$$

Since b is odd and k is a digit, it follows $k \in \{1, 3, 5, 7, 9\}$. But

$$k \equiv 111k \equiv 3b^2 + 8 \equiv 2 \pmod 3$$

implies $k = 5$; hence $b = 43$. Answer: $41^2 + 43^2 + 45^2 = 5555$.

336. (a) Let $3N = x^4 + x^2y^2 + y^4$ and assume that $x \geq y \geq 0$. If $x = y$ or $y = 0$, then N is not a prime number; therefore $x > y \geq 1$.

If $y = 1$, then $3N = (x^2 + x + 1)(x^2 - x + 1)$; therefore $x = 2$; otherwise N is the product of two integers greater than 1 or is not an integer.

If $y > 1$, then $3N = (x^2 + xy + y^2)(x^2 - xy + y^2)$; therefore N is not a prime or is not an integer.

Answer: $(x, y) = (\pm 2, \pm 1), (\pm 1, \pm 2)$.

(b) Let $5N = x^4 + x^2y^2 + y^4$. If $x \equiv 0 \pmod 5$, then $y \equiv 0 \pmod 5$ and N is not a prime, since it is divisible by 5^3. If $xy \not\equiv 0 \pmod 5$, then N is not an integer, since the nonzero squares modulo 5 are 1 and 4, so that

$$x^4 + x^2y^2 + y^4 \equiv 1 + (xy)^2 + 1 \equiv 1, 3 \pmod 5.$$

337. Answer: $k = 0, 1, 2$.

If a prime $p | n! + 1$, then $p > n$; therefore $n + 1 = p$ is a prime, so that $n! + 1 = (n + 1)^m$, with m positive integer. For $n = 1, 2, 4$ the number $n! + 1$ is a perfect power. Let $p > 5$ such that $(p - 1)! + 1 = p^m$; then

$$(p-2)! = \sum_{i=0}^{m-1} p^i \Rightarrow (p-2)! \equiv m \pmod{p-1}$$

and since $p - 1 > 4$ is composite, we have $m \equiv (p - 2)! \equiv 0 \pmod{p-1}$. From

$$p^m - 1 = (p-1)! < (p-1)^{p-1} < p^{p-1},$$

it follows that $m < p - 1$, contradiction.

338. 1. If there is a prime $p \neq 3$ that divides D_n, then

$$p \mid a^n + (a+1)^n + (a+2)^n, \quad p \mid (a-1)^n + a^n + (a+1)^n,$$

imply that p divides their difference $(a + 2)^n - (a - 1)^n$. For $a = 1$, we get $p | 3^n$, contradiction.

2. Let $x, y \in \{a, a+1, a+2\}$ be not divisible by 3. Then $3 | x + y$ and lifting the exponent lemma implies $\nu_3(x^n + y^n) = \nu_3(x + y) + \nu_3(n)$. For $n = 3^{k-1}$, we have

$$\nu_3(a^n + (a+1)^n + (a+2)^n) \geq k$$

for all positive integers a. Taking $a = 1$, we get $\nu_3(3) + \nu_3(n) = k$; therefore $D_n = 3^k$.

339. If $m \geq 2015$, then by the AM–GM inequality we have

$$(ab)^{2015} = (a^2 + b^2)^m \geq (a^2 + b^2)^{2015} > (ab)^{2015},$$

Contradiction; hence $m < 2015$. Take $g = \gcd(a, b)$, $a = gx$, $b = gy$; then $\gcd(x, y) = 1$, and the equation becomes

$$g^{2(2015-m)} (xy)^{2015} = (x^2 + y^2)^m.$$

If $x > 1$, then there it has a prime factor p. Since $\gcd(x, y) = 1$, we must have $p \nmid y$. Then p is a divisor of LHS, but not a divisor of RHS, contradiction. Then $x = 1$ and similarly, $y = 1$ and thus

$$g^{2(2015-m)} = 2^m.$$

Then $g = 2^n$; whence $2n(2015 - m) = m$, where n is a positive integer. This equation rearranges as $(2n + 1)(2015 - m) = 2015$; hence $2015 - m$ is a divisor of 2015. From $2015 = 5 \cdot 13 \cdot 31$, we find the divisors of 2015: 1, 5, 13, 31, 65, 155, 403, and 2015; whence

$$m \in \{0, 1612, 1860, 1950, 1984, 2002, 2010, 2014\}.$$

340. If $n = dx$, then $dn + 1 = d^2x + 1$ and $d^2 + n^2 = d^2(x^2 + 1)$. Note also that $\gcd(d, dn + 1) = 1$. Since $d^2x + 1 | d^2(x^2 + 1)$, we must have $d^2x + 1 | x^2 + 1$; hence $d^2x + 1 \leq x^2 + 1$ and thus $d^2 \leq x$. Note also that

$$d^4(x^2 + 1) - (d^2x + 1)(d^2x - 1) = d^4 + 1$$

is a multiple of $d^2x + 1$; hence $d^2x + 1 \leq d^4 + 1$ and thus $x \leq d^2$. Since $x \leq d^2 \leq x$, we must have $x = d^2$ and so $n = d^3$. Now we check that all cube numbers work: if $n = d^3$, then $dn + 1$ will always be a divisor of $d^2 + n^2$, because

$$d^2 + n^2 = d^2(1 + d^4) = d^2(dn + 1).$$

341. Answer: $(a, b) = (1, 1), (-2, 1), (-1, 0), (5, 3)$.

From $a^2 + a \geq -\frac{1}{4}$, we deduce $b^3 + b \geq 0$, whence $b \geq 0$. Write the equation As $a(a + 1) = b(b^2 + 1)$ and deduce $a|b^2 + 1$ and $b|a + 1$, since a, b are coprime. Multiplying these relations, we get $ab|(a + 1)(b^2 + 1)$, whence $ab|a + b^2 + 1$.

Suppose that $a \geq 0$, and find the negative solutions from the quadratic equation $a^2 + a - b(b^2 + 1) = 0$.

Case 1. $a \leq b$. Then $b^3 + b \geq a^2 + a$, with equality for $a = b = 1$. For $b = 1$ we have two solutions: $(a, b) = (1, 1), (-2, 1)$.

Case 2. $b = 0$. Then $(a, b) = (-1, 0)$, since the pair $(a, b) = (0, 0)$ doesn't satisfy the condition of coprimality.

Case 3. $a > b > 1$. Then $ab \leq a + b^2 + 1$, so that

$$b < a < \frac{b^2 + 1}{b - 1} = b + 1 + \frac{2}{b - 1}.$$

If $b = 2$, the equation $a^2 + a - 10 = 0$ is without integer solutions.

If $b = 3$, the equation $a^2 + 1 - 30 = 0$ has the solutions 5 and -6. But a, b are coprime, so that $(a, b) = (5, 3)$.

If $b > 3$, from $b < a < b + 2$, we deduce $a = b + 1$. But $b|a + 1$; hence $b|2$, contradiction.

342. The number $m = \varphi(6^{43})$ satisfies the desired condition. Multiplying 6^{43} by any number from the following list yields a solution of the equation $\varphi(n) = m$:

$$\frac{5}{4}, \frac{7}{6}, \frac{13}{12}, \frac{17}{16}, \frac{19}{18}, \frac{37}{36}, \frac{73}{72}, \frac{109}{108}, \frac{163}{162}, \frac{257}{256}, \frac{65537}{65536}.$$

Hence there are at least $2^{11} = 2048 > 2015$ solutions.

2.44 2016

343. (a) Consider the set $S = \{a_{i+1} - a_i, i \in \mathbb{N}^*\}$. Let d be the minimal element of S and let k be minimal such that $a_{k+1} - a_k = d$. Then $d < p$, since $d \leq a_2 - a_1 = a_1 < p$. Also, since the sequence is incresing, we must have $d > 0$.
Now, for large n such that $2^n > p$, consider $a_{2^n k}, a_{2^n k+1}, \ldots, a_{2^n(k+1)}$. Note that

$$\sum_{i=2^n k}^{2^n(k+1)-1} (a_{i+1} - a_i) = a_{2^n(k+1)} - a_{2^n k} = 2^n d.$$

Also,

$$\sum_{i=2^n k}^{2^n(k+1)-1} (a_{i+1} - a_i) \geq \sum_{i=2^n k}^{2^n(k+1)-1} d = 2^n d,$$

so we must have equality everywhere, that is, the sequence is an arithmetic sequence with common difference d. Since $2^n > p$ and $0 < d < p$, then some term of this arithmetic sequence must be divisible by p.
(b) Consider $a_i = pi + 2^{\lfloor \log_2 i \rfloor}$, which clearly satisfies $a_{2n} = 2a_n$ for all positive integers n. Note that for all i, we have $a_i \equiv 2^{\lfloor \log_2 i \rfloor} \not\equiv 0 \pmod{p}$. The first terms of this sequence are $p+1, 2p+2, 3p+4, 4p+4, 5p+8, 6p+8, 7p+8, 8p+8, 9p+16, \ldots$

344. Let $k \geq 3$ be an integer. The following representations hold:

$$6k = 2+3+(6k-5), 6k+2 = 3+4+(6k-5), 6k+4 = 2+3+(6k-1),$$
$$12k+1 = 9+(6k-1)+(6k-7), 12k+3 = 3+(6k-1)+(6k+1),$$
$$12k+5 = 9+(6k-5)+(6k+1), 12k+7 = 3+(6k-1)+(6k+5),$$
$$12k+9 = 9+(6k-1)+(6k+1), 12k+11 = 3+(6k+1)+(6k+7).$$

The number 17 does not have the required property. Indeed, if $17 = a+b+c$ with coprime positive integers a, b, and c, then a, b, and c are odd. For $a = 3$ we have

$$3+5+7 < a+b+c < 3+5+11,$$

which is impossible. If $a \geq 5$, then $b \geq 7$, $c \geq 9$, and $a+b+c \geq 21 > 17$, contradiction.

345. **Lemma.** There exist only finitely many positive integers b such that $j+b$ and $k+b$ have all prime factors in P.

Proof. Let

$$j+b = \prod p_i^{\alpha_i}, k+b = \prod p_i^{\beta_i};$$

whence $|j-k| = \left|\prod p_i^{\alpha_i} - \prod p_i^{\beta_i}\right|$, the number of solutions of which is bounded.

Back to the problem, we will inductively construct a set B that satisfies the required condition. Put one element of A in B. Suppose that $B = \{b_1, \ldots, b_n\}$. If we cannot put another element in B, then there exists an integer k and an infinite set S_k of integers a in A such that $k+a$ has all prime factors in P, which contradicts the lemma, so we can choose another element of A to put in B and we are done by induction.

346. If $p = q$, then $p^2 - p - 1 = 2p + 3$; whence $p \in \{-1, 4\}$, which are not solutions.

If $p \neq q$, then $p | 2q + 3$, so that $2q + 3 = kp$, for some positive integer k. The given equation is equivalent to

$$2p^2 - (k^2 + 2)p + 3k - 2 = 0;$$

and the discriminant $\Delta = (k^2 + 2)^2 - 8(3k - 2)$ must be a square. It is easy to see that for $k \geq 11$ we have $(k^2 + 1)^2 < \Delta < (k^2 + 2)^2$. Checking all cases $1 \leq k \leq 10$, we find the solution $k = 5$, $(p, q) = (13, 31)$, which is unique.

347. Note that $10^{4p} \equiv 1 \pmod{4p+1}$ implies that the order of 10 modulo $4p+1$ is either 1, 2, 4, p, $2p$, $4p$. Since $p > 10^9$, the order is restricted to $\{p, 2p, 4p\}$.

Let S be the reduced set of residues of the form 10^k modulo $4p+1$. Note that for any $s \in S$, $\frac{s}{4p+1}$ and $\frac{1}{4p+1}$ have the same repeating block in their decimal expansions, since they are shifts by some 10^k.

If 10 has order $4p$, $2p$, or p, respectively, then 10 can be written in the form g, g^2, and g^4, respectively, for some primitive root g. In all cases, the set $\{10^k\}$ will cover all residues of the form $g^{4k} \bmod 4p+1$, so every quartic residue is an element of S.

Consider all fourth powers $1^4, 2^4, \ldots,$ mod $4p+1$, which are at most $4p+1$. We claim that for each decimal digit d, there exists some k such that $\frac{d}{10} \leq \frac{k^4}{4p+1} < \frac{d+1}{10}$. This will imply that the first digit of $\frac{k^4}{4p+1}$ is d, and since $\frac{k^4}{4p+1}$ and $\frac{1}{4p+1}$ are cyclic shifts of each other, we will be able to conclude that $\frac{1}{4p+1}$ contains every digit d.

In order to prove the above claim, we need to show that the sequence $\{k^4\}$ never has a gap greater than $\frac{4p+1}{10}$. In other words, it suffices to prove that $(k+1)^4 - k^4 < \frac{4p+1}{10}$ for $(k+1)^4 < 4p+1$, which follows from $(k+1)^4 - k^4 \approx 4k^3$ and $p > 10^9$.

348. Answer: if $p = 2$, then $(2, 1, 1)$, $(2, 1, -1)$, $(2, 2, -1)$, and $(-2, 2, -1)$, and cyclic permutations are all solutions; if $p > 2$, then $(p, 1, 1)$ and $(-p, 1, -1)$ and cyclic permutations are all solutions.

If a, b, and c satisfy the equation, then $|a|^b |b|^c |c|^a = p$ and none of a, b, c can be zero. Furthermore, $\gcd(a, b, c) = 1$. Indeed, if $d = \gcd(a, b, c)$, then the exponent of p in the canonical representation of each positive rational number $|a|^b$, $|b|^c$, and $|c|^a$ is divisible by d. Since their product is p, it follows that $d = 1$.

Let $q \neq p$ be a prime, and let α, β, γ be the exponents of q in the canonical representation of $|a|$, $|b|$, and $|c|$, respectively. Then $\alpha b + \beta c + \gamma a = 0$ and not all exponents α, β, and γ are positive, since $\gcd(a, b, c) = 1$. Consequently, there are exactly two positive exponents among α, β, and γ. Assume that $\alpha > 0$, $\beta > 0$, $\gamma = 0$. Then $\alpha b + \beta c = 0$ implies $\alpha |b| = \beta |c|$; hence $|b|$ divides $\beta |c|$. Since q^β divides $|b|$ and is relatively prime to $|c|$, it follows that $q^\beta | \beta$ and $q^\beta \leq \beta$, contradiction. It follows that $\alpha = \beta = \gamma = 0$ and $|a|$, $|b|$, and $|c|$ are powers of p.

Then the equation becomes $p^{\alpha b} p^{\beta c} p^{\gamma a} = p$, where α, β, and γ are now the exponents of p in the canonical representation of $|a|$, $|b|$, and $|c|$, respectively; whence it follows that $\alpha b + \beta c + \gamma a = 1$. By $\gcd(a, b, c) = 1$, one of α, β, and γ must be zero, and, clearly, one of the summands αb, βc, and γa must be positive. Suppose that $\alpha b > 0$, i.e., $\alpha > 0$, $b > 0$.

Case 1. $\beta = \gamma = 0$. Then $b = |c| = 1$; therefore $\alpha = 1$; whence $|a| = p$. If $p > 2$, then the exponents of a and c in the given equation are both odd; whence a and c must have the same sign to make the product $a^b b^c c^a$ positive. Triples $(p, 1, 1)$ and $(-p, 1, -1)$ satisfy the given equation. If $p = 2$, then c^a is positive; hence $a > 0$. Triples $(2, 1, 1)$ and $(2, 1, -1)$ satisfy the given equation.

Case 2. $\beta = 0$, $\gamma > 0$. Then $b = 1$; hence $\alpha + \gamma a = 1$, so that $a < 0$. We get

$$p^\alpha \leq \gamma p^\alpha = \gamma |a| = \alpha - 1 < \alpha,$$

contradiction.

Case 3. $\beta > 0$, $\gamma = 0$. Then $|c| = 1$, $\alpha b + \beta c = 1$; whence $c = -1$, $\alpha p^\beta = 1 + \beta$. If $p > 2$, this leads to a contradiction, as above. If $p = 2$, then $\alpha = \beta = 1$; whence we get triples $(2, 2, -1)$ and $(-2, 2, -1)$ which satisfy the given equation.

349. We shall prove first that $kl \geq n$. Let $f(i)$ be the length of the longest increasing subsequence ending with a_i, and let $g(i)$ be the length of the longest decreasing subsequence ending with a_i, $1 \leq i \leq n$. For $i < j$, if $a_i < a_j$, then $f(i) < f(j)$ and $g(i) > g(j)$. Hence pairs $(f(i), g(i))$, $1 \leq i \leq n$, are distinct. The largest number of the form $f(i)$ is k and the largest number of the form $g(i)$ is l. Thus the number of pairs $(f(i), g(i))$ is at most kl; hence $n \leq kl$.

It follows that $k+l \geq 2\sqrt{kl} \geq 2\sqrt{n}$. At most one number can belong to an increasing and a decreasing subsequence simultaneously. Thus subsequences

$$(a_{i_1}, \ldots, a_{i_k}), (a_{j_1}, \ldots, a_{j_l})$$

contain together at least $2\sqrt{n} - 1$ different numbers. By assumptions, the number n has at most $\lfloor\sqrt{n}\rfloor - 1$ divisors that are not larger than $\sqrt{n}$, the total number of divisors $\tau(n) \leq 2\lfloor\sqrt{n}\rfloor - 2$. Consequently, subsequences $(a_{i_1}, \ldots, a_{i_k})$ and $(a_{j_1}, \ldots, a_{j_l})$ together contain at least one number that does not divide n.

350. Answer: $n = 1, 2$.

Denote $a_n = (n^2 + 11n - 4) \cdot n! + 33 \cdot 13^n + 4$. If $n \geq 4$, then 8 divides $n!$; hence

$$a_n \equiv 33 \cdot 13^n + 4 \equiv 5^n + 4 \pmod{8}.$$

From $5^2 \equiv 1 \pmod{8}$, it follows that $5^n \equiv 1 \pmod 8$ for all even n; therefore $a_n \equiv 5 \pmod 8$ and a_n cannot be a square, since the squares modulo 8 are 0, 1, and 4.

If $n \geq 7$, then $7 | n!$, so that $a_n \equiv 33 \cdot 13^n + 4 \equiv 5 \cdot (-1)^n + 4 \equiv -1 \pmod 7$ if n is odd and a_n cannot be a square, since the squares modulo 7 are 0, 1, 2, and 4.

Now we are left possible candidates $n \in \{1, 2, 3, 5\}$. From

$$a_1 = (1 + 11 - 4) \cdot 1 + 33 \cdot 13 + 4 = 441 = 21^2,$$
$$a_2 = (4 + 22 - 4) \cdot 2 + 33 \cdot 169 + 4 = 5625 = 75^2,$$
$$a_3 = (9 + 33 - 4) \cdot 6 + 33 \cdot 13^3 + 4 \equiv 3 + 3^4 - 1 \equiv 3 \pmod 5,$$
$$a_5 \equiv 33 \cdot 13^5 + 4 \equiv 3^6 - 1 \equiv 3 \pmod 5,$$

we conclude that $n = 1$ and $n = 2$ are the only solutions, since the squares modulo 5 are 0, 1, and 4.

351. Answer: $n = 7$.

Observe that $n^2 + 10 \equiv (n-2)(n+2) \pmod 7$, $n^2 - 2 \equiv (n-2)(n+2) \pmod 7$, $n^3 + 6 \equiv (n-1)(n^2+n+1) \pmod 7$, $n^5 + 36 \equiv (n+1)(n^4 - n^3 + n^2 - n + 1) \pmod 7$. Since any seven consecutive integers must contain a multiple of 7, it follows that the given numbers cannot be simultaneously prime for $n > 7$. By checking the values $n \leq 7$, we find that the only possibility is $n = 7$, for which the given numbers are the primes 7, 59, 47, 349, and 16,843.

352. We see that in the decomposition

$$2^m p^2 = q^7 - 1 = (q-1)(q^6 + q^5 + q^4 + q^3 + q^2 + q + 1)$$

the second factor of RHS is odd; therefore $q - 1 = 2^m$ or $q - 1 = 2^m p$. The latter is impossible, for otherwise we will have $2^m p^2 + 1 = q^7 = (2^m p + 1)^7 > 2^m p^7$.

Hence we have $q = 2^m + 1$ and so $2^m p^2 + 1 = (2^m + 1)^7 = 2^{7m} + 7 \cdot 2^{6m} + 21 \cdot 2^{5m} + 35 \cdot 2^{4m} + 35 \cdot 2^{3m} + 21 \cdot 2^{2m} + 7 \cdot 2^m + 1$. After some simplifications, we get

$$p^2 = 2^{6m} + 7 \cdot 2^{5m} + 21 \cdot 2^{4m} + 35 \cdot 2^{3m} + 35 \cdot 2^{2m} + 21 \cdot 2^m + 7.$$

If $m \geq 3$, then $p^2 \equiv 7 \pmod 8$, which is impossible since the quadratic residues modulo 8 are 0, 1, and 4. If $m = 2$, then $2^{12} + 7 \cdot 2^{10} + 21 \cdot 2^8 + 35 \cdot 2^6 + 35 \cdot 2^4 + 21 \cdot 2^2 + 7 = 19531$, which is not a perfect square. If $m = 1$, then $2^6 + 7 \cdot 2^5 + 21 \cdot 2^4 + 35 \cdot 2^3 + 35 \cdot 2^2 + 21 \cdot 2 + 7 = 1093$, which is not again a perfect square. We conclude that such triples do not exist.

353. Answer: $(p, q) \in \{(2, 3), (3, 2)\}$.
If $p = 2$, then $q^2 | 2^6 - 1$ implies $q = 3$.
If $q = 2$, then $p^2 | 2^3 + 1$ implies $p = 3$.
If $p \geq 3$, $q \geq 3$, then clearly $p \neq q$.
Case 1. $p < q$. Then $q \geq p + 2$ and

$$q^2 | p^6 - 1 \Rightarrow q^2 | (p-1)(p+1)(p^2 - p + 1)(p^2 + p + 1) \Rightarrow$$
$$\Rightarrow q^2 | (p^2 - p + 1)(p^2 + p + 1),$$

since $q \geq p + 2$. The only possibility is $q | p^2 - p + 1$, $q | p^2 + p + 1$; whence $q | 2p$, which is impossible.
Case 2. $p > q$. Then $p \geq q + 2$ and

$$p^2 | q^3 + 1 \Rightarrow p^2 | (q+1)(q^2 - q + 1) \Rightarrow p^2 | q^2 - q + 1,$$

since $p \geq q + 2$. But $p^2 \geq (q+2)^2 > q^2 - q + 2$, so that $p^2 \nmid q^2 - q + 2$, contradiction.

354. There is k such that $k^2 \equiv -5 \pmod p$. By the Thue's lemma, there exist x, y such that $0 < x, y < \sqrt{p}$ and $x \equiv ky \pmod p$. This implies

$$x^2 \equiv k^2 y^2 \equiv -5y^2 \pmod p,$$

that is, $p | x^2 + 5y^2$. Since $x^2 + 5y^2 < 6p$, it follows that $x^2 + 5y^2 = tp$, with $1 \leq t \leq 5$.
Case 1. $t = 5$. From $x^2 + 5y^2 = 5p$, it follows that $5 | x$, $5x'^2 + y^2 = p$ and we use the Brahmagupta's identity.
Case 2. $t = 4$. If both x and y are odd, then $2 \equiv 4p \pmod 4$, contradiction. Therefore both x and y are even and $x'^2 + 5y'^2 = p$.
Case 3. $t = 3$. We find x, y not divisible by p such that $x^2 + 5y^2 = 9p^2$. If one of x, y is divisible by p, then so is the other. Otherwise, looking mod 9, we see that at least one of $x' = 2x \pm 5y$ is divisible by 9, and the same holds for $y' = -x \pm 2y$. Then

$$xt^2 + 5yt^2 = 9(x^2 + 5y^2) = 81p^2.$$

But x' and y' are divisible by 9 and are not divisible by p, as is easy to see.

Case 4. $t = 2$. We find x, y are not divisible by p such that $x^2 + 5y^2 = 4p^2$. Looking in mod 4, we see that both x and y are even.

Case 5. $t = 1$. Obvious.

355. Assume that $x_1, \ldots, x_{n+1}$ are integers, and define the positive integers

$$a_k = x_k - 1 = \frac{m-n}{n+k}, 1 \leq k \leq n+1.$$

Let $P = x_1 x_2 \ldots x_{n+1} - 1$. We need to prove that P is divisible by an odd prime, i.e., P is not a power of 2. Let 2^d be the largest powe of 2 dividing $m - n$, and let 2^c be the largest power of 2 not exceeding $2n + 1$. Then $2n + 1 \leq 2^{c+1} - 1$; therefore $n + 1 \leq 2^c$. We conclude that 2^c is one of the numbers $n + 1, n + 2, \ldots, 2n + 1$ and that it is the only multiple of 2^c among these numbers. Define $l = 2^c - n$. Since $\frac{m-n}{n+l}$ is an integer, it follows that $d \geq c$. Therefore $a_l = \frac{m-n}{n+l}$ is not divisible by 2^{d-c+1}, but 2^{d-c+1} divides a_k for all $k \neq l, 1 \leq k \leq n+1$.

Then we have

$$P = (a_1 + 1)(a_2 + 1) \ldots (a_{n+1} + 1) - 1 \equiv (a_l + 1) \cdot 1^n - 1 \not\equiv 0 \pmod{2^{d-c+1}};$$

hence P is not divisible by 2^{d-c+1}. On the other hand, $P \geq a_k \geq 2^{d-c+1}$, so that P is not a power of 2.

356. Answer: 17.

It suffices to compute $\gcd((2+8)^{16} - 1, (1+8)^{16} - 1) = 17$.

Suppose that

$$2a^2 \cdot 19^{b^2} \cdot 53^{c^2} + 8 \equiv 2^{a^2+b^2+c^2} + 8 \equiv 0 \pmod{17}.$$

Examining modulo 4, we see that the only problem is the case $a^2 + b^2 + c^2 = 4k + 3$ with odd k, but this case is impossible.

357. (a) The numbers a, b, and c are not divisible by p. Suppose, without loss of generality, that $a < b < c$. From $p|(bc + 1) - (ab + 1)$, we deduce $p|b(c - a)$, and since $p \nmid b$, it follows that $p|c - a$ and similarly, $p|b - a$. It follows that $b = a + mp, c = a + np$ for some positive integers $m < n$. Then

$$\frac{a+b+c}{3} = a + \frac{(m+n)p}{3} \geq a + p,$$

and the result is true, unless $a = 1, b = p + 1, c = 2p + 1$, but then $ab + 1 = p + 2$ is not divisible by p.

(b) The same argument shows that equality holds only when $a = 2$, $b = p + 2$, $c = 2p + 2$. Then the odd prime p divides $ab + 1 = 2p + 5$ only for $p = 5$.

358. Let (x, y) be a solution of the equation for some n. From $xy \leq x^2 + y^2$, we deduce $n \leq 2016$. If $d = (x, y)$, then there are positive integers a, b such that $x = da$, $y = db$, $(a, b) = 1$. The equation becomes

$$(a^2 + b^2)^n = (ab)^{2016} d^{4032 - 2n}.$$

Since a and b are divisors of the left side and coprime to it, we deduce $a = b = 1$ and $2^n = d^{4032 - 2n}$.

Conversely, if $2^n = d^{4032 - 2n}$, then $x = y = d$ is a solution of the given equation. It follows that $d = 2^k$ for some k; whence

$$n = \frac{4032k}{2k + 1}.$$

Since $(k, 2k + 1) = 1$, it follows that $2k + 1$ divides $4032 = 2^6 \cdot 63$. Analyzing all cases, we find $n \in \{1344, 1728, 1792, 1920, 1984\}$.

359. We shall prove by induction on k. Since $a \equiv 3 \pmod{8}$, $m = 1$ works for $k = 1$, 2, 3. Let $k \geq 3$ and let m be a positive integer such that $2^k | a^m + a + 2$.

If $(a^m + a + 2)/2^k$ is even, then clearly $2^{k+1} | a^m + a + 2$.

If $(a^m + a + 2)/2^k$ is odd, then we shall prove that $2^{k+1} | a^{m+2^{k-2}} + a + 2$. To this end, write $a^{m+2^{k-2}} + a + 2 = a^{2^{k-2}}(a^m + a + 2) - (a + 2)\left(a^{2^{k-2}} - 1\right)$. The first term is an odd multiple of 2^k, and it suffices to prove that so is the second.

Induction on $k \geq 3$ shows that $a^{2^{k-2}} - 1$ is an odd multiple of 2^k. Since $a \equiv 3 \pmod{8}$, this is clearly the case for $k = 3$, and the induction step follows from the identity

$$a^{2^{k-1}} - 1 = \left(a^{2^{k-2}} - 1\right)\left(\left(a^{2^{k-2}} - 1\right) + 2\right).$$

360. We write $2^r \| x$ when $2^r | x$ and $2^{r+1} \nmid x$. We call (a, b) a (k, l) − pair if $a + w(b + 1)$ and $b + w(a + 1)$ are powers of two and $2^k \| a + 1$, $2^l \| b + 1$. Let (a, b) be a (k, l)-pair, and let

$$a + 1 = 2^k c, b + 1 = 2^l d, a + w(b + 1) = 2^k c + d - 1 = 2^m,$$

$$b + w(a + 1) = 2^l d + c - 1 = 2^n. \quad (*)$$

We see that $c = 1$ if and only if $d = 1$ and in this case $(a, b) = (2^k - 1, 2^l - 1)$.

If $c, d > 1$, then $(*)$ implies that $2^k \| d - 1 = 2^k b'$, $2^l \| c - 1 = 2^l a'$ for some odd a', b'; when substituted in $(*)$, these yield $2^l a' + b' + 1 = 2^{m-k}$, $2^k b' + a' + 1 = 2^{n-l}$. From this we get $2^k \| a' + 1$, $2^l \| b' + 1$, so the above relations become

$$a' + w(b' + 1) = a' + \frac{b' + 1}{2^l} = 2^{m-k-l}, \quad b' + w(a' + 1) = 2^{n-k-l}.$$

Thus (a', b') is a (k, l)–pair with $a' < a$, $b' < b$. Define

$$a_1 = a, \quad b_1 = b, \quad a_{n+1} = \frac{a_n + 1 - 2^k}{2^{k+l}}, \quad b_{n+1} = \frac{b_n + 1 - 2^l}{2^{k+l}}.$$

By the above considerations, each pair (a_i, b_i) is a (k, l)-pair and $a_n = 2^k - 1$, $b_n = 2^l - 1$ for some n. Going backward, we find

$$a = \frac{\left(2^{n(k+l)} - 1\right)\left(2^k - 1\right)}{2^{k+l} - 1}, \quad b = \frac{\left(2^{n(k+l)} - 1\right)\left(2^l - 1\right)}{2^{k+l} - 1}.$$

Since $\gcd(a, b) = 1$, we finally get $n = 1$, $a = 2^k - 1$, $b = 2^l - 1$.

361. We shall use mathematical induction. For $n = 0$, the number $7^{7^0} + 1 = 2^3$ has $2 \cdot 0 + 3 = 3$ prime factors. For the induction step, it suffices to show that if $x = 7^{2m-1}$ for a positive integer m; then $(x^7 + 1)/(x + 1)$ is composite. We have

$$\frac{x^7 + 1}{x + 1} = (x+1)^6 - \frac{7x(x^5 + 3x^4 + 5x^3 + 5x^2 + 3x + 1)}{x + 1} =$$
$$= (x+1)^6 - 7x(x^4 + 2x^3 + 3x^2 + 2x + 1) =$$
$$= (x+1)^6 - 7^{2m}(x^2 + x + 1)^2 =$$
$$= \left[(x+1)^3 - 7^m(x^2 + x + 1)\right]\left[(x+1)^3 + 7^m(x^2 + x + 1)\right].$$

Now it suffices to show that the first factor is greater than 1. Indeed, $\sqrt{7x} \leq x$ implies

$$(x+1)^3 - 7^m(x^2+x+1) = (x+1)^3 - \sqrt{7x}(x^2+x+1) \geq$$
$$\geq x^3 + 3x^2 + 3x + 1 - x(x^2+x+1) = 2x^2 + 2x + 1 \geq 113 > 1.$$

It follows that $x^7 + 1$ has at least two prime factors more than $x + 1$, which ends the proof.

362. Case 1. n even. Clearly, $n > 2$, since $n = 2$ is not a solution. Then $1, 2, n/2$ are divisors of n. From

$$1^2 + 2^2 + \left(\frac{n}{2}\right)^2 \leq 5(n+1),$$

we deduce $n \leq 20$. Direct calculations show that $n = 16$ is a solution.

Case 2. n odd. Then all the divisors of n are odd and their number is also odd; therefore n must be a square. But n cannot be the power of a prime p, since then

$$\text{LHS} \equiv 1 \pmod{p} \text{ and RHS} \equiv 5 \pmod{p},$$

which is impossible. It follows that $n = (ps)^2$, with p prime and s having at least a prime divisor different from p. From $p, s > 2$, we deduce

$$5 + 9(ps)^2 = 1 + 4 + 9(ps)^2 \leq 1 + p^2 + s^2(ps)^2 \leq 5\left((ps)^2 + 1\right),$$

contradiction. In this case there are no solutions.

363. Answer: $(a, b) = (4, 2)$.

If $a - b = 1$, then $a = b = 1$, contradiction.
If $a - b \geq 2$, then

$$\ln(a - b) = \frac{\ln a}{a} + \frac{\ln b}{b}.$$

For $b \geq e$, we have

$$\ln 2 \leq \ln(a - b) < \frac{2 \ln b}{b};$$

hence $b < 4$, that is, $1 \leq b \leq 3$. Observe that a, b must be both even, so that $b = 2$. This gives $(a - 2)^{2a} = a^2 \cdot 2^a$. If a prime $p \neq 2$ divides a, then p divides $a - 2$, contradiction. So a and $a - 2$ are powers of 2; whence $a = 4$.

364. The following implications hold:

$$2^n - n^2 \mid a^n - n^a \Rightarrow 2^n - n^a \mid a^{2n} - n^{2a} \Rightarrow 2^n - n^a \mid a^{2n} - 2^{na},$$

since $2^n - n^a \mid 2^{na} - n^{2a}$. Choose a prime $p > \max\{a^2, 2^a\}$ and take $n = p - 1$. Then

$$2^{p-1} - (p-1)^2 \mid a^{2(p-1)} - 2^{(p-1)a} \Rightarrow p \mid 2^a - a^2 \Rightarrow 2^a - a^2 = 0 \Rightarrow a \in \{2, 4\},$$

by the Fermat theorem. Checking, we find that both values are solutions.

365. By the substitutions $u = x + y$, $v = x - y$, the equation $u^2 + v^2 = 2p^2$ becomes $x^2 + y^2 = p^2$, that is, (x, y, p) is a Pythagorean triple. Then there exist positive integers $m > n$ such that $x = m^2 - n^2$, $y = 2mn$, $p = m^2 + n^2$. We have

$$2p - u - v = 2(p - x) = (2n)^2 \text{ or } 2p - u - v = 2(p - y) = 2(m - n)^2,$$

as desired.

366. The function $\varphi(x)$ decreases $\nu_2(x)$ by at most 1, and only when x is a power of 2. Let $f(x)$ be the maximal possible value of ν_2 upon repeatedly applying φ to x. It follows that $k \geq f(x)$, with equality only when x is a power of 2. Now, we will prove that $f(x) \geq \log_3 x$, using induction on the largest prime dividing x.

The base cases $x = 1$ and $x = 2^k$ are trivial. Suppose that the result holds for all x with only prime divisors $p_i < p$; then for any x with largest prime divisor p, we have

$$f(x) = f(p^m) + f\left(\frac{x}{p^m}\right) \geq mf(p-1) + \log_3\left(\frac{x}{p^m}\right) \geq \log_3 x,$$

where $m = \nu_p(x)$.

367. Clearly, $(a, 3^{2016})$ is a solution for all a.

If $n > 3^{2016}$, then Zsigmondy's theorem implies that there is a prime divisor p of $a^n - 1$, not dividing $a^{3^{2016}} - 1$, contradiction.

If $n < 3^{2016}$, then let $d = \gcd(n, 3^{2016})$. For $n \nmid 3^{2016}$, we have $d < n$ and Zsigmondy's theorem yields a contradiction. Then $n | 3^{2016}$, which is a contradiction for all a.

2.45 2017

368. Rewrite the equation As $20a^3 = b^3 + 1 = (b+1)(b^2 - b + 1)$. Then $b^3 \equiv -1$ (mod 20), so that $b = 20k - 1$; whence $a^3 = k(b^2 - b + 1)$, with k integer. If $k = 1$, then $(a, b) = (7, 19)$. If $k = a$, then $k^2 = (20k - 1)^2 - (20k - 1) + 1$, without real solutions.

369. Case 1. $d_2 = p$, $d_3 = p^2$, with p prime. Then d_5 divides $d_2 d_3$; therefore $d_5 = p^3$, $d_4 = p^2 + p + 1 = q$ is prime, $n = p^3 q$, so that $k = 8$.

Case 2. $d_2 = p$, $d_3 = q$, p, q primes. Then d_5 divides pq; therefore $d_5 = pq$ and $n = pq(p + q + 1)$. If $(p, q) = (2, 3)$, then $n = 36$, $k = 9$. If $(p, q) \neq (2, 3)$, then

2.1. $d_4 = p^2$, $q = p^2 - p - 1$, $d_5 = pq$, $n = p^3 q$, $k = 8$.
2.2. $d_4 = p + q + 1 = r$ is prime, $n = pqr$; therefore $k = 8$.

370. Suppose that the assertion is false. Then there is a positive integer N such that $ap + b$ is prime for every prime $p > N$. It follows that

$$a(ap+b)+b, a(a(ap+b)+b)+b, \ldots, a^t p + b \cdot \frac{a^t - 1}{a-1} = P_t$$

is a prime for every positive integer t. Taking $t = ap + b$, we deduce that $ap + b | P_t$, contradiction.

371. Answer: 2 and all composite numbers.

Let $m = 3^x p_1^{a_1} \ldots p_s^{a_s} q_1^{b_1} \ldots q_s^{b_s}$ be the prime factorization of m, where $p_i \equiv 1 \pmod{3}$, $1 \leq i \leq s$, $q_j \equiv 2 \pmod{3}$, $1 \leq j \leq t$. Then

$$\tau_1(m) = \prod_{i=1}^{s}(a_i + 1)\left\lceil \frac{1}{2}\prod_{j=1}^{t}(b_j) + 1 \right\rceil. \quad (1)$$

Indeed, a divisor of m congruent to 1 modulo 3 does not contain 3 as prime divisor and contain some of p_i and an even number of q_j (counted with multiplicity).

If $\prod_{j=1}^{t}(b_j + 1)$ is even, then we may assume wlog that $b_1 + 1$ is even. The factors $q_2, q_3, \ldots, q_t$ can be chosen in $\prod_{j=2}^{t}(b_j + 1)$ ways. Then there are $\frac{1}{2} \times (b_1 + 1)$ ways to choose the exponent of q_1; hence (1) is verified.

If $\prod_{j=1}^{t}(b_j + 1)$ is odd, we use the induction on t to count the number of choices. When $t = 1$, there are $\left\lceil \frac{1}{2}(b_1 + 1) \right\rceil$ choices for which the exponent is even and $\left\lfloor \frac{1}{2}(b_1 + 1) \right\rfloor$ choices for which the exponent is odd. For the inductive step, we find that there are

$$\left\lceil \frac{1}{2}\prod_{j=1}^{t-1}(b_j + 1) \right\rceil \cdot \left\lceil \frac{1}{2}(b_t + 1) \right\rceil + \left\lfloor \frac{1}{2}\prod_{j=1}^{t-1}(b_j + 1) \right\rfloor \cdot \left\lfloor \frac{1}{2}(b_t + 1) \right\rfloor$$

$$= \left\lceil \frac{1}{2}\prod_{j=1}^{t}(b_j + 1) \right\rceil$$

choices with an even number of prime factors and hence $\left\lfloor \frac{1}{2}\prod_{j=1}^{t}(b_j + 1) \right\rfloor$ choices with an odd number of prime factors; hence (1) is also true.

Let $n = 3^x 2^y 5^z p_1^{a_1} \ldots p_s^{a_s} q_1^{b_1} \ldots q_t^{b_t}$ be the prime factorization of n. Using a known formula and (1), we have

$$\tau(10n) = (x+1)(y+2)(z+2)\prod_{i=1}^{s}(a_i+1)\prod_{j=1}^{t}(b_j+1), \qquad (2)$$

$$\tau_1(10n) = \prod_{i=1}^{s}(a_i)+1\left[\frac{1}{2}(y+2)(z+2)\prod_{j=1}^{t}b_j+1\right]. \qquad (3)$$

If $c = (y+2)(z+2)\prod_{j=1}^{t}(b_j+1)$ is even, then (2) and (3) imply

$$\frac{\tau(10n)}{\tau_1(10n)} = 2(x+1).$$

In this case, $\frac{\tau(10n)}{\tau_1(10n)}$ can be any even positive integer.

If c is odd, which means y, z odd and each b_j even, then (2) and (3) imply

$$\frac{\tau(10n)}{\tau_1(10n)} = \frac{2(x+1)c}{c+1}, \qquad (4)$$

which is an integer only when $c+1$ divides $2(x+1)$, since $\gcd(c, c+1) = 1$. For $2(x+1) = k(c+1)$, equality (4) becomes

$$\frac{\tau(10n)}{\tau_1(10n)} = kc = k(y+2)(z+2)\prod_{j=1}^{t}(b_j+1). \qquad (5)$$

The integers $y + 2$, $z + 2$ are at least 3, since y and z are odd, which shows that kc must be composite. On the other hand, for any odd composite number ab, with $a, b \geq 3$, we can take $n = 3^{\frac{ab-1}{2}} \cdot 2^{a-2} \cdot 5^{b-2}$ so that $\frac{\tau(10n)}{\tau_1(10n)} = ab$ from (5).

372. **Answer:** $n + \nu_p(n!)$.

The condition $m \geq n + \nu_p(n!)$ is necessary: take $f(x) = (x+p)(x+2p)\ldots(x+np)$ and $k = 1$. Whenever $p | f(k')$, then $p | k'$. Let $k' = pr$, then

$$f(k') = p^n(r+1)(r+2)\ldots(r+n)$$

implies $p^n \cdot n! \,|\, f(k')$, so $m \geq n + \nu_p(n!)$, as desired.

Now, we show that $m = n + \nu_p(n!)$ is the answer. The following lemma is a special case of the problem.

Lemma. If $n, a_1, a_2, \ldots, a_n$ are positive integers, then there exists a positive integer x such that $f(x) = (x+a_1)(x+a_2)\ldots(x+a_n)$ is divisible by p, but not by $p^{n+\nu_p(n!)+1}$.

Proof. We can choose x (mod p^i) such that p^i divides at most $\left\lfloor \frac{n}{p^{i-1}} \right\rfloor$ numbers. Induction on i: $x \equiv -a_1$ (mod p) proves the base case. Suppose that we have x (mod p^{i-1}) such that $x + b_1, x + b_2, \ldots, x + b_k$ are divisible by p^{i-1}, with $k \leq \left\lfloor \frac{n}{p^{i-2}} \right\rfloor$. By the Pigeonhole principle, there exists a residue class y (mod p) which contains at most $\left\lfloor \frac{k}{p} \right\rfloor = \left\lfloor \frac{n}{p^{i-1}} \right\rfloor$ of numbers

$$\frac{x+b_1}{p^{i-1}}, \frac{x+b_2}{p^{i-1}}, \ldots, \frac{x+b_k}{p^{i-1}};$$

hence $yp^{i-1} + x$ (mod p^i) satisfies the required condition. This process repeats until there are no factors divisible by p^i, so

$$\nu_p(f(x)) \leq n + \left\lfloor \frac{n}{p} \right\rfloor + \left\lfloor \frac{n}{p^2} \right\rfloor + \cdots = n + \nu_p(n!).$$

Back to the main problem, fix a positive integer k and let $c_i = \nu_p(x + a_i)$. We can suppose that $c_1 \geq c_2 \geq \cdots \geq c_n$. By suitable translation, we can assume that $a_1 = 0$. Take $k' = p^{c_1} r$. We see that

$$f(k') = p^{\nu_p(f(k))} r \left(p^{c_1 - c_2} r + \frac{a_2}{p^{c_2}} \right) \cdots \left(p^{c_1 - c_n} + \frac{a_n}{p^{c_n}} \right).$$

If $c_1 > c_i$ for some i, then $p^{c_j} \| a_j$ for any $j \geq i$, so the later $n - i + 1$ brackets never divisible by p; thus we can ignore them. Otherwise, $c_1 = c_2 = \cdots = c_n = c$, which implies

$$f(k') = p^{\nu_p(f(k))} r \left(r + \frac{a_2}{p^c} \right) \cdots \left(r + \frac{a_n}{p^c} \right).$$

By the lemma, we can choose r such that ν_p of the right product not exceed $n + \nu_p(n!)$ and is greater than 0, which ends the proof.

373. We shall use the well-known result: $\tau(n) = n^{o(1)}$, that is, $\frac{\tau(n)}{n^\varepsilon}$ is bounded, for any $\varepsilon > 0$. It follows that $|D_n| = n^{o(1)}$; therefore $|D_n - D_n| = n^{o(1)}$, as well; hence $f(n) = n^{o(1)}$.

374. Answer: n.
Consider $a = 1, b = 2$. If $2n | ax + by$, then

$$2n \leq x + 2y \leq 2(x + y) \leq 2m;$$

hence $m \geq n$.

It remains to show that $m \leq n$. Fix $a, b \in \{1, 2, \ldots, 2n - 1\}$.
Case 1. $\gcd(a, 2n) > 1$ or $\gcd(b, 2n) > 1$. If $\gcd(a, 2n) > 1$, then choose $x = \frac{2n}{\gcd(a, 2n)}$, $y = 0$, so

$$x+y = \frac{2n}{\gcd(a,2n)} \le \frac{2n}{2} = n.$$

Case 2. $\gcd(a,2n) = \gcd(b,2n) = 1$. Let $0 \le c \le 2n-1$ such that $c \equiv ba^{-1} \pmod{2n}$. The congruence $ax + by \equiv 0 \pmod{2n}$ is equivalent to $x + cy \equiv 0 \pmod{2n}$. Choose $y = \lfloor \frac{2n}{c} \rfloor, x = 2n - c\lfloor \frac{2n}{c} \rfloor$. Since $a \ne b$, we have $c \ge 2$. Also, $c \nmid 2n$, since $c > 1$ and $\gcd(c,2n) = 1$. In particular, $c \notin \{1, 2, n\}$.

If $2 < c < n$, then

$$x + y \le c - 1 + y < c - 1 + \frac{2n}{c} = \frac{(c-2)(c-n)}{c} + n + 1 < n + 1.$$

If $c \ge n+1$, then

$$x + y = 2n - (c-1)\left\lfloor \frac{2n}{c} \right\rfloor = 2n - (c-1) \le 2n - (n+1-1) = n.$$

375. Answer: no.

Suppose that $5^x \cdot 10^n + 5^y = 5^z$, where 5^y has n digits. Then

$$5^{x+n} \cdot 2^n = 5^y(5^{z-y} - 1);$$

whence $2^n = 5^{z-y} - 1$. Case $n = 1$ does not work. For $n = 2$, we get $z - y = 1$. Since 5^y has two digits, the only possibility is $y = 2, z = 3$; whence $x = 0$, which is not positive. If $n > 2$, then $5^{z-y} \equiv 1 \pmod 8$; therefore $z - y = 2k$, with k positive integer. Now $2^n = 25^k - 1 = 24(25^{k-1} + \cdots + 1)$, impossible, since $3 | 24$.

376. Answer: (a) $6^{n(n-1)}$; (b) 1, if $n = 1$; 6, if $n = 2$; $6^{(n+1)(n-2)}$, if $n \ge 3$.

(a) Let's fill the top left $(n-1) \times (n-1)$ subtable arbitrarily; this can be done in $6^{(n-1)^2}$ ways. Now there are three ways to fill each of the top $n-1$ cells of the rightmost column and two ways to fill each of the left $n-1$ cells of the bottom row to satisfy the requirements. The value of the last empty cell in the bottom right is then uniquely determined (mod 2 by the bottom row and mod 3 by the rightmost column). In conclusion, there are $6^{(n-1)^2} \cdot 3^{n-1} \cdot 2^{n-1} = 6^{n(n-1)}$ ways to fill the table.

(b) For $n = 1$, the only solution is writing 0 into the single cell. For $n = 2$, let a be the top left number. The bottom right must then be $(6-a) \bmod 6$. Using the conditions for rows and columns, for the top right number x, we get the equations $x \equiv -a \pmod 2$ and $x \equiv a \pmod 3$, and for the bottom left number y, $y \equiv a \pmod 2$ and $y \equiv -a \pmod 3$. The Chinese remainder theorem determines x and y uniquely, and we see from the equations that their sum is also divisible by 6. Thus there are 6 ways to fill the table in this case, one for each value of a.

Consider now $n \geq 3$. Fill the top left $(n-1) \times (n-1)$ subtable arbitrarily; this can be done in $6^{(n-1)^2}$ ways. The bottom right cell's value is uniquely determined by other values on the failing diagonal. Denote the value in the top left cell by a, the sum of the second to n-first cells (inclusive) in the top row by b, the sum of the second to n-first cells in the leftmost column by c, and the sum of the second to n-first cells on the rising diagonal by d.

Using the Chinese remainder theorem, fill the top right cell with the unique value x such that $x \equiv -a - b \pmod{2}$ and $x \equiv a + c - d \pmod{3}$, and the bottom left cell with the unique value y such that $y \equiv a + b - d \pmod{2}$ and $y \equiv -a - c \pmod{3}$. The divisibility conditions are now fulfilled for the top row, the leftmost column and both diagonals (the rising diagonal is verified by summing mod 2 and mod 3 separately).

Now, we leave one cell both in the rightmost column and in the bottom row empty for the time being. For the other $n-3$ empty cells in the rightmost column, there are three possible values for each, and for the other $n-3$ empty cells in the bottom row, two values for each. Having made all those choices (which can be done in $3^{n-3} \cdot 2^{n-3}$ ways), the values for the two remaining cells are now uniquely determined (mod 2 by the values in the respective row and mod 3 by the column). The total number of ways to fill the table is $6^{(n-1)^2} \cdot 3^{n-3} \cdot 2^{n-3} = 6^{(n+1)(n-2)}$.

377. The equality

$$A = \binom{4n}{2n} \cdot \binom{2n}{n} \cdot n!$$

shows that A is an integer. For the second part, we shall use the formula $\nu_2(n!) = n - s_2(n)$, where $s_2(n)$ is the sum of binary digits of n. Then

$$\nu_2(A) = \nu_2((4n)!) - [\nu_2((2n)!) + \nu_2(n!)] =$$
$$= 4n - s_2(4n) - [2n - s_2(2n) + n - s_2(n)] = n + s_2(n) \geq n + 1,$$

since $s_2(4n) = s_2(2n) = s_2(n)$. Equality holds when $s_2(n) = 1$, that is, n is a power of 2.

378. Answer: 6.

Define

$$a_n = \frac{1}{n+1} + \frac{1}{n+2} + \cdots + \frac{1}{2n}.$$

Then

2.45 2017

$$a_n - a_{n-1} = \frac{1}{2n(2n-1)},$$

so that

$$a_n = \sum_{k=1}^{n} \frac{1}{2k(2k-1)} < \sum_{k=1}^{\infty} \frac{1}{2k(2k-1)} = \sum_{k=1}^{\infty} \frac{(-1)^{k-1}}{k} = \ln 2 < 0,7.$$

On the other hand,

$$a_{1008} = \sum_{k=1}^{1008} \frac{1}{2k(2k-1)} > \frac{1}{2} + \frac{1}{12} + \frac{1}{30} > 0,6.$$

379. We will prove that n satisfies the required condition if and only if $n = 2^k$ for some positive integer k. Indeed,

$$2^n - 1 = 2^{2^k} - 1 = \left(2^{2^{k-1}} + 1\right)\left(2^{2^{k-2}} + 1\right) \cdots \cdot 5 \cdot 3,$$

and it is well known that $p \mid 2^{2^l} + 1$ implies $p \equiv 1 \pmod{2^{l+1}}$; therefore we can find an integer m, by quadratic reciprocity.

Conversely, $n' \mid n$ if and only if $2^{n'} - 1 \mid 2^n - 1$ with $n' > 2$ odd. It follows that there is a prime $p \equiv 3 \pmod 4$, $p \neq 3$, so that $p \mid 2^{n'} - 1 \mid 2^n - 1$.

380. Suppose that $p \geq q \geq r > 1$. If $r \geq 3$, then $r \mid p! + q! + r!$, contradiction; therefore $r = 2$. If $q \geq 4$, then $4 \mid p! + q!$, but $4 \nmid 2!$, so that $q = 3$.

Now, $p \leq 5$, since $\nu_2(2^s) \geq \min(\nu_2(p!), \nu_2(q! + r!))$.

At the end, we find the solutions $(p, q, r, s) = (2, 3, 4, 5), (2, 3, 5, 7)$.

381. Denote

$$X = \sum_{cyc} \frac{a}{a+b}, \quad Y = \sum_{cyc} \frac{b}{a+b}.$$

Then

$$X > \sum \frac{a}{a+b+c+d} = 1$$

and similarly, $Y > 1$. From $X + Y = 4$, it follows that X is an integer if and only if $X = 2$. We have

$$X - Y = \frac{2(ac-bd)(a-c)(b-d)}{\prod\limits_{cyc}(a+b)} = 0$$

if and only if $ac = bd$. By the four numbers theorem, there exist integers p, q, r, and s such that $a = pq$, $b = qr$, $c = rs$, and $d = sp$. We deduce that

$$a + b + c + d = (p + r)(q + s)$$

is not a prime number.

382. We have $a_k a_{k+2} - a_{k+1}^2 = 5^k(4mn - 5m^2 - n^2)$. Since $p \equiv 1 \pmod{4}$, there exist m, n such that $m^2 \equiv -1 \pmod{p}$, $n \equiv 2m + 1 \pmod{p}$. Then

$$a_k a_{k+2} - a_{k+1}^2 \equiv 0 \pmod{p}.$$

If $p | a_{k+2}$, then $p | a_{k+1}$ and finally, $p | m$, contradiction

383. Assume $x, y > 0$. Then $c = x + y$, $d = x - y$ have the same parity and $c > d$. It suffices to show that a solution with $d > 0$ exists if and only if a solution with $d < 0$ exists. The equation rewrites as

$$4a = ky^2 - (k-1)x^2 = k\left(\frac{c-d}{2}\right)^2 - (k-1)\left(\frac{c+d}{2}\right)^2 = c^2 + d^2 - (4k-2)cd \Leftrightarrow$$
$$\Leftrightarrow c^2 - (4k-2)d \cdot c + d^2 - 4a = 0.$$

It follows that a solution (c, d) generates another solution:

$$(c, d) \to \left(d, \frac{d^2 - 4a}{c}\right).$$

If we take a minimal possible solution $c > d > 0$, then we have

$$\frac{d^2 - 4a}{c} < \frac{d^2}{c} < c.$$

If we take a maximal negative solution $c > 0 > d$, then

$$|c| > |d| \Rightarrow \frac{d^2 - 4a}{c} < \frac{d^2}{c} < c, |c| < |d| \Rightarrow \frac{c^2 - 4a}{d} > \frac{c^2}{d} > d.$$

384. Answer: no.

Consider an arbitrary sequence $n_1 < n_2 < \ldots$ with at least an integer i for which $2^{n_i} \equiv 2 \pmod{p_i}$, and let x be a solution of all the equations. Then we have $x \neq 2$. If p_k is a prime factor of $x - 1$, then $x^{n_k} \equiv 2 \pmod{p_k}$. But $x^{n_k} \equiv 1 \pmod{p_k}$, contradiction.

385. The identity
$$n(n-1) = \frac{n^2(n^2-1)}{n(n+1)}, n > 1,$$
proves the result.

386. By substitutions $a = dpqx$, $b = dpry$, $c = dqrz$, we get
$$n = \frac{ab}{\gcd(a,b)} + \frac{bc}{\gcd(b,c)} + \frac{ca}{\gcd(c,a)} = dpqr(xy + yz + zx).$$

We see that n cannot be a power of 2, if x, y, and z are pairwise coprime. On the other hand, choosing $y = z = 1$, we obtain $n = M(2x+1)$, with $M = dpqr$, so that any number which is not a power of 2 admits such a representation.

387. The given number is $a_p = (2-\sqrt{5})^p + (2+\sqrt{5})^p - 2^{p+1}$, since a_p is an integer (by the binomial expansion) and $2-\sqrt{5}$ is a negative number, in absolute value less than 1. The sequence (a_n) defined by
$$a_0 = a_1 = 0, a_2 = 10, a_{n+3} = 6a_{n+2} - 7a_{n+1} - 2a_n, n \geq 0,$$
satisfies $20 | a_n$ for all odd n. In particular, $20 | a_p$.

Now, checking the case $p = 5$ separately, it suffices to prove that $p | a_p$. But
$$a_p = 2 \sum_{k=1}^{\frac{p-1}{2}} \binom{p}{2k} 5^k 2^{p-2k}$$
is clearly divisible by p, as each of the binomial coefficients is, hence the result.

Generalization. The number $\lfloor (\sqrt{n^2+1} + n)^p - 2n^p \rfloor$ is divisible by $2n(n^2+1)p$.

388. Two consecutive Fibonacci numbers x and y always satisfy $x^2 - xy - y^2 = \pm 1$ (*Cassini*). The converse statement is also true: all solutions of the above equation are consecutive Fibonacci numbers (*Wasteels*).

Note that $x^4 - x^2y^2 + y^4 + 2x^3y - 2xy^3 = (x^2 + xy - y^2)^2 = 1$.

389. We have $p^2 + 5pq + 4q^2 = k^2 \iff (k - p - 2q)(k + p + 2q) = pq$. Since $k + p + 2q > p$ and $k + p + 2q > q$, we must have $k - p - 2q = 1$, yielding $(p-4)(q-2) = 9$, with solutions $(p,q) = (13,3), (7,5), (5,11)$.

390. (a) If $a^2 \equiv b^2 \pmod{p}$, then $a - b \equiv 0 \pmod{p}$ or $a + b \equiv 0 \pmod{p}$, which is impossible, since $0 < |a-b| < p$ and $0 < a+b < p$.
(b) If $a^3 \equiv b^3 \pmod{p}$, then $a^2 + ab + b^2 \equiv 0 \pmod{p}$, so that
$$(a+b)(a+b+c) = a^2 + ab + b^2 + ab + bc + ca \equiv 0 \pmod{p},$$
which is impossible, since $0 < a+b < p$, $0 < a+b+c < p$.

391. It is easy to see that integers $n = 1, 2, 3, 4$ satisfy the given condition. We shall prove that if $n > 4$ satisfies the given condition, then $2n - 5$ is a prime number. Suppose the contrary and study two cases.

Case 1. $2n - 5$ has an odd prime divisor $p > 3$. We have $p \leq \frac{2n-5}{3} < n$, so that $p | n!$, $p \nmid n! + 1$ and $p \nmid 3(n! + 1)$. We deduce that $2n - 5 \nmid 3(n! + 1)$.

Case 2. $2n - 5 = 3^k$, with $k > 1$ a positive integer. Then $3 \nmid n! + 1$; therefore $2n - 5$ cannot satisfy the given condition.

If $q = 2n - 5$ is a prime number, then $q | n! + 1$, that is $n! \equiv -1 \pmod{q}$. The Wilson theorem states $(q - 1)! \equiv -1 \pmod{q}$, so that

$$-1 \equiv (2n-6)! \equiv (2n-6)(2n-7)\ldots(n+1)n! \equiv (-1)(-2)\ldots(-n+6)n! \equiv$$
$$\equiv (-1)^{n-6}(n-6)! \cdot n! \equiv (-1)^{n-6}(n-6)!(-1) \pmod{q};$$

whence $(n - 6)! \equiv (-1)^{n-6} \pmod{q}$. Since $n! \equiv -1 \pmod{q}$, we deduce

$$n(n-1)\ldots(n-5) \equiv (-1)^{n-1} \pmod{q}.$$

Multiplying by 2^6, we get

$$2n(2n-2)\ldots(2n-10) \equiv 5 \cdot 3 \cdot 1 \cdot (-1)(-3)(-5) \equiv -225$$
$$\equiv (-1)^{n-1} \cdot 64 \pmod{q}.$$

If n is odd, then q divides $225 + 64 = 289 = 17^2$; therefore $q = 17$ and $n = 11$.

If n is even, then q divides $225 - 64 = 161 = 7 \cdot 23$; therefore $q = 7, n = 6$ or $q = 23, n = 14$. We shall check these values:

$(q, n) = (17, 11)$: $11! = 1 \cdot (2 \cdot 9) \cdot (3 \cdot 6) \cdot (5 \cdot 7) \cdot 4 \cdot 8 \cdot 10 \cdot 11 \equiv 4 \cdot 8 \cdot 10 \cdot 11 \equiv$
$\equiv 88 \cdot 40 \equiv 3 \cdot 6 \equiv 1 \pmod{17}$,

so that $n = 11$ doesn't satisfy the condition.

$(q, n) = (23, 14)$: $14! = 1 \cdot (2 \cdot 12) \cdot (3 \cdot 8) \cdot (4 \cdot 6) \cdot (5 \cdot 14) \cdot (7 \cdot 10) \cdot 9 \cdot 11 \cdot 13 \equiv$
$\equiv 9 \cdot 11 \cdot 13 \equiv -1 \pmod{23}$,

so that $n = 14$ satisfies the condition.

$(q, n) = (7, 6)$: $6! \equiv -1 \pmod{7}$ by the Wilson theorem, so that $n = 6$ satisfies the condition. Finally, the integers n that satisfy the condition are 1, 2, 3, 4, 6, and 14, and their product is 2016.

392. If there is a positive integer n for which $16p + 1 = n^3$, then

$$16p = (n-1)(n^2 + n + 1).$$

Since $n^2 + n + 1$ is odd, we deduce that p divides $n^2 + n + 1$ and 16 divides $n - 1$. Then we have $16 = n - 1, p = n^2 + n + 1 = 17^2 + 17 + 1 = 307$, which is a prime number.

393. If $\max\{a,b,c,d\} = a$, then $\min\{a,b,c,d\} = b$. If $p = ac + bd$ is a prime number, then $ac \equiv -bd \pmod{p}$, so that $(ac)^4 \equiv (bd)^4 \pmod{p}$. We have also

$$(bd)^4 + (bc)^4 \equiv (ac)^4 + (bc)^4 \equiv c^4(a^4 + b^4) \equiv c^4(c^4 + d^4)$$
$$\equiv c^8 + (cd)^4 \pmod{p};$$

whence it follows $(c^4 + d^4)(c^4 - b^4) \equiv 0 \pmod{p}$. Then p must divide at least one of the numbers $c - b$, $c + b$, $c^2 + b^2$, $c^4 + d^4$. From

$$0 < c - b < c + b < c^2 + b^2 < ac + bd = p,$$

it follows that p divides $c^4 + d^4 = e^5$. But then $p^5 = (ac + bd)^5$ divides $c^4 + d^4$, contradiction, since $(ac + bd)^5 > c^4 + d^4$.

394. Answer: all composite numbers.

It is clear that a prime number n doesn't have the required property.

Let n be a composite positive integer, and denote by S_i the number of the board after i exclamation signs ($S_0 = n!$). Let p_i be the largest prime such that $\nu_{p_i}(S_i)$ is odd. The following algorithm will make the number of the board a perfect square:

1. If p_i doesn't exist, then the algorithm terminates;
2. Otherwise, apply an exclamation sign after $p_i + 1$.

We see that p_i decreases after the application of the second step, therefore after a finite number of steps, S_i becomes a perfect square.

395. Suppose that such k exists. Let $n = \prod_{i=1}^{m} p_i^{\alpha_i}$ be the prime decomposition of n, with primes p_i in increasing order, $\alpha_m > 0$, $k = (\alpha_1 + 1)\ldots(\alpha_m + 1)$ and $3 \mid \alpha_i$ for all i. The minimality of n implies $\alpha_1 \geq \alpha_2 \geq \cdots \geq \alpha_m > 0$. Now we need a lemma.

Lemma. If $\alpha_r + 1 = ab$, with a and b integers greater than 1 and $p_s < p_r^a < p_{s+1}$, then $\alpha_s \geq b - 1 \geq \alpha_{s+1}$.

Proof. The number $n_1 = p_r^{(a_s+1)a-1} p_s^{b-1} \prod_{i \neq r,s} p_i^{r_i}$ has k divisors; therefore $n_1 \geq n$.

This reduces to $(p_r^a/p_s)^{\alpha_s - b + 1} \geq 1$; whence $\alpha_s \geq b - 1$. Similarly,

$$n_1' = p_r^{(a_{s+1}+1)a-1} p_{s+1}^{b-1} \prod_{i \neq r, s+1} p_i^{r_i} \geq n \text{ yields } \alpha_{s+1} \leq b - 1.$$

Consider the largest r such that $\alpha_r + 1 = ab$ for some $a \equiv b \equiv 2 \pmod{3}$, and let s and t be such that $p_s < p_r^a < p_{s+1}$ and $p_t < p_r^b < p_{t+1}$. Observe that the Bertrand postulate implies $\frac{1}{2}p_r^a < p_s$ and $p_{s+1} < 2p_r^a$; hence $p_r^{a-1} < p_s < p_r^a < p_{s+1} < p_r^{a+1}$ and analogously $p_r^{b-1} < p_t < p_r^b < p_{t+1} < p_r^{b+1}$. It follows that $s, t > r$. We similarly obtain $|s - t| \neq 1$. By the lemma we have $\alpha_s > b - 1 > \alpha_{s+1}$ and $\alpha_t > a - 1 > \alpha_{t+1}$. Number $n_2 = p_r^{(\alpha_s+1)(\alpha_{t+1}+1)-1} p_s^{b-1} p_{t+1}^{a-1} \prod_{i \neq r,s,t+1} p_i^{r_i}$ also has k divisors, so $n_2 \geq n$, i.e.

$$1 \leq \frac{n_2}{n} = \frac{p_r^{(\alpha_s+1)(\alpha_{t+1}+1)-ab} p_{t+1}^{a-1-\alpha_{t+1}}}{p_s^{\alpha_s-b+1}} < \frac{p_r^{(\alpha_s+1)(\alpha_{t+1}+1)-ab+(b+1)(a-1-\alpha_{t+1})}}{p_r^{(a-1)(\alpha_s-b+1)}} =$$
$$= p_r^{1-(\alpha_s-b)(a-2-\alpha_{t+1})};$$

whence $(\alpha_s - b)(a - 2 - \alpha_{t+1}) < 1$. Since $3 \mid \alpha_s$ and $\alpha_s \neq b$, we must have $\alpha_{t+1} = a - 2$. From the hypothesis, $\alpha_i + 1$ has no divisors of the form $3j + 2$ if $i > r$, so it must be odd. In particular, $2 \mid \alpha_{t+1}$; therefore $2 \mid a$. Analogously, 2; whence $\alpha_r = 4c - 1$ for some positive integer c. Since $2 \mid \alpha_m$ implies $\alpha_m > 3 > 1 > \alpha_{m+1} = 0$, the lemma for $(a, b) = (2, 2c)$ and $(a, b) = (4, c)$, respectively, gives us $p_m < p_r^{2c} < p_{m+1}$ and $p_m < p_r^c < p_{m+1}$. The interval (p_r^c, p_r^{2c}) contains at least one prime by the Bertrand postulate, contradiction.

396. Case 1. $c = p^n$, with p prime. Then there is a sequence $x_1, x_2, \ldots, x_n$ such that for $1 \leq k \leq n$ holds $a^{x_k} + x_k \equiv b \pmod{p^k}$ and $x_k \equiv x_{k-1} \pmod{(p-1)p^{k-1}}$.

If $p \mid a$, then $p \mid a^x$ and we can take $x_1 \equiv b \pmod{p}$. For $x_{k+1} = x_k + \lambda(p-1)p^k$, we have

$$a^{x_{k+1}} + x_{k+1} = a^{\lambda(p-1)p^k} a^{x_k} + \lambda(p-1)p^k + x_k \equiv -\lambda p^k + x_k \pmod{p_{k+1}},$$

since $a^{\lambda(p-1)p^k} \equiv 0 \pmod{p^{k+1}}$. Now it suffices to choose $0 \leq \lambda < p$ such that

$$\lambda \equiv \frac{x_k - b}{p^k} \pmod{p}$$

in order to get the desired sequence. Moreover, $x_k \equiv x_1 \pmod{p-1}$ for every k, therefore x_1 can be arbitrarily chosen mod $p - 1$.

If $p \nmid a$, then choose x_0 among $1, 2, \ldots, p - 1$ and after defining x_{k-1}, take

$$x_k = \lambda(p-1)p^{k-1} + x_{k-1}.$$

Then we have

$$a^{x_k} + x_k = a^{\lambda(p-1)p^{k-1}} a^{x_{k-1}} + \lambda(p-1)p^{k-1} + x_{k-1}$$
$$\equiv a^{x_{k-1}} - \lambda p^{k-1} + x_{k-1} \pmod{p^k},$$

and we can choose

$$\lambda = \frac{a^{x_{k-1}} + x_{k-1} - b}{p^{k-1}}$$

in order to get the desired sequence. Since $x_n \equiv x_0 \pmod{p-1}$, it follows that x_0 can be arbitrarily chosen mod $p - 1$.

Case 2. c has at least two different prime factors. We shall use mathematical induction on the number k of distinct prime divisors of c. Let $c = p_1^{\alpha_1} \ldots p_k^{\alpha_k}$ be the prime factorization of c, with p_i in increasing order. We shall find a solution x mod $d_k = \operatorname{lcm}(c, \varphi(c))$.

If $k = 0$, i.e., $c = 1$, then x is arbitrary.

If $k = 1$, i.e., $c = p^n$, then we have find a solution mod c, which get us a solution mod $(p-1)p^n$.

If $k \geq 2$, suppose that x is a solution for $c_{k-1} = p_1^{\alpha_1} \ldots p_{k-1}^{\alpha_{k-1}}$, and let d_{k-1} be defined as above. Take $d = \gcd(d_{k-1}, p_k \varphi(p_k^{\alpha_k}))$, and choose a solution y of the equation mod $p_k^{\alpha_k}$ such that $y \equiv x \pmod{d}$, which is always possible. Then $d_{k-1} = f'g$ and $(p_k - 1)p_k^{\alpha_k} = f''h$, where factors g and h are those of d. In particular, $\gcd(g, h) = 1$ and we define $f = \operatorname{lcm}(f', f'')$. By the Chinese remainder theorem, there is a solution z mod fgh of the system

$$z \equiv x \pmod{f}, z \equiv x \pmod{g}, z \equiv y \pmod{h}.$$

We deduce $z \equiv x \pmod{d_{k-1}}$, $z \equiv y \pmod{(p_k-1)p_k^{\alpha_k}}$; therefore z is the desired solution mod $p_1^{\alpha_1} \ldots p_k^{\alpha_k}$. Moreover, we have $fgh = \operatorname{lcm}(d_{k-1}, (p_k-1)p_k^{\alpha_k}) = d_k$.

397. Answer: $(1, 2, 3), (2, 1, 3)$.

Suppose that $m^3 + n = p^2$, $n^3 + m = p$. Subtracting these equalities, we get

$$m(m-1)(m+1) - n(n-1)(n+1) = p(p-1);$$

therefore $3 | \text{LHS}$, which implies $3 | \text{RHS}$; whence $p = 3$. Solving the system of equations $m^3 + n = 9$, $n^3 + m = 3$, we obtain the solution $(m, n) = (2, 1)$. The equation from the enounce is symmetric, so that $(1, 2)$ is also a solution.

398. For $n \leq 4$ there are solutions

$$2^2 + 3^0 + 5^0 = 2^1 + 3^1 + 5^0 = 3!, \ \ 2^4 + 3^1 + 5^1 = 4!.$$

For $n \geq 5$ there are no solutions. Observe that

$$2^a \pmod{120} \in \{1, 2, 4, 8, 16, 32, 64\},$$
$$3^b \pmod{120} \in \{1, 3, 9, 27, 81\},$$
$$5^c \pmod{120} \in \{1, 5, 25\}$$

and that no three elements have vanishing sum mod 120.

399. **Lemma.** If $n \equiv 1 \pmod 3$, then the homogenous polynomial

$$P_n(a,b) = (a+b)^n - a^n - b^n \in \mathbb{Z}[a,b]$$

is divisible by $(a^2 + ab + b^2)^2$.
Proof. Evaluate $P_n(x,1)$ and its derivative at a primitive cubic root of unity ε.
Corollary. If p is a prime, $p \equiv 1 \pmod 3$ and a, b are integers such that

$$p^k \mid a^2 + ab + b^2;$$

then $p^{2k+1} \mid P_n(a,b)$.
Proof. Observe that $P_p \in p\mathbb{Z}[a,b]$ and use the lemma.

400. (a) For any odd prime p,

$$\frac{p^k}{2} < n < p^k \Longrightarrow p \mid \binom{2n}{n}.$$

Indeed, the inequality

$$\left\lfloor \frac{2n}{p^i} \right\rfloor \geq 2 \left\lfloor \frac{n}{p^i} \right\rfloor$$

is true, with equality only when $i = k$. Now apply the Legendre formula.
(b) h cannot have any odd prime divisor p, since x_n is a multiple of p for $\frac{p^k}{2} < n < p^k$, but not for $n = p^k$. For any value of T, we can choose k such that $p^k > T$. It is known that $\nu_2\left(\binom{2n}{n}\right) = 1$ if n is a power of 2 and $\nu_2\left(\binom{2n}{n}\right) \geq 2$ otherwise. The only power of 2 that works is $h = 2$.

2.46 2018

401. Answer: all k such that there exists a prime p with $\nu_p(k) \geq 1009$, excepting $k = 2^{1009}$.
Call a number which satisfies the arithmetic progression condition *good*; otherwise call it bad. If a number is good, then so is any multiple of it, so if a number is bad, then all divisors of it are bad. Observe that a number k is bad if and only if every residue class of k contains arbitrarily large uninteresting numbers. Note that 1009 is a prime number.
We will prove first that all the above numbers are good. If $p > 2$, then it suffices to show that p^{1009} is good. Let r be a quadratic nonresidue mod p. Then every number $n \equiv rp^{1008} \pmod{p^{1009}}$ has $\nu_p(n) = 1008$, so $1009 \mid \tau(n)$. Also, $\frac{n}{p^{1008}}$ is never a square; hence n isn't a square, so $2 \mid \tau(n)$. It follows that $2018 \mid \tau(n)$.

All $k = 2^{1009}p$, $p \neq 2$, are good. Take r a quadratic nonresidue mod p and

$$n \equiv 2^{1008} \pmod{2^{1009}}, n \equiv r \pmod{p}.$$

Then n is not a square, $\nu_2(n) = 1008$, so $2018 \mid \tau(n)$.
The number $k = 2010$ is good, taking $n \equiv 3 \cdot 2^{1008} \pmod{2^{2010}}$. Then.

$$\nu_2(n) = 1008, n = 2^{1008}(4t+3)$$

and $4t + 3$ is not a square, so $2 \mid \tau(n)$.
Now we need a lemma.

Lemma. For any distinct primes $p_1, \ldots, p_k$, the number $N = (p_1 \ldots p_k)^{1008}$ is bad.

Proof. Let $1 \leq r \leq N$ and consider the residue class $r \pmod N$. Let $M = (p_1 \ldots p_k)^{1010}$, and consider the residue class $c \pmod M$, which exists by the Chinese remainder theorem, satisfying $c \equiv r \pmod{p_i^{1010}}$ for all p_i such that $\nu_{p_i}(r) < 1008$; $c \equiv p_i^{1009} \pmod{p_i^{1010}}$ for all p_i such that $\nu_{p_i}(r) \geq 1008$.

Then $c \equiv r \equiv 0 \pmod{p_i^{1008}}$, so that the residue class $c \pmod M$ is a subset of the residue class $r \pmod N$. By the Dirichlet's theorem, there exist arbitrarily large elements in this residue class that are $n \equiv q \cdot \gcd(c, M)$, where $q \nmid M$ is a prime. Since $\nu_{p_i}(c) \in \{0, \ldots, 1007, 1009\}$, $\nu_{p_i}(M) = 1010$ and there are no other primes dividing M, it follows that any prime $p \mid n$ satisfies $\nu_p(n) \not\equiv -2 \pmod{1009}$; hence $1009 \nmid \tau(n)$. It follows that n is uninteresting, so that N is bad.

It remains to show that 2^{1009} is bad: for $n \equiv 0 \pmod{2^{1009}}$, take $n = 2^{1009i}$, so $1009 \nmid \tau(n) = 1009i + 1$; for $n \equiv 2^{1008} \pmod{2^{1009}}$, we have $n = 2^{1008}(2k+1)$. Taking $2k+1$ to be some arbitrarily large square, we have $2 \nmid \tau(n)$; if $\nu_2(n) = a \leq 1008$, let $n' = \frac{n}{2^a}$, then $n = 2^a(2^{1009-a}k + n')$, and for $a \geq 1$, we can apply the above claim; for $a = 0$, we can choose n arbitrarily large prime, so $1009 \nmid \tau(n) = 2$.

Thus we obtain the solution set as described.

402. Let $p^2 \mid k-1$ and let $d = \frac{k-1}{p}$. Consider the number

$$\alpha = N + \overline{0.dd \ldots d_k}$$

in base k. Then

$$\lfloor \alpha k^n \rfloor = k^n N + d \cdot \frac{k^n - 1}{k - 1} = \left(N + \frac{1}{p}\right) k^n - \frac{1}{p}.$$

If we choose N such that $p \nmid N$, then the middle-hand side expression is not divisible by p, since d is divisible by p. We can select N such that $q \mid N + p^{-1}$ for every $q \mid M, q \neq \neq p$. The Chinese remainder theorem completes the proof.

403. Answer: no solutions.

The equation $(xy + 1)(xy + x + 2) = k^2$ is equivalent to

$$(y-1)^2 = [2y(y+1)x + 3y + 1]^2 - 4y(y+1)k^2.$$

Set $u = 2y(y+1)x + 3y + 1$, $v = k$, so it suffices to show that there is no solution to the Pell-like equation

$$(y-1)^2 = u^2 - 4y(y+1)v^2$$

with $u \equiv 3k + 1 \pmod{2y(y+1)}$. Assume for contradiction that such a solution exists. Define $N(u, v) = u^2 - 4y(y+1)v^2$ as the norm of $u + v\sqrt{4y(y+1)}$. The fundamental unit in $\mathbb{Z}[\sqrt{4y(y+1)}]$ is $2y + 1 + \sqrt{4y(y+1)}$. Then the solutions of the Pell-like equation are given by the recurrence

$$(u_{i+1}, v_{i+1}) = ((2y+1)u_i + 4y(y+1)v_i, u_i + (2y+1)v_i).$$

One solution is $(2y^2 - y - 1, y - 1)$, with $\equiv -(3y+1) \pmod{2y(y+1)}$. Then there is a t such that

$$-(3y+1)(2y+1)^t \equiv 3y + 1 \pmod{2y(y+1)}.$$

Note that $(2y+1)^2 \equiv 1 \pmod{2y(y+1)}$; hence $(2y+1)^t$ can have only the values 1 and $2y + 1$, neither satisfying the above congruence.

404. We may assume $p \geq 7$. Let $S = \{g^0, \ldots, g^{d-1}\} \pmod{p}$, $d \mid p - 1$, $d \neq p - 1$. Then

$$\sum_{s \in S} s = \frac{g^d - 1}{g - 1} = 0, \quad \sum_{s \in S} s^2 = \frac{g^{2d} - 1}{g^2 - 1} = 0,$$

as $g \neq \pm 1$. For $S = \{a, a+k, \ldots, a + (d-1)k\} \pmod{p}$, it follows

$$\sum_{s \in S} s = da + \frac{d(d-1)}{2}k = 0, \quad \sum_{s \in S} s^2 = da^2 + d(d-1)ak + \frac{d(d-1)(2d-1)}{6}k^2 = 0.$$

Setting $a = -\frac{d-1}{2}k$ in the second expression, we get

$$k^2\left(\frac{(d-1)^2}{4} - \frac{(d-1)^2}{2} + \frac{(d-1)(2d-1)}{6}\right) = 0;$$

whence $d = -1$, contradiction.

405. If $m^2 = n^4 + n^3 + n^2 + n + 1$ with m a positive integer, then

$$(2n^2 + n)^2 < (2m)^2 < (2n^2 + n + 2)^2;$$

therefore $(2m)^2 = (2n^2 + n + 1)^2$; whence $n^2 - 2n - 3 = 0$, so that $n = 3$.

406. Answer: 146250.

First, we show by induction on the length of n that if $10n$ is bright, and then n is bright as well. Assume that for one digit numbers, the statement holds. If after deleting 0, we obtain a bright divisor, and then the statement holds trivially. If the bright divisor of $10n$ is obtained after deleting some other digit, then in the end of this divisor, we still have 0, i.e., it can be written as $10d$. By the induction hypothesis, d is bright. But then after deleting from n the corresponding digit, we get a bright divisor d, which means that also n is bright.

Next we show that any bright divisor of at least two-digit bright number not ending with 0 can be obtained by deleting the first or the second digit.

Assume the contrary that a bright divisor d of a bright number n not ending with 0 is obtained by deleting the third or a further digit. This means that $n = (10a + x) \cdot 10^k + b$ and $d = 10^k a + b$, where $a \geq 10$, $0 \leq x < 10$ and $b < 10^k$. Since $9d = 9 \cdot 10^k a + 9b < 9 \cdot 10^k a + 9 \cdot 10^k = 9 \cdot 10^k(a + 1)$ $< 10a \cdot 10^k \leq (10a + x) \cdot 10^k + b = 10a \cdot 10^k + 10^k x + b$ $< 10a \cdot 10^k + 10^k a + 11b = 11d$, the only possibility is $n = 10d$. Then n ends with 0, contradiction.

Let now n be a five-digit bright number not ending with 0, and let d be its bright divisor. We consider two cases depending on which digit is deleted to obtain d.

Case 1. d is obtained by deleting the first digit of n. Then $d \mid n - d = 10^4 x$, where x is the deleted digit. As d does not end with 0, it is not divisible either by 2 or 5. Then $d \leq 2^4 x \leq 2^4 \cdot 9 < 10^3$, contradiction. d is odd and $d \mid 5^4 x$. Since d is a four-digit number not ending with 0, it must divide one of the numbers 1875, 3125, 4375, and 5625. We may leave out 3125, because in this case the first digit of n must be $x = 5$, but the last digit is 5 as well. The four-digit divisors of the remaining numbers are 1125, 1875, 4375, and 5625. The first and the last numbers contain equal digits; from the other two numbers, we cannot obtain a divisor by deleting the first or the second digit.

Case 2. d is obtained by deleting the second digit of n. Then $d \mid n - d = 10^3 z$, where z is at most two-digit. Since d does not end with 0, it is not divisible either by 2 or 5, implying that $\frac{n-d}{d}$ is divisible either by 2^3 or 5^3. Since n and d start with the same digit a, we have

$$\frac{n}{d} < \frac{10^4(a+1)}{10^3 a} = 10 + \frac{10}{a} \leq 20,$$

implying that $\frac{n-d}{d}$ can be only 8 or 16, in both cases $5^3|d$. We can write $d = 1000a + 125r$, where $r \in \{1, 3, 5, 7\}$. If $\frac{n-d}{d} = 16$, equivalently $n = 17d$, then $n = 17{,}000a + 2125r = 1000(17a + 2r) + 125r$. Since n starts with a and $125r < 1000$, we get $17a + 2r < 10(a + 1)$ yielding $7a + 2r < 10$. This gives $a = r = 1$ and $d = 1125$, which is not bright. If $\frac{n-d}{d} = 8$, equivalently $n = 9d$, then $n = 9000a + 1125r = 1000(9a + r) + 125r$. Since n starts with a, we get $9a + r + \frac{r}{8} > 10a$ yielding $r > \frac{8}{9}a \geq a - 1$, i.e., $r \geq a$. Leaving out numbers with repeated digits, we get $d \in \{1375, 1625, 1875, 2375, 2875, 3625, 3875, 4625, 4875, 6875\}$. Among these numbers only 1625 is bright; deleting the first or the second digit from other candidates does not give a divisor. A check shows that $9d = 14625$ is bright as well.

Therefore 14625 is the only five-digit bright number not ending with 0. Let now n be an arbitrary six-digit bright number. If n ends with 0, then deleting 0 we obtain a five-digit bright number not containig 0, whence $n = 146250$. If n is not ending with 0, then after deleting the first or the second digit, we obtain a bright divisor d not ending with 0. Thus $d = 14625 = 117 \cdot 125$ yielding $117|n - d = 10^4 z$, where z is at most two-digit. Since 117 and 10 are coprime, this is not possible. Consequently, 146250 is the only six-digit bright number. If n is a bright seven-digit number, then by deleting its some digit we obtain 146250. Then also n ends with 0, which means that $\frac{n}{10}$ is a bright six-digit number not containing 0. But there are no such numbers. Since there are no seven-digit bright numbers, there cannot be longer bright numbers either.

407. Answer: 2.

A single nonzero coefficient is not sufficient for any $n > 1$, as the only simple polynomial functions with a single nonzero coefficient are $P(x) = x^n$ and $P(x) = -x^n$, but in both cases $n \nmid P(1)$. Let us show that the values of the polynomial function $P_n(x) = x^n - x^{n - \varphi(n)}$ at all integral arguments are divisible by n. (Here φ is the Euler's totient function.) This shows that having 2 nonzero coefficients is sufficient.

Let k be an integer. Let the canonical form of n be $p_1^{\alpha_1} \ldots p_m^{\alpha_m}$, and let us assume without loss of generality that k is divisible by primes $p_1, \ldots, p_l$ and is not divisible by primes $p_{l+1}, \ldots, p_m$. Define $u = p_1^{\alpha_1} \ldots p_l^{\alpha_l}$ and $v = p_{l+1}^{\alpha_{l+1}} \ldots p_m^{\alpha_m}$. Let us now show that $u|k^{n - \varphi(n)}$ and $v|k^{\varphi(n)} - 1$. Having $uv = n$, we can conclude that $n|P_n(k)$, as $P_n(k) = k^n - k^{n - \varphi(n)} = k^{n - \varphi(n)}(k^{\varphi(n)} - 1)$.

To prove that $u|k^{n - \varphi(n)}$, it is sufficient to prove for all that $i = 1, \ldots, l$ that $p_i^{\alpha_i}|k^{n - \varphi(n)}$. It is sufficient to prove that $\alpha_i \leq n - \varphi(n)$, as by the assumption $p_i|k$. Inequality $\alpha_i \leq n - \varphi(n)$ holds as $p_i, p_i^2, \ldots, p_i^{\alpha_i}$ are α_i positive integers which are not greater than n and not coprime with n.

To prove the statement $v|k^{\varphi(n)} - 1$, we derive from Euler's theorem that $v|k^{\varphi(v)} - 1$, as k and v are coprime. Also, u and v are coprime; therefore $\varphi(n) = \varphi(uv) = \varphi(u)\varphi(v)$, from which $k^{\varphi(v)} - 1|k^{\varphi(n)} - 1$. Consequently, $v|k^{\varphi(n)} - 1$.

408. The equality $n^2 + 32n + 8 = k^2$ implies $(n+16)^2 - k^2 = 16^2 - 8 = 2^3 \cdot 31$. Then there are divisors d_1, d_2 of $2^3 \cdot 31$ such that $d_1 < d_2$, $d_1 d_2 = 2^3 \cdot 31$ and $2(n+16) = d_1 + d_2$. It is easy to see that the only solutions are $n = 17$ and $n = 47$, for which we find $k = 29$ and $k = 61$, respectively.

409. By Dirichlet's theorem, there are infinitely many prime numbers of the form $m = 13k + 1$. From $8191 = 2^{13} - 1$, it follows that $8191 | 2^{m-1} - 1$. The Fermat's theorem implies $m | 2^{m-1} - 1$. Since $m \neq 8191$, we deduce that $8191m | 2^{m-1} - 1$.

410. For $p \leq 5$ the solutions (p, a, m) are

$(2,1,1), (2,3,2), (2,7,3), (2,15,4), (3,1,1), (3,7,2), (3,25,3), (5,1,2), (5,101,3).$

For $p > 5$ there are no solutions. Indeed, from

$$a - 1 = p^m - 1 - (p-1)!,$$

it follows that $p(p-1)$ divides $a - 1$, so that $a = kp(p-1) + 1$. The condition $a \leq 5p^2$ implies $0 \leq k \leq 5$. Since $p - 1$ is even, it follows that $p - 1$ divides $(p-2)!$. Then the equality

$$kp = p^{m-1} + \cdots + p + 1 - (p-2)!$$

implies $k \equiv m \pmod{(p-1)}$. We deduce $k = m$ and

$$p^2 - 2p > mp = p^{m-1} + \cdots + p + 1 - (p-2)!$$
$$> 1 + 2(p-2)(p-3),$$

contradiction.

411. We claim that S_k is finite if and only if k is a power of 2.

First, we show that S_k is infinite when an odd prime p divides k. By the lifting the exponent lemma, since n is odd, we have

$$\nu_p\left(1^{p^l} + (k-1)^{p^l}\right) = \nu_p(1 + k - 1) + \nu_p(p^l) \geq l + 1,$$

so $n = p^l$ divides $a^n + b^n$; hence $(n, a, b) = (p^l, 1, k-1) \in S_k$. Since l is arbitrary, it follows that S_k is infinite.

Suppose that k is a power of 2 and assume that S_k is infinite. Then we find an odd value $n > 1$. Let p be the smallest prime divisor of n. Then $p \nmid a$, otherwise $p | b$, so $p | k$, contradiction. Now, $a^n \equiv -b^n \pmod{p}$ implies $(ab^{-1})^n \equiv -1 \pmod{p}$; hence $(ab^{-1})^{\gcd(2n, p-1)} \equiv 1 \pmod{p}$. If an odd prime $q | \gcd(2n, p-1)$, then $q | 2n, q \leq p - 1$, so $q | n$, with $q < p$, contradiction. Hence $\gcd(2n, p-1) = 2$, as n, p are odd. But then $p | a^2 - b^2 = k(a-b)$, since $\gcd(p, k) = 1$; hence $p | a - b$. But then $b^n \equiv a^n \equiv -a^n \pmod{p}$; therefore $p | a$, contradiction.

412. Let $n > 4$ divisible by 4 and $a_k = 2^{k-1}$, $1 \leq k \leq n-1$, $a_n = n-2$. Then

$$\sum_{k=1}^{n} a_k = \sum_{1 \leq k < l \leq n} \gcd(a_k, a_l) = 2^{n-1} + n - 3.$$

413. If a has more than 11 divisors, then $a < b$ implies $a_{11} + b_{11} = b \leq \frac{a}{2} + \frac{b}{2} < b$, contradiction. It follows that a has 11 divisors, $a = p^{10}$. From the first equation, we get $b_{10} = p^9(p-1)$, so that $p^{10} = a < 2b_{10}$. If b has more than 12 divisors, then

$$a + b_{11} = b \leq 2b_{10} + b_{11} < \frac{2b}{3} + \frac{b}{3} = b,$$

contradiction. Finally, $p^9 | b$ implies $b = p^{11}$. Putting these values in the equations, we obtain $p = 2$.

414. We will apply the well-known result

$$\binom{p}{k} \equiv p \cdot \frac{(-1)^{k-1}}{k} \pmod{p^2}.$$

If $p = 6n + 1$, then

$$\sum_{i=1}^{\lfloor 2p/3 \rfloor} \binom{p}{i} \cdot \frac{1}{p} = \frac{1}{1} - \frac{1}{2} + \cdots - \frac{1}{4n} = 1 + \frac{1}{2} + \cdots + \frac{1}{4n} - 2\left(\frac{1}{2} + \frac{1}{4} + \cdots + \frac{1}{4n}\right) =$$

$$= \frac{1}{2n+1} + \cdots + \frac{1}{4n} \equiv 0 \pmod{p},$$

which is obvious, as we take pairs of fractions with denominators $2n + 1 + i$, $4n - i$.

If $p = 6n + 5$, then

$$\sum_{i=1}^{\lfloor 2p/3 \rfloor} \binom{p}{i} \cdot \frac{1}{p} = \frac{1}{1} - \cdots + \frac{1}{4n+3} = \frac{1}{1} + \cdots + \frac{1}{4n+3} - 2\left(\frac{1}{2} + \cdots + \frac{1}{4n+2}\right) =$$

$$= \frac{1}{2n+2} + \cdots + \frac{1}{4n+3} \equiv 0 \pmod{p},$$

again by pairings.

415. Divisibility by 3: the four integers a, b, c, and d are at most three distinct mod 3; therefore, by the pigeonhole principle, there are at least two whose difference is divisible by 3.

Divisibility by 4. Case 1. Exactly two of the numbers a, b, c, and d, say a and b, are even and c and d are odd. Then $a - b$ and $c - d$ are both even; therefore 4 divides $(a - b)(c - d)$.

Case 2. Three or four of the numbers a, b, c, and d have the same parity. In this case there are at least two pairs with the same parity, and, as above, the product of the differences is divisible by 4.

416. The given equation is equivalent to $(a - 1)(a + b + 1) = 2017$. Since 2017 is a prime number, it follows that $a - 1 \in \{\pm 1, \pm 2017\}$, whence the solutions

$$(a, b) \in \{(2018, -2018), (2, 2014), (0, -2018), (-2016, 2014)\}.$$

417. Suppose that n is not a power of p, and choose a solution (x, y, p, n, k) with the smallest $x + y + p + n + k$. If n is composite, then $n = mr$, with $1 < m < n$, and m is not a multiple of p. From

$$x^n + y^n = (x^m + y^m)\left(x^{(r-1)m} - \cdots + y^{(r-1)m}\right) = p^k,$$

We deduce $x^m + y^m = p^{k'}$, and we have a solution (x, y, p, m, k') with smaller sum of components, contradiction.

If $n \neq p$ is a prime number, then we have

Case 1. $x \equiv y \equiv 0 \pmod{p}$. We have a solution $\left(\frac{x}{p}, \frac{y}{p}, p, n, k - n\right)$ with smaller sum of components, contradiction.

Case 2. $x \not\equiv 0 \pmod{p}$ and $y \not\equiv 0 \pmod{p}$. The decomposition

$$x^n + y^n = (x + y)\left(x^{n-1} - \cdots + y^{n-1}\right)$$

and the condition $n > 1$ show that the two factors are powers of p; whence

$$x^{n-1} - x^{n-2}y + \ldots + y^{n-1} \equiv x^{n-1} - x^{n-2}(-x) + \ldots + (-x)^{n-1}$$
$$\equiv nx^{n-1} \equiv 0 \pmod{p},$$

contradiction.

418. Assume the contrary. Let M be a positive integer such that for all $n \geq M$, both numbers are primes. Fix a large n and let $p = 2^{2^n} + 1$. Let $d = ord_p(2018)$. Then $d | p - 1$, so that $d = 2^j$, $j \geq 2$. Hence $p \nmid 2018^{2^{j-1}} - 1$; therefore $p | 2018^{2^{j-1}} + 1$, so that $2018^{2^{j-1}} + 1$ is not a prime. This implies $j \leq M + 1$. Choosing n large, we get a contradiction.

419. Answer: $n = 2^k - 2$, $k \geq 2$.

For $j = i + 1$, $i < n$, we have

$$\binom{n+1}{i+1} = \binom{n}{i} + \binom{n}{i+1} \equiv i + i + 1 \equiv 1 \pmod{2}.$$

Since $\binom{n+1}{0} = 1$, it follows that $\binom{n+1}{i}$ is odd, for every $0 \leq i \leq n+1$. For every $0 \leq i \leq n+1$, we can write

$$\binom{n+1}{i} = \frac{(n+1)n(n-1)\ldots(n-i+2)}{1 \cdot 2 \cdot \ldots \cdot i}.$$

that is, $i - 1$ becomes i we need multiply by $q = \frac{n-i+2}{i}$ and viewing that $\binom{n+1}{i-1}$ and $\binom{n+1}{i}$ are odd, we conclude that the irreducible fraction associated to q is the quotient of two odd numbers. In other words, for every $1 \leq i \leq n+1$, the numbers i and $n - i + 2$ have the same exponent of 2 in the decomposition in prime factors.

We shall prove that $n + 2$ is a power of 2. Let 2^l be the greatest power of 2 less than $n + 2$. Then $2^l \geq \frac{n+2}{2}$. Take $i = 2^l$, and observe that 2^l divides $n - 2^l + 2$ only if $n + 2 = 2^{l+1}$, since $n - 2^l + 2 \leq 2^l$.

It remains to prove that the numbers $n = 2^k - 2$ are solutions. It suffices to prove that $\binom{n}{i} \equiv i + 1 \pmod{2}$ for $0 \leq i \leq n$, since then

$$\binom{n}{i} + \binom{n}{j} \equiv i + j + 2 \equiv i + j \pmod{2}.$$

We shall prove this property by induction on i. For $i = 0$ and $i = 1$, clearly we have $\binom{n}{i} \equiv i + 1 \pmod{2}$. Suppose that the property is true for $0, 1, \ldots, i-1$. If i is odd, then we shall multiply by $\frac{n-i+1}{i}$, and since the numerator is even and the denominator is odd, it follows that $\binom{n}{i}$ is even. If i is even, then we shall multiply $\binom{n}{i-2}$ by $\frac{n-i+1}{i-1} \cdot \frac{n-i+2}{i}$. The first fraction is a quotient of two odd numbers, and since $n + 2 = 2^k$, the second fraction can be represented as a quotient of two odd numbers. But $\binom{n}{i-2}$ is odd, so that $\binom{n}{i}$ is also odd.

420. Suppose that $P(n) = n^2 + an + b$ has m prime divisors. Observe that $P(n)$ divides $P(n + kP(n))$ for all positive integers k. Since $P(n + kP(n))$ is a nonconstant polynomial in k, we can find a prime p and a positive integer k_0 such that $p \nmid P(n)$ and $p | P(n + k_0 P(n))$. We deduce that $P(n + k_0 P(n))$ has at least $m + 1$ prime divisors,

In conclusion, the number of prime divisors of $P(n)$ is unbounded.

421. Suppose that $d | pn^2$ and $d = a^2 - n^2$ for some positive integer $a > n$. Then $a^2 - n^2 | pa^2$. Let $k(a^2 - n^2) = pa^2$. If $a + n$ has a prime factor $q \neq p$, and $q \nmid a - n$, then $q \nmid a^2$, contradiction. By the Zsigmondy's theorem, it follows that $a + n = 2^j$ or $a + n = p^j$ for some j. In the first case, we get a contradiction analyzing the exponents of 2. In the second case, $k(p^j - 2n)p^j = p(p^j - n)^2$. Since

$$\nu_p(n) = \nu_p(p^j - 2n) = \nu_p(p^j - n),$$

we get $\nu_p(p^j - n) = j - 1$; hence $n = xp^{j-1}$ for some $x < \frac{p}{2}$. If x satisfies the condition, then

$$p - 2x | (p - x)^2 \Rightarrow p - 2x | p \Rightarrow x = \frac{p-1}{2}.$$

Therefore there exists at most one such d.

422. Answer: no.

We have

$$4\left[(m+1)^3 + \cdots + (2m)^3\right] = [2m(2m+1)]^2 - [m(m+1)]^2$$
$$= m^2(5m+3)(3m+1),$$

so that the given sum is a square if and only if $(5m + 3)(3m + 1)$ is a square. Suppose that this is the case.

Case 1. m even. Then $5m + 3$ and $3m + 1$ are both odd and gcd $(5m + 3, 3m + 1) = 1$, since $3(5m + 3) - 5(3m + 1) = 4$. We deduce that there are positive integers u, v such that $5m + 3 = u^2$, $3m + 1 = v^2$; whence $3u^2 - 5v^2 = 4$. But this equality is impossible, since $u^2 \equiv v^2 \equiv 1 \pmod 4$.

Case 2. m odd. Then $\gcd(5m + 3, 3m + 1) = d \in \{2, 4\}$.

If $d = 2$, then $5m + 3 = 2u^2$, $3m + 1 = 2v^2$, whence $3u^2 - 5v^2 = 2$, which implies $v^2 \equiv 2 \pmod 3$, contradiction.

If $d = 4$, then $5m + 3 = 4u^2$, $3m + 1 = 4v^2$; whence $3u^2 - 5v^2 = 1$, which implies $u^2 + v^2 \equiv 3 \pmod 4$, contradiction.

423. Answer: $b \neq 2^k - 1$.

If $b + 1 \neq 2^k$, then some odd prime $p_1 | b + 1$. By the lifting the exponent lemma, it follows that $p_1^2 | b^{p_1} + 1$. Since $b > 2$, and $b + 1 \neq 2^k$, the Zsigmondy's theorem implies that there is a prime p_2 that divides $b^{p_1} + 1$, so $(p_1 p_2)^2 | b^{p_1 p_2} + 1$, by the lifting the exponent lemma. This process can continue indefinitely.

If $b + 1 = 2^k$, then let q be the smallest prime divisor of n. For $q = 2$ we get a contradiction. So $q | b + 1$, $b^n \equiv -1 \pmod{q}$; hence $\gcd(q - 1, n) > 1$, but q is the smallest prime divisor of n, contradiction.

424. If $d | n$, then any d consecutive prime divisors contain at most $\varphi(d)$ numbers coprime to n. If there exists a number in that interval coprime to d, but not coprime to n, then the equality does not hold.

Using this result, we see that among any $\sigma(n)$ positive integers, there are at most $\sum_{d|n} \varphi(d) = n$ numbers coprime to n. If there exist prime numbers $p < q$ that both divide n, then any q consecutive integers contain a multiple of p, and the inequality is strict. If n is a prime power, then the equality holds.

2.47 2019

425. (a) If $a = 1$, then $b = 1$.
If $a > 1$ and $p | a$, p prime, then

$$b \nu_p(a) + c \nu_p(b) = a \nu_p(c).$$

If $p \nmid b$, then

$$c \nu_p(a) = a \nu_p(c) \Rightarrow p^{\nu_p(a)} | \nu_p(a) \Rightarrow \nu_p(a) \geq 2^{\nu_p(a)} \Rightarrow \nu_p(a) = 0,$$

contradiction.

(b) If $b \geq a$, then $a^a | a^b | c^a$, which implies $a | c$, so that $c = ka$, $k \geq 1$. We have

$$a^b b^{ka} = (ka)^a \Leftrightarrow a^{b-a} b^{ka} = k^a \Rightarrow k^a \geq b^{ka} \Rightarrow k \geq b^k \Rightarrow k$$
$$= a = b = c = 1.$$

(c) $a = c = n^n$, $b = n^{n-1}$ is a solution, for every positive integer n.

426. Answer: all pairs (m, n) such that $m \nmid n$, $m \nmid n$.

Clearly, if $m | n$ or $n | m$, there exists no such s, so it suffices to construct a s for all other pairs (m, n). We can delete pairs in the prime decompositions of m, n with equal powers. Assume that

$$m = d_1^{p_1} \ldots d_i^{p_i} b_1^{q_1} \ldots b_j^{q_j}, n = a_1^{x_1} \ldots a_i^{x_i} b_1^{y_1} \ldots b_j^{y_j}, p_k > x_k, q_l < y_l,$$

for $1 \leq k \leq i$, $1 \leq l \leq j$. Choose a sufficiently large positive integer S, and define

$$s_k = (iS+k)(p_k - x_k) - 1 - x_k, \quad t_l = (jS+l)(y_l - q_l) - 1 - q_l, \quad 1 \leq k \leq i, \, 1 \leq l \leq j.$$

Let $s = a_1^{s_1} \ldots a_i^{s_i} b_1^{t_1} \ldots b_j^{t_j}$. Then

$$\frac{\tau(sm)}{\tau(sn)} = \prod_{k=1}^{i} \frac{p_k + s_k + 1}{x_k + s_k + 1} \cdot \prod_{l=1}^{j} \frac{q_l + t_l + 1}{y_l + t_l + 1} = \prod_{k=1}^{i} \frac{iS + k + 1}{iS + k} \cdot \prod_{l=1}^{j} \frac{jS + l}{jS + l + 1} =$$

$$= \frac{S+1}{S} \cdot \frac{S}{S+1} = 1.$$

An example is $m = 2^3 \cdot 3^5 \cdot 5^1 \cdot 11^0$, $n = 2^2 \cdot 3^6 \cdot 5^{10} \cdot 11^1$, $s = 2^0 \cdot 3^0 \cdot 5^{124} \cdot 11^{14}$.

427. **Answer: no.**

Assume that $xy = a^2$, $zt = c^2$, $a, c > 0$. If $x + y = z + t$ is odd; then x, y and z, t have opposite parity, so that xy and zt are both even; hence $xy - zt = x + y$ is even, contradiction. We deduce that $x + y = z + t = 2s$, with s a positive integer. Denote $|x - y| = 2b$, $|z - t| = 2d$. The given conditions yield

$$s^2 = a^2 + b^2 = c^2 + d^2, \quad 2s = a^2 - c^2 = d^2 - b^2.$$

We shall use the fact that a, d, and s are positive integers, and b and c are nonnegative integers, at most one of which may be zero. We can assume, wlog, due to symmetry, that $b \geq c$; hence $b > 0$. It follows that $d^2 = 2s + b^2 > c^2$; whence

$$d^2 > \frac{c^2 + d^2}{2} = \frac{s^2}{2}.$$

Since $d^2 - b^2$ is even, the numbers b and d have the same parity, so that $0 < b \leq d - 2$; therefore

$$2s = d^2 - b^2 \geq d^2 - (d-2)^2 = 4(d-1) \Rightarrow d \leq \frac{s}{2} + 1 \Rightarrow$$

$$\Rightarrow 2s^2 < 4d^2 \leq 4\left(\frac{s}{2} + 1\right)^2 \Rightarrow (s-2)^2 < 8 \Rightarrow s \leq 4.$$

Finally, each number s^2, $1 \leq s \leq 4$ has a unique representation as a sum of two squares, $s^2 = s^2 + 0^2$; therefore $b = c = 0$, contradiction.

Note. All quadruples (x, y, z, t) satisfying the given equalities are

$$x = 2kl - k + l, \, y = 2kl + k - l, \, z = 2kl + k + l, \, t = 2kl - k - l,$$

with $l \geq k > 0$, $(k, l) \neq (1, 1)$. For $D = 4(kl)^2 - k^2 - l^2$, it follows

$$(D-1)^2 < xyzt < D^2,$$

so that xy and zt cannot be both squares.

428. First of all, n must be odd. Let p be the greatest odd divisor of n. The binomial theorem implies $p|n|2^n + 2 - 2n$, contradiction.

429. We have

$$(x+1)^3 - x^3 = y^2 \implies 3(2x+1)^2 = (2y-1)(2y+1).$$

Since $\gcd(2y - 1, 2y + 1) = 1$, there are two cases:
Case 1. $2y - 1 = 3a^2$, $2y + 1 = b^2$. Then $b^2 - 3a^2 = 2$, which is impossible, since squares modulo 3 are 0 and 1.
Case 2. $2y - 1 = c^2$, $2y + 1 = 3d^2$. Then c and d are odd, $2y = c^2 + 1 = (2k+1)^2 + 1$; whence $y = k^2 + (k+1)^2$, with k a positive integer.

430. Denote $x = m - 1009$, $y = n + 1009$. The given equation becomes

$$x^2 + y^2 = 2 \cdot 1009^2.$$

Since RHS is not a multiple of 4, it follows that x and y are odd. Take $x = a + b$, $y = a - b$ where $a = \frac{x+y}{2}$, $b = \frac{x-y}{2}$ are integers, since the sum and the difference of two odd numbers are even. The equation becomes

$$a^2 + b^2 = 1009^2.$$

From $1009 \equiv 1 \pmod{4}$, we deduce that the prime number 1009 is the sum of two nonzero squares, and this representation is unique: $1009 = 15^2 + 20^2$; whence it follows $1009^2 = 559^2 + 840^2$. We have also $1009^2 = 0^2 + 1009^2$, so that the given equation has two solutions in positive integers: $(m, n) = (728, 390), (1290, 390)$.

431. From $n^2 - m^2 = 2018^2 - 2017^2 = 4035 = 1 \cdot 3 \cdot 5 \cdot 269$ and $n - m < n + m$, we deduce $(n - m, n + m) = (1, 4035), (3, 1345), (5, 807), (15, 269)$; whence it follows $(m, n) = (2017, 2018)$, which doesn't satisfy the given conditions or

$$(m, n) = (671, 674), (401, 406), (127, 142).$$

432. If there are m_1, n_1 dividing $p - 1$, then $5^{m_1} \equiv 7^{n_1} \equiv 1 \pmod{p}$.
Suppose that such m_1, n_1 do not exist. Then $5 \cdot 7^k$, $1 \leq k \leq p - 2$ are different modulo p. Also, $5 \cdot 7^n \not\equiv 0, 1, 5 \pmod{p}$, so that there exist $n_1 < n_2 < p - 1$ such that $5 \cdot 7^{n_1} \equiv 5 \cdot 7^{n_2} \pmod{p}$, which implies $7^{n_2 - n_1} \equiv 1 \pmod{p}$, contradiction.

433. Observe that

$$a_1 + a_2 + \cdots + a_m \geq 2 + 4 + \cdots + 2m = m^2 + m,$$
$$b_1 + b_2 + \cdots + b_n \geq 1 + 3 + \cdots + (2n-1) = n^2;$$

hence $n^2 + m^2 + m \leq 2019$. By the Cauchy-Schwarz inequality,

$$169 \cdot 2019 \geq 169(n^2 + m^2 + m) = (5^2 + 12^2)(n^2 + m^2) + 169m \geq$$
$$\geq (5m + 12n)^2 + 169m.$$

Then

$$(5m + 12n)^2 \leq 169(2019 - m) \leq 169 \cdot 2019,$$

so that $5m + 12n \leq 584$. It remains to verify that $5m + 12n \notin \{582, 583, 584\}$.

434. Answer: n square-free.

The condition is necessary, since $S_{k^2 n} \subseteq S_n$ for every positive integers k, n. We will show that the condition is sufficient. This verifies easy for $n = 1, 2$.

Lemma 1. Let p be a prime number, and let n be a positive integer such that $\left(\frac{-n}{p}\right) = -1$. If a positive integer k is such that $\nu_p(k) = 1$, then $k \notin S_n$.

Proof. If $k = x^2 + ny^2$ and $p \nmid y$, then $\left(\frac{x}{y}\right)^2 \equiv -n \pmod{p}$, which contradicts $\left(\frac{-n}{p}\right) = -1$. Thus $p|y$, so $p|x$; hence $p^2|k$, contradiction with $\nu_p(k) = 1$.

Lemma 2. Let $p_1, \ldots, p_t$ be odd primes, and let n be a positive integer such that $\left(\frac{-n}{p_i}\right) = 1$ for each i. Then there exists $k \in S_n$ with $\nu_{p_i}(k) = 1$ for each i.

Proof. It suffices to prove the lemma for $t = 1$; the general case follows from the Chinese remainder theorem. Let $p = p_1$; as $\left(\frac{-n}{p}\right) = 1$, there exists a positive integer x such that $p|x^2 + n$. Now we can choose $x^2 + n$ or $(x + p)^2 + n$; otherwise p^2 divides both $x^2 + n$ and $(x + p)^2 + n$; whence $p|x$, contradiction.

Applying these two lemmas, we can take odd primes $p_1, \ldots, p_{n-1}$ such that $\left(\frac{-i}{p_i}\right) = -1$ and $\left(\frac{-n}{p_i}\right) = 1$ for each i. Then if $k \in S_n$ and $\nu_{p_i}(k) = 1$ for each i, then k satisfies the required condition.

We prove now that such p_i exist.

Lemma 3. Let a and b be positive integers such that $b > \min\{2, a\}$ is square-free. Then there exists an odd prime p such that $\left(\frac{-a}{p}\right) = -1$ and $\left(\frac{-b}{p}\right) = 1$.

Proof. We can suppose that a is also square-free.

Case 1. $a = 1$. Choose $p \equiv 7 \pmod{8}$, so that $\left(\frac{-1}{p}\right) = -1, \left(\frac{2}{p}\right) = 1$. Dividing by 2, we can suppose that $b > 1$ is odd. We have to find p such that $\left(\frac{b}{p}\right) = -1$. Using the Jacobi's symbol, we find

$$\left(\frac{b}{p}\right) = (-1)^{\frac{p-1}{2}} \left(\frac{p}{b}\right),$$

which is possible choosing appropriate residue $p \pmod{b}$ and applying the Dirichlet's theorem.

Case 2. $a > 1$. Choose $p \equiv 1 \pmod{8}$, so that $\left(\frac{-1}{p}\right) = 1$ and $\left(\frac{2}{p}\right) = 1$. Then the required condition is equivalent to $\left(\frac{a}{p}\right) = -1, \left(\frac{b}{p}\right) = 1$. Cancelling by 2, if necessary, we can suppose that a and b are odd. Using the Jacobi's symbol, we have

$$\left(\frac{p}{a}\right) = -1, \left(\frac{p}{b}\right) = -1.$$

Take a prime $q|b, q \nmid a$. Then by the Dirichlet's theorem, we find $p \equiv 1 \pmod{\frac{ab}{q}}, \left(\frac{p}{q}\right) = -1$, which ends the proof.

435. For an integer $n \geq 2$, denote

$$g(n) = \sum_{r=2}^{n} s(r),$$

with the convention $g(1) = 0$. Note that

$$s(6k-4) = 2, s(6k-3) = 3, s(6k-2) = 2, s(6k-1)$$
$$\geq 5, s(6k) = 2, s(6k+1) \geq 5$$

for any $k \geq 1$. Then for $n \geq 1$, we have

$$g(n+6) - g(n) \geq 2 + 3 + 2 + 5 + 2 + 5 = 19.$$

Suppose that for some $n \geq 1$, $g(n) \geq 3n - 5$ holds. Then

$$g(n+6) \geq 3n - 5 + 19 > 3(n+6) - 5.$$

It remains to check the result for $n \leq 6$:

n	$s(n)$	$g(n)$	$3n-5$
1	–	0	-2
2	2	2	1
3	3	5	4
4	2	7	7
5	5	12	10
6	2	14	11

436. If $3 \nmid p$, then $x^2 + x + 1 | x^{2p} + x^p + 1$. For $n = 2^{2^k}$, we have

$$4^n + 2^n + 1 = \left(2^{2^k}\right)^2 + 2^{2^k} + 1.$$

Since $2^k | 2^{2^k}$, the result follows.

437. If 6 or both 2 and 3 are not primitive roots modulo p, then the result is obvious.

If 6 and 2 are primitive roots modulo p, then $6^{\frac{p+1}{2}} \equiv 2^t \pmod{p}$ for some t. Clearly, $t \neq 1, \frac{p+1}{2}$. Now we can choose

$$(m,n) = \begin{cases} \left(\frac{p+1}{2} - t, \frac{p+1}{2}\right), \frac{p+1}{2} > t \\ \left(t - \frac{p+1}{2}, \frac{p-3}{2}\right), \frac{p+1}{2} < t \end{cases}.$$

438. Let $T_k = \frac{k(k+1)}{2}$ be the kth triangular number.

Lemma. Every positive integer n has a unique representation as $T_m + i$, where $|i| \leq \lfloor \frac{m}{2} \rfloor$. Moreover, $m = \lfloor \sqrt{2n} \rfloor$.

Proof. It is easy to verify that the representation is unique. We prove that $m = \lfloor \sqrt{2n} \rfloor$ is the desired m. It suffices to show that $\left|\frac{m(m+1)}{2} - n\right| \leq \lfloor \frac{m}{2} \rfloor$. Note that m satisfies

$$m^2 \leq 2n < (m+1)^2.$$

This implies $-2m \leq m^2 - 2n \leq 0$, with equality if and only if m is even, which proves the desired inequality.

Back to the problem, suppose $a_0 = k$; denote $m = \lfloor \sqrt{2k} \rfloor$ and $k = T_m + i$. Note that

$$a_0 + a_1 > 2a_0 = m(m+1) + 2i \geq m^2.$$

Then $a_0 + a_1 = (m+1)^2$; hence $a_1 = T_{m+1} - i$. Continuing in this manner, we get $a_2 = T_{m+2} + i$, $a_3 = T_{m+3} - i$ and so on. Define $b_n = a_{n+1} - a_n$, and then

$$b_0 = (m+1) - 2i, \; b_1 = (m+2) + 2i, \; b_2 = (m+3) - 2i, \ldots$$

In particular, $b_{n+2} = b_n + 2$. Denote $b = \min\{b_0, b_1\}$. Since b_0, b_1 have different parity, it follows that b_n cover all differences, except

$$1, 2, \ldots, b-1 \text{ and } b+1, b+3, \ldots, \max\{b_0, b_1\} - 1.$$

This comprises m numbers which cannot be written as differences of the a_n, which ends the proof.

439. Direct calculations show that there are no solutions for $n \leq 5$. Assume $n > 5$.
If n is odd, then

$$\left(n^{\frac{n+1}{2}}\right)^2 < n^{n+1} + n - 1 < \left(n^{\frac{n+1}{2}} + 1\right)^2.$$

If $n \equiv 2 \pmod{3}$, then

$$\left(n^{\frac{n+1}{3}}\right)^3 < n^{n+1} + n - 1 < \left(n^{\frac{n+1}{3}} + 1\right)^3.$$

If $n \equiv 0 \pmod 6$, then $n^{n+1} + n - 1 \equiv 2 \pmod 3$, which is not a square.
If $n = 6k - 2$, observe that

$$x^{n+1} + x - 1 = (x^2 - x + 1)\left(x^3 + x^2 - 1 + \sum_{l=1}^{k-1}(x^{6l+3} + x^{6l+2} - x^{6l} - x^{6l-1})\right).$$

Modulo $x^2 - x + 1$, we have

$$x^3 + x^2 - 1 + \sum_{l=1}^{k-1}(x^{6l+3} + x^{6l+2} - x^{6l} - x^{6l-1}) \equiv$$

$$\equiv x - 3 + \sum_{l=1}^{k-1}(x^3 + x^2 - 1 - x^5) \equiv k(x-3),$$

so,

$$\gcd\left(n^3+n^2-1+\sum_{l=1}^{k-1}(n^{6l+3}+n^{6l+2}-n^{6l}-n^{6l-1}), n^2-n+1\right)=$$
$$=\gcd(k(n-3),n^2-n+1)=1.$$

It follows that n^2-n+1 must be a sixth power. But it cannot even be a square: if $n^2-n+1=a^2$, then $(2n-1)^2-(2a)^2=5$, which is not possible for $n>5$.

440. Define $S_n = a_0 + \cdots + a_n$, and then

$$S_n = 2^{n+1} + 2^{\lfloor\frac{n+1}{2}\rfloor} + 2^{\lceil\frac{n+1}{2}\rceil} - 3 = a_{n+1} + 2^{\lceil\frac{n+1}{2}\rceil} - 3.$$

Call a positive integer *expressible* if it can be written as the sum of distinct terms of the sequence $(a_n)_{n \geq 0}$. If $x < S_n$ is expressible, then $S_n - x$ is expressible, by taking all terms not included in the expression of x and vice versa. Write $x \sim y$ for "x is expressible if and only if y is expressible." Then $x \sim S_n - x$, for $S_n > x$. Thus, $a_n \sim 2^{\lceil\frac{n}{2}\rceil} - 3$, so it suffices to find infinitely many n such that $2^n - 3$ is expressible and infinitely many n such that it's not. Since $S_{2n-3} = 2^{2n-2} + 2^n - 3$, we have $2^n - 3 \sim 2^{2n-2}$. Then, $S_{4n-7} = 2^{4n-6} + 2^{2n-2} - 3$ implies $2^{2n-2} \sim 2^{4n-6} - 3$.

We see that $2^3 - 3 = 5$ is expressible and $2^7 - 3 = 125$ is not. The infinite chains

$$2^3 - 3 \sim 2^6 - 3 \sim 2^{18} - 3 \sim \ldots \text{ and } 2^7 - 3 \sim 2^{22} - 3 \sim 2^{82} - 3 \sim \ldots$$

solve the problem.

441. Answer: $(x,y) = (2,0), (3,1), (7,3)$.

Assume $x \geq 8$ and let $p = 2^8 + 1$, which is a prime number. Then $5^y \equiv -4 \pmod{p}$ implies $5^{8y} \equiv 1 \pmod{p}$. It follows that the order of 5 modulo p divides $8y$ and 256. But $5^8 \equiv 111^2 \pmod{p}$, so y must be even. Looking at the equation mod 8, we have a contradiction, so there are no solutions for $x \geq 8$.

442. Answer: $(x,y) = (3,1), (7,3)$.

Assume $x \geq 8$. Then $5^y \equiv -4 \pmod{257}$ implies $5^{8y} \equiv 1 \pmod{257}$. So the order of 5 mod 257 divides $8y$ and 256. But $5^8 \equiv 111^2 \pmod{257}$ shows that y must be even. Considering the equation mod 8, we get $0 \equiv 4 \pmod 8$, contradiction.

443. Answer: $p = 5$, $(a,b) = (39, 15)$.

Denote $d = \gcd(a,b)$; then $a = da'$, $b = db'$, and $\gcd(a', b') = 1$. We have

$$p = \frac{db'}{2}\sqrt{\frac{a'-b'}{a'+b'}}$$

and $\gcd(a'-b', a'+b') = \gcd(a'-b', 2b') = \gcd(a'-b', 2) = \delta$.

Case 1. $\delta = 1$. Then $a' - b' = r^2$, $a' + b' = s^2$, so that $2b' = s^2 - r^2$. Moreover, $a' - b'$, $a' + b'$ are odd; therefore r and s are odd; whence

$$ps = dr \cdot \frac{s-r}{2} \cdot \frac{s+r}{2}.$$

Since s is coprime to r, it is also coprime to $s - r$ and $s + r$; hence $s|d$. This means that p is divisible by $r \cdot \frac{s-r}{2} \cdot \frac{s+r}{2}$, which is possible only when $r = 1$, $s - r = 2$; therefore $s + r = 4$ and $p = 2$.

Case 2. $\delta = 2$. Then $a' - b' = 2r^2$, $a' + b' = 2s^2$, so that $b' = s^2 - r^2 = (s-r)(s+r)$ and

$$2ps = dr(s-r)(s+r).$$

As above, s divides d and $2p$ is divisible by $r(s - r)(s + r)$.
If $r = s - r = 1$, then $s + r = 3$, $p = 3$.
If $r = 1$, $s - r = 2$, then $s + r = 4$ and 4 divides p, impossible.
If $r = 2$, $s - r = 1$, then $s + r = 5$, $p = 5$.

Now it remains to verify that $p = 5$ is a solution, and it is easy to see that the pair $(a, b) = (39, 15)$ satisfies the equality.

444. (a) Let $b_n = a_n^2 + na_n$; then

$$a_{n+2} - a_{n+1} = b_{n+1} - b_n, \quad \sum_{i=1}^{n-1}(a_{i+2} - a_{i+1}) = \sum_{i=1}^{n-1}(b_{i+1} - b_i), \quad a_{n+1} = b_n.$$

Let $\mathbb{P}$ be the set of prime numbers that divide at least one term of the sequence. Assume that $\mathbb{P}$ is finite, $\mathbb{P} = \{p_1, \ldots, p_k\}$. We know that $a_n | a_{n+1}$, so there exists N such that for all $n > N$ and $i \in \{1, 2, \ldots, k\}$, we have $p_i | a_n$. Then there exists t sufficiently large such that $tp_1 \ldots p_k + 1 > N$. For $n = tp_1 \ldots p_k + 1$, $a_n + n$ is not divisible by p_i, so there is a prime $p \notin \mathbb{P}$ which divides a_{n+1}, contradiction.

(b) Modulo 3, 5, and 17, the sequence is periodic and all terms are nonzero.

445. We shall use the fact: $\gcd(m^a - 1, m^b - 1) = m^{\gcd(a,b)} - 1$.

We will prove that n is a power of p. First, we show that $p|n$. Assume $p \nmid n$. If $n = 1$; then we have the condition trivially. If not, let $m^{pn} - 1 = (m^n - 1)q$, for some prime q. Note that if $\gcd(p, n) = 1$, then $\gcd(m^p - 1, m^n - 1) = m - 1$; hence

$$m^{pn} - 1 = l \cdot \frac{m^p - 1}{m - 1} \cdot (m^n - 1),$$

for some $l > 1$; therefore $q = l \cdot \frac{m^p - 1}{m - 1}$ is not a prime.

Now, let $n = p^k n_1$, with $n_1 > 1$, $p \nmid n_1$. We have $m^{p^{k+1}n_1} - 1 = \left(m^{p^k n_1} - 1\right) q$, for some prime q. Observe that $\gcd\left(m^{p^{k+1}} - 1, m^{p^k n_1} - 1\right) = m^{p^k} - 1$; hence

$$m^{p^{k+1}n_1} - 1 = l' \cdot \frac{m^{p^{k+1}} - 1}{m^{p^k} - 1} \cdot \left(m^{p^k n_1} - 1\right) = \left(m^{p^k n_1} - 1\right) q,$$

for some $l' > 1$, so $q = l' \cdot \frac{m^{p^{k+1}} - 1}{m^{p^k} - 1}$ is never a prime. Hence, $n = p^k$.

By the lifting the exponent lemma,

$$\nu_p\left((p-1)^{p^k} + 1\right) = \nu_p(p) + \nu_p(p^k) = 1 + k;$$

hence $p^{k+1} = pn | (p-1)^n + 1$.

446. Answer: $(x, y, z) = (3, 1, 1), (6, 2, 4)$.

Denote by C_m the congruence $2^x + 1 \equiv 7^y + 2^z \pmod{m}$.

Case 1. $z = 1$. Then $2^x = 7^y + 1$. If $x \geq 4$, then $16 | 7^y + 1$, contradiction. The only solution is $(x, y) = (3, 1)$.

Case 2. $z = 2$. Then $2^x = 7^y + 3$. C_8 implies $0 \equiv 7^y + 3 \pmod{8}$, contradiction.

Case 3. $z = 3$. Then $2^x = 7^y + 7$, so that $7 | 2^x$, contradiction.

Case 4. $z = 4$. Then $2^x = 7^y + 15$. If $x = 4$, then $1 = 7^y$, contradiction. If $x = 5$, then $17 = 7^y$, contradiction. If $x = 6$, then $y = 2$. If $x \geq 7$, then C_{128} implies

$$7^y + 15 \equiv 0 \pmod{128};$$

whence $y \equiv 10 \pmod{16}$

Case 5. $z \geq 5$. Then C_{32} implies $2^5 | 7^y - 1$; hence $4 | y$. C_5 implies $x \equiv z \pmod{4}$, C_7 implies $x \equiv 0 \pmod{7}$, and $y \equiv 1 \pmod{3}$. C_9 implies $x \equiv 3 \pmod{6}$, $y \equiv 4 \pmod{6}$, $z \equiv 1 \pmod{6}$ or $x \equiv 0 \pmod{6}$, $y \equiv 2 \pmod{6}$, and $z \equiv 4 \pmod{6}$. C_{13}, respectively, C_{19} lead to contradictions.

2.48 2020

447. Let $p_1, \ldots, p_m$ be the prime factors of $\mathrm{lcm}(S)$ and $t_j = \frac{x}{p_j}$. Observe that $t_j \in [0, 1]$. Let $g_j(A) = \nu_{p_j}(\mathrm{lcm}(A))$, and denote $f(S) = f_S(x)$. Then

$$\frac{x^{\Omega(\mathrm{lcm}(A))}}{\mathrm{lcm}(A)} = \prod_{p_j} \frac{x^{\nu_{p_j}(\mathrm{lcm}(A))}}{p_j^{\nu_{p_j}(\mathrm{lcm}(A))}} = \prod_{p_j} \left(\frac{x}{p_j}\right)^{\nu_{p_j}(\mathrm{lcm}(A))} = \prod_{j=1}^m t_j^{g_j(A)}.$$

The given sum is

$$\sum_{A \subseteq S} (-1)^{|A|} \prod_{j=1}^{m} t_j^{g_j(A)}.$$

We will prove the result by induction on $|S| + \sum_{j \in S} j$. The base case $|S| = 1$ is clear. For the inductive step, we first show that $f(S) \leq 0$. Let M be the set of elements in S not divisible by p_m. If $M \neq \emptyset$, then multiply all elements of M by p_m, which changes the sum by adding 1 to all $g_m(A)$ and get a new set S'. The sum becomes

$$\sum_{\emptyset \neq A \subseteq M} (-1)^{|A|} (t_m - 1) \prod_{j=1}^{m-1} t_j^{g_j(A)} = (t_m - 1) \left(\sum_{\emptyset \neq A \subseteq M} (-1)^{|A|} \prod_{j=1}^{m-1} t_j^{g_j(A)} \right),$$

which is the product of two nonpositive factors by the induction hypothesis, so $f(S') \geq f(S)$. Now we can divide each element in S' by p_m. Note that for all $A \neq \emptyset$, $g_j(A)$ is smaller by exactly 1, therefore $f(S) \leq f(S') = t_m f\left(\frac{S'}{p_m}\right) \leq 0$.

It remains to show that the sum is ≥ -1. Add the empty set into the calculations for this part, and denote this sum by $g(S) = f(S) + 1$. For each $A \subseteq M$ (including the empty set), we add 1 to $g_j(A)$, whose initial value is 0. By the indiction hypothesis, the difference in sum is

$$(t_m - 1) \sum_{A \subseteq M} (-1)^{|A|} \prod_{j=1}^{m-1} t_j^{g_j(A)},$$

which is nonpositive, since $t_m - 1 \leq 0$, and the other sum is ≥ 0 by the induction hypothesis. Therefore

$$g(S) \geq \sum_{A \subseteq S} (-1)^{|A|} \prod_{j=1}^{m-1} t_j^{g_j(A)} \cdot t_m^{\max(g_m(A),1)} =$$

$$= t_m \sum_{A \subseteq S} (-1)^{|A|} \prod_{j=1}^{m-1} t_j^{g_j(A)} \cdot t_m^{\max(g_m(A) - 1, 0)}$$

$$= t_m f(M \cup (S - M)/p_m) \geq 0.$$

448. Suppose without loss of generality that $x \leq y \leq z$ and consider the equation

$$t(1 - A^{-2}) = \frac{(x+1)(y+1)(z+1)}{(x-1)(y-1)(z-1)} := f(x, y, z).$$

(a) $t \leq 36$. Indeed, for $A = 1$, there is no solution and neither does if $x = 1$ or $y = 1$ or $z = 1$. Then $1 - A^{-2} \geq \frac{3}{4}$, and as f is decreasing in each variable, we have

$$f(x, y, z) \leq f(2, 2, 2) = 27.$$

There are only finitely many solutions with $y \geq 22$. Indeed, if $y \geq 22$, then

$$\frac{(y+1)(z+1)}{(y-1)(z-1)} < \left(\frac{23}{21}\right)^2 < \frac{6}{5}.$$

If $x = 2$, then we can see that for $t > 6$ there is no solution; if $t \in \{4, 5\}$ there are only finitely many solutions, as A is bounded above. If $x = 3$, then we can see that $t > 5$ yields no solution and $t \in \{3, 4\}$ yields finitely many solutions. Otherwise,

$$\frac{x+1}{x-1} \leq \frac{5}{3}$$

and there are no solution for $t > 3$, and there are only finitely many solutions for $t \in \{2, 3\}$.

For a fixed triple (t, x, y), there are only finitely many solutions z. Indeed, for

$$\frac{(x-1)(y-1)t}{(x+1)(y+1)} \leq 1$$

there are no solution z, else

$$\frac{3}{4} \leq \frac{(x+1)(y+1)}{(x-1)(y-1)t} \cdot \frac{z+1}{z-1},$$

which shows that z is bounded, as desired.

449. Answer: yes.

Call a sequence $(b_k)_{k \geq 0}$ *good* if $\gcd(b_k, b_{k+2}) = 1$ for all k. For a good sequence $(b_k)_{k \geq 0}$, let $a_0 = b_0$, $a_k = b_{k-1} b_k$ for $k > 0$. Then $\gcd(a_0, a_1) = b_0$, $\gcd(a_k, a_{k+1}) = b_k$ for all $k > 0$. It suffices to find a good sequence $(b_k)_{k \geq 0}$ such that it contains every positive integer exactly once.

Let p_k be the kth prime number and q_k the kth composite number (including 1). It is easy to see that $p_k \geq 2k - 1$, $q_k \leq 2k$; hence $p_k \geq q_k - 1$. It follows that

$$q_k \leq p_k + 1 < 2p_k < 2p_{k+2}, \gcd(q_k, p_k) = \gcd(q_k, p_{k+2}) = 1.$$

Define $b_{4k} = p_{2k+1}$, $b_{4k+1} = p_{2k+2}$, $b_{4k+2} = q_{2k+1}$, $b_{4k+3} = q_{2k+2}$ for all $k \geq 0$. This sequence contains every positive integer exactly once. Moreover,

$$\gcd(b_{4k}, b_{4k+2}) = \gcd(p_{2k+1}, q_{2k+1}) = 1, \gcd(b_{4k+1}, b_{4k+3}) = \gcd(p_{2k+2}, q_{2k+2}) = 1,$$
$$\gcd(b_{4k+2}, b_{4k+4}) = \gcd(p_{2k+3}, q_{2k+1}) = 1, \gcd(b_{4k+3}, b_{4k+5}) = \gcd(p_{2k+4}, q_{2k+2}) = 1.$$

450. We shall prove that the function

$$f(m) = \begin{cases} p, & m = p^k, \ p \text{ prime}, k \geq 1, \\ 1, & \text{otherwise} \end{cases}$$

is the unique solution. Let f be an arbitrary solution. Putting $m = 1$ in the given condition, we obtain $f(1) = 1$. Setting $m = p$, where p is a prime number, in the given condition, we get $f(p) = f(1)f(p) = p$. Induction on k shows that $f(p^k) = p$ for $k \geq 1$. Finally, let m have at least two distinct prime divisors, say

$$m = p_1^{k_1} p_2^{k_2} \ldots p_l^{k_l}, l \geq 2, k_j \geq 1 \text{ for } 1 \leq j \leq l \text{ and } p_1, \ldots, p_l \text{ distinct primes.}$$

Let D be the set of those divisors of m that have at least two distinct prime divisors. Then

$$m = f(1) \left(\prod_{\substack{1 \leq j \leq l, \\ 1 \leq i \leq k_j}} f(p_j^i) \right) \left(\prod_{d \in D} f(d) \right) = \left(\prod_{1 \leq j \leq l} p_j^{k_j} \right) \left(\prod_{d \in D} f(d) \right) = m \prod_{d \in D} f(d);$$

hence

$$\prod_{d \in D} f(d) = 1.$$

It follows that $f(d) = 1$ for all $d \in D$. Since $m \in D$, we conclude that $f(m) = 1$.

451. Consider all pairs $(m, n) \in \mathbb{N}^2$ such that $m \equiv 99 \pmod{1000}$ and $n = m + 9$. Then the first condition is satisfied. Now we will show that there are infinitely many pairs as above, satisfying the second condition.

Call k *good* if there are at least one such pair (m, n) and *very good* if there are infinitely many such pairs. We have to show that all positive integers are very good.

Lemma 1. There are arbitrarily small good numbers.

Proof. Consider

$$m = \frac{10^{2t+1} + 98198}{198}, n = m + 9,$$

for sufficiently large t. Then the corresponding k is as large as needed.

Lemma 2. If k is good, then $k+1$ is very good.

Proof. Let $k \in \mathbb{Z}$ good, and let (m, n) be a pair such that $n = m + 9$, $m \equiv 99$ (mod 1000) and $kn + s(n^2) = km + s(m^2)$. Then choosing the pairs $(10^N + m, 10^N + n)$ for sufficiently large N, we conclude that $k+1$ is very good.

From these two lemmas, it follows that all integers are very good.

452. Denote $n = \frac{p-1}{2}$. For $p = 3$ we have $n = 1$, and any value of x_1 satisfies the condition. For $p \geq 5$ there are 2^n solutions with $x_i \in \{0, 1\}$ for all i.

Lemma 1. For any x in the tuple and $t \neq 0, 1$ a quadratic residue, it follows that $(t-1)x + 1$ is a quadratic nonresidue.

Proof. Let s be the common value of the given sums. Then

$$\sum_{i=1}^{\frac{p-1}{2}} \left(\frac{x_i + r}{p}\right) \equiv \sum_{i=1}^{\frac{p-1}{2}} (x_i + r)^n \equiv s + \binom{n}{1} sr^1 + \ldots + \binom{n}{n-1} sr^{n-1} + \binom{n}{n} nr^n \equiv$$
$$\equiv s\left[\left(\frac{1+r}{p}\right) - \left(\frac{r}{p}\right)\right] + n\left(\frac{r}{p}\right) \pmod{p},$$

which implies

$$\sum_{i=1}^{\frac{p-1}{2}} \left(\frac{r^{-1}x_i + 1}{p}\right) \equiv s\left[\left(\frac{1+r^{-1}}{p}\right) - 1\right] + n \pmod{p}.$$

Choosing $r = (t-1)^{-1}$, we have

$$\sum_{i=1}^{\frac{p-1}{2}} \left(\frac{(t-1)x_i + 1}{p}\right) \equiv n \pmod{p}.$$

Since there are n Legendre symbols, each in $\{-1, 0, 1\}$ and $p > 2n$, it follows that each symbol is 1, whence the conclusion.

Lemma 2. For each i, $x_i \in \{0, 1\}$.

Proof. Let $Q = \{a^2 \bmod p\} \setminus \{0, 1\}$ be the set of quadratic residues mod p, other than 0, 1. Then $t \mapsto (t-1)x + 1$ is a bijection from Q to itself. From

$$\sum_{t \in Q} t \equiv -1 \pmod{p},$$

it follows by summation

$$-1 \equiv (-1)x - nx + n \pmod{p};$$

whence $x \equiv 1 \pmod{p}$, since $n + 1 \not\equiv 0 \pmod{p}$.

These two lemmas prove the assertion.

453. If $n = 1$, then the equality holds only when $(n, m) = (1, 1), (1, 2), (1, 3)$.

For each prime p, $n < p \leq m$, write $p = nk + r$, $0 \leq r \leq n - 1$. Clearly, $r \neq 0$ and $\gcd(r, n) = 1$; therefore we have $\varphi(n)$ values for r. For each fixed r, we have $k \geq 1$ and $nk + 1 \leq p \leq m$; whence $k \leq \frac{m-1}{n}$. There are $\lfloor \frac{m-1}{n} \rfloor$ possible choices for k, which proves the desired inequality.

Equality holds only when $n | m - 1$. Let $m = nt + 1$; for $n \geq 3$, there is $1 < r_0 < n$ such that $\gcd(r_0, n) = 1$. Equality holds only if $nt + r_0$ is a prime not greater than $m = nt + 1$, contradiction.

For $n = 2$, we want $\pi(m) = \frac{m+1}{2}$; therefore m is odd and $3, 5, 7, \ldots, m$ must be primes, which implies $m = 3, 5, 7$.

454. Let $(br^n)_{n \geq 0}$ be the geometric progression, with $b = b_1 \in \mathbb{Z}$. There are a, c such that $\frac{br^n - c}{a}$ is an integer for all n. Define $c_k = \frac{b}{a}(1 - r^k)$, and then

$$c_k r^n = \frac{br^n - c}{a} - \frac{br^{n+k} - c}{a}$$

is an integer for all nonnegative integers n and positive integers k. Thus $r = \frac{c_k r^{n+1}}{c_k r^n}$ is rational, so c_k is rational for all k.

Fix k and suppose that $\nu_q(r) < 0$ for some prime q. Then $\nu_q(c_k r^n) \to -\infty$ as $n \to \infty$, contradiction, since $c_k r^n$ is an integer. We deduce $\nu_q(r) \geq 0$ for all primes q; hence r is an integer.

455. We will first show that it is possible to choose c, d such that $\frac{a}{b} - \frac{c}{d} = \frac{1}{bd}$. Because $\gcd(a, b) = 1$, there exists an inverse b^{-1} of b modulo a. Let c with $1 \leq c \leq a - 1$ be such that $c \equiv -b^{-1} \pmod{a}$. Then we have $bc \equiv -1 \pmod{a}$; hence $a | bc + 1$. Define $d = \frac{bc+1}{a}$, which is a positive integer. We have $d = \frac{bc+1}{a} < b + \frac{1}{a}$. Because $a \geq 2$ and d is an integer, we get $d \leq b$. Hence we have

$$\frac{a}{b} - \frac{c}{d} = \frac{ad - bc}{bd} = \frac{1}{bd}.$$

If this is the smallest possible outcome, then we are done, because $\frac{1}{r} = bd$ is an integer. We will show that no smaller positive outcome is achievable. Let c, d be as above, and suppose that there are positive integers $c' \leq a$, $d' \leq b$ such that

$$0 < \frac{a}{b} - \frac{c'}{d'} < \frac{1}{bd}.$$

For $x = ad' - bc'$, we have $\frac{a}{b} - \frac{c'}{d'} = \frac{x}{bd'}$; hence $xd < d'$. Then $0 < xd < d' \leq b$, which means that xd and d' are distinct numbers whose difference is less than b. From $x = ad' - bc'$, we get $x \equiv ad' \pmod{b}$. We know that $ad - bc = 1$; hence $ad \equiv 1 \pmod{b}$; hence $xad \equiv x \pmod{b}$. Combining this, we get $ad' \equiv xad \pmod{b}$. Because $\gcd(a, b) = 1$, we may divide by a; hence $d' \equiv xd \pmod{b}$. However, d' and xd are distinct positive integers whose difference is smaller than b, contradiction.

456. If $a = 1$, then $\varphi(1) = 1$ and $\gcd(a,b) = 1$ for all positive integers b. The equation becomes

$$1 + b = 1 + \varphi(b) + 1 \Leftrightarrow \varphi(b) = b - 1.$$

This equation has a solution if and only if b is a prime number. It follows that the solutions are $(1, p)$ and similarly $(p, 1)$, with p prime number.

If $a, b \geq 2$, as $\gcd(b, b) > 1$, we have $\varphi(b) \leq b - 1$; therefore

$$\gcd(a, b) = a + b - \varphi(a) - \varphi(b) \geq a - \varphi(a) + 1.$$

Let p be the smallest prime divisor of a. Since for all t such that $tp \leq a$ we have $\gcd(tp, a) > 1$, it follows that $a - \varphi(a) \geq \frac{a}{p}$; hence

$$\gcd(a, b) \geq a - \varphi(a) + 1 \geq \frac{a}{p} + 1.$$

The two largest divisors of a are a and $\frac{a}{p}$. Since $\gcd(a, b)$ is a divisor of a greater than $\frac{a}{p}$, it must be equal to a, that is, $\gcd(a, b) = a$ and similarly, $\gcd(a, b) = b$; hence $a = b$. The equation becomes $2a = 2\varphi(a) + a$, that is, $a = 2\varphi(a)$. Then a is even, $a = 2^k m$, with $k \geq 1$ and m odd. Applying the multiplicative function φ, we get $\varphi(a) = 2^{k-1}\varphi(m)$, and the equation becomes $2^k m = 2 \cdot 2^{k-1}\varphi(m)$, or equivalently $m = \varphi(m)$; whence $m = 1$ and $(a, b) = (2^k, 2^k)$.

457. For $i \geq 0$, take $a = 2 \cdot 5^i$, $b = 3 \cdot 5^i$, then $2a + 3b = 13 \cdot 5^i$ and

$$\tau(a) = \tau(b) = \tau(2a + 3b) = 2(i + 1);$$

hence every even k satisfies the condition.

If k is odd, then $a, b, 2a + 3b$ have an odd number of divisors; hence they are squares, $a = x^2, b = y^2, 2a + 3b = z^2$; whence $2x^2 + 3y^2 = z^2$. This equation has no integer solutions, as we see easy, considering it modulo 3.

458. Answer: no solutions.

Case 1. $101 \nmid xy$. The Fermat's theorem implies $x^{100} \equiv y^{100} \equiv 1 \pmod{101}$; hence $100! \equiv 0 \pmod{101}$, contradiction.

Case 2. $101|x, 101|y$. Then $101|100!$, contradiction.

Case 3. $101 \nmid x, 101|y$. Then $100! \equiv 1 \pmod{101}$, which contradicts the Wilson's theorem.

Case 4. $101|x, 101 \nmid y$. Then $x = 101k$, with $k \geq 1, y \leq x - 1$, so that

$$100! = x^{100} - y^{100} \geq x^{100} - (x - 1)^{100}$$
$$= x^{99} + x^{98}(x - 1) + \ldots + (x - 1)^{99} > 101^{99},$$

contradiction.

459. By induction on k, we will show that all such sequences can be written as

$$\{0, 1, \ldots, x\} \cup \{0, 1, \ldots, y\}, x + y = k - 1 \text{ or } \{0, 1, \ldots, k\}.$$

The base case is trivial. Suppose that the statement is true for $k = n$ and prove it for $k = n + 1$. Note that if $x < y$, then

$$\binom{k+1}{0} + \cdots + \binom{k+1}{x} + \binom{k+1}{0} + \cdots + \binom{k+1}{y} =$$
$$= \binom{k+1}{0} + \cdots + \binom{k+1}{x} + \binom{k+1}{x+2} + \cdots + \binom{k+1}{k+1}$$
$$= 2^{k+1} - \binom{k+1}{x+1}.$$

Now we have $a_{k+1} = x + 1$ or $a_{k+1} = y + 1$, which finishes the proof.

460. Answer: $(1, 2, 3)$ and its permutations.

Assume wlog that $a \geq b \geq c$. Then

$$3a^3 \geq a^3 + b^3 + c^3 = (abc)^2 > a^3;$$

hence $3a > (bc)^2 > a$. Now $b^3 + c^3 = a^2(b^2c^2 - a) \geq a^2$ and so

$$18b^3 \geq 9(b^3 + c^3) \geq 9a^2 \geq b^4c^4 \geq b^3c^5,$$

so $18 \geq c^5$; therefore $c = 1$.

We must have $a > b$; otherwise $2a^3 + 1 = a^4$, which has no positive integer solutions. So

$$a^3 - b^3 \geq (b+1)^3 - b^3 > 1 \text{ and } 2a^3 > 1 + a^3 + b^3 > a^3,$$

which implies $2a^3 > a^2b^2 > a^3$ and so $2a > b^2 > a$. Therefore

$$4(1 + b^3) = 4a^2(b^2 - a) \geq 4a^2 > b^4,$$

so $4 > b^3(b-4)$; whence $b \leq 4$.

For each value $2 \leq b \leq 4$ we get a cubic equation for a:

$$b = 2: a^3 - 4a^2 + 9 = 0,$$
$$b = 3: a^3 - 9a^2 + 28,$$
$$b = 4: a^3 - 16a^2 + 65 = 0.$$

2.48 2020

The only case with an integer solution for a such that $b \leq a$ is $b = 2$; whence

$$(a, b, c) = (3, 2, 1).$$

461. We first show that b is a-good if and only if b is even and $p|a$ for all primes $p \leq b$. The condition $an + 1 \mid \binom{an}{b} - 1$ can be rewritten as

$$\frac{an(an-1)\ldots(an-b+1)}{b!} \equiv 1 \pmod{(an+1)}.$$

Suppose that there is a prime $p \leq b$ with $p \nmid a$. Take $t = \nu_p(b!)$. Then there exist positive integers c such that $ac \equiv 1 \pmod{p^{t+1}}$. If we take c big enough, and $n = (p-1)c$, then $an = a(p-1)c \equiv p - 1 \pmod{p^{t+1}}$ and $an \geq b$. Since $p \leq b$, one factor of the numerator is $an - p + 1$, which is divisible by p^{t+1}; hence the exponent of p in the numerator is at least $t + 1$, whereas the exponent of p in the denominator is exactly t. This means that $p \mid \binom{an}{b}$, so that $p \nmid \binom{an}{b} - 1$. As $p \mid an + 1$, we get that $an + 1 \nmid \binom{an}{b}$, so b is not a-good.

On the other hand, if for all primes $p \leq b$, we have $p|a$, and then every factor of $b!$ is coprime with $an + 1$, hence invertible modulo $an + 1$. Then

$$an(an-1)\ldots(an-b+1) \equiv b! \pmod{(an+1)}.$$

We can rewrite the LHS as

$$an(an-1)\ldots(an-b+1) \equiv (-1)(-2)\ldots(-b) \equiv (-1)^b b! \pmod{(an+1)}.$$

If b is even, we deduce $(-1)^b b! = b!$, as needed. If b is odd and we take $an + 1 > 2(b!)$, we have $(-1)^b b! \not\equiv b!$; therefore b is not a-good.

Suppose that b is a-good, but $b + 2$ is not. Then b is even and $p|a$ for all primes $p \leq b$, but there is a prime $q \leq b + 2$ for which $q \nmid a$, so $q = b + 1$ or $q = b + 2$. We cannot have $q = b + 2$, since b is even, so that $q = b + 1$ is a prime number.

462. Let k be a product of distinct primes congruent to 5 mod 6. Then nonnegative integers a, b satisfy

$$\frac{a^3 + b^3}{ab + 4} = 4k$$

if and only if $\{a, b\} = \{2k - 1, 2k + 1\}$.

We need a

Lemma. Let p be a prime number congruent to 2 mod 3. If integers a, b satisfy

$$a^2 + ab + b^2 \equiv 0 \pmod{p},$$

then $a \equiv b \equiv 0 \pmod{p}$.

Proof. If $p = 2$, then both a and b are even. Suppose that $p > 2$ and assume that $p \nmid b$. We can take an integer c such that $bc \equiv 1 \pmod{p}$, so we have $x^2 + x + 1 \equiv 0 \pmod{p}$ for $x = ac$; hence $(2x+1)^2 \equiv -3 \pmod{p}$. But then $\left(\frac{-3}{p}\right) = 1$, so that $p \equiv 1 \pmod{6}$, contradiction.

Back to the problem, suppose that $a^3 + b^3 = 4k(ab+4)$, $a \geq b$. If $a = b$, then

$$a^3 = 2k(a^2 + 4),$$

which implies $a^2 | 8k$. Since k is a product of distinct odd primes, we must have $\gcd(a, k) = 1$, so $a^2 | 8$; whence $a \in \{1, 2\}$. There is no solution in this case.

If $a > b$, then $(a, b) = (ds, dt)$, where $d = \gcd(a, b)$, so we have

$$d^3(s+t)(s^2 - st + t^2) = 4k(d^2st + 4).$$

Then $d^2 | 16k$. Since $a > b$, we have $s > t$. Also, $\gcd(d, k) = 1$, since k is a product of odd primes, so we must have $d^2 | 16$, which implies $d \in \{1, 2, 4\}$.

If $d = 4$, then $64 | 4k(16st + 4)$, impossible.

If $d \in \{1, 2\}$, then $2 | s + t$. By the lemma, we have $k | s + t$, so we can take a nonnegative integer u such that $s + t = 2ku$; thus,

$$d^3 u(s^2 - st + t^2) = 2(d^2 st + 4). \quad (*)$$

Since $s > t$ and $2 | s + t$, we must have $s - t \geq 2$. If $d = 2$, then $(*)$ implies

$$u(s^2 - st + t^2) = st + 1.$$

But $(s-t)^2 > 1$, so that $s^2 - st + t^2 > st + 1$; hence $u < 1$, impossible. If $d = 1$, then $(*)$ implies

$$u(s^2 - st + t^2) = 2(st + 4).$$

Since $2 | u$, we have $u \geq 2$, so $s^2 - st + t^2 \leq st + 4$, which implies $(s-t)^2 \leq 4$. Since $s - t \geq 2$, we have $s - t = 2$ and $u = 2$. From $s - t = 2$ and $s + t = 2ku = 4k$, we have $(s, t) = (2k+1, 2k-1)$; therefore $(a, b) = (2k+1, 2k-1)$. On the other hand, if $(a, b) = (2k+1, 2k-1)$, then $a^3 + b^3 = 4k(ab + 4)$.

Note that $2020 = 4 \cdot 5 \cdot 101$. By the above result, $a^3 + b^3 = 2020(ab + 4)$ if and only if $\{a, b\} = \{1009, 1011\}$.

463. For $n = 1$ is trivial: $f(1) = 1$. Let n be the smallest positive integer such that $f(n) \neq n$, and then $f(i) = i$ for $i \in \{1, 2, \ldots, n - 1\}$.

Case 1. $f(n) < n$. Let $f(n) = c$. Since $c < n$, we have $f(c) = c$, so

$$\prod_{l=1}^{2020} f_l(n) = c \prod_{l=1}^{2019} f_l(c) = c^{2020}, (n-1)^{2020} < c^{2020}$$
$$< n^{2020} + n^{2019} < (n+1)^{2020},$$

which implies $c = n$, contradiction.

Case 2. $f(n) > n$. Let $f(n) = c \geq n + 1$. Then

$$\prod_{l=1}^{2020} f_l(n) = c \prod_{l=1}^{2019} f_l(c) < n^{2020} + n^{2019} \Rightarrow \prod_{l=1}^{2019} f_l(c) < n^{2019}, (*)$$

so we must have $f_i(c) < n$ for some $i \in \{1, 2, \ldots, 2019\}$. Let q be the smallest among them. Since $f_q(c) < n$, we have $f_r(c) = f_q(c)$ for $r \geq q$. Let $f_q(c) = a < n$. We can write $(*)$ as

$$\prod_{l=1}^{2019} f_l(c) = \left(\prod_{l=1}^{q-1} f_l(c)\right) \cdot (f_q(n))^{2020-q} = \left(\prod_{l=1}^{q-1} f_l(c)\right) \cdot a^{2020-q} < n^{2019}. (**)$$

The inequality must hold for $n = c$, which yields

$$(c-1)^{2020} < \prod_{l=1}^{2020} f_l(c) < c^{2020} + c^{2019}.$$

Again we can rewrite it as

$$(c-1)^{2020} < \left(\prod_{l=1}^{q-1} f_l(c)\right) \cdot (f_q(c))^{2021-q} = \left(\prod_{l=1}^{q-1} f_l(c)\right) \cdot a^{2021-q} <$$
$$< c^{2020} + c^{2019}. (***)$$

From $(**)$, $(***)$ we get

$$\frac{(c-1)^{2020}}{a^{2021-q}} < \prod_{l=1}^{q-1} f_l(c) < \frac{n^{2019}}{a^{2020-q}} \Rightarrow (c-1)^{2020} < n^{2019} \cdot a < n^{2020} \Rightarrow$$
$$\Rightarrow c - 1 < n \Rightarrow c < n + 1,$$

contradiction, so such n doesn't exist, that is, $f(n) = n$ for all n.

464. For $1 \leq n \leq 8$, direct calculations show that only $n = 5$ is a solution. If both $n^2 + 2$ and $n^2 + 5$ are composite, then LHS> RHS.

For every composite n, we have $f(n) \leq \sqrt{n}$. Indeed, if n is a prime power, this verifies easy. If n is not a prime power and $f(n) > \sqrt{n}$, then the $p(m)$th prime divisor of n is larger than $\sqrt{n}$, contradiction.

Now, if $f(n^2 + 2) = n$, then $n|n^2 + 2$ implies $n|2$; if $f(n^2 + 2) = n - 1$, then $n - 1|n^2 + 2$; thus $n - 1|3$; if $f(n^2 + 2) = n - 2$, then $n - 2|n^2 + 2$; thus $n - 2|6$. If $f(n^2 + 5) = n$, then $n|n^2 + 5$; thus $n|5$; if $f(n^2 + 5) = n - 1$, then $n - 1|n^2 + 5$; thus $n - 1|6$.

We conclude that if $n > 8$, then $f(n^2 + 2) \leq n - 3$ and $f(n^2 + 5) \leq n - 2$; hence

$$f(n^2 + 2) + f(n^2 + 5) < 2n - 4.$$

2.49 2021

465. Let $m := \frac{n(n-1)}{2}, M = \{1, 2, \ldots, m\}$ and $D := \{a_i - a_j | a_i > a_j, 1 \leq i, j \leq n\} = \{d_1, \ldots, d_m\}$ be the multiset of differences. Let $A := M \cap D, B := M \cap A = \{b_1, \ldots, b_t\}, C := D - A = \{c_1, \ldots, c_t\}$, where t is the number wanted to be larger than $\left\lceil \frac{n(n-6)}{19} \right\rceil$.

Consider the generating function

$$F(z) = z^{a_1} + \cdots + z^{a_n} \text{ and } G(z) = F(z)F\left(\frac{1}{z}\right) = n + \sum \left(z^{d_k} + z^{-d_k}\right),$$

$$H(z) := z^{-m} + z^{-m+1} + \cdots + ! + z + \cdots + z^m.$$

We have

$$G(z) = n - 1 + H(z) - \sum \left(z^{b_i} + z^{-b_i}\right) + \sum \left(z^{c_i} + z^{-c_i}\right).$$

Take $|z| = 1$, and then $G(z) = |F(z)|^2 \geq 0$ and $|G(z) - (n - 1) - H(z)| \leq 4t$. Write $z = e^{2i\theta}$ and then $H(z) = \frac{\sin(2m+1)\theta}{\sin \theta}$. Take $(2m + 1)\theta = \frac{3\pi}{2}$, and then

$$H(z) = -\frac{1}{\sin \theta} < -\frac{n^2 - n + 1}{\frac{3\pi}{2}};$$

hence

$$|G(z) - (n - 1) - H(z)| \geq \frac{2}{3\pi}\left(n^2 - \left(1 + \frac{3\pi}{2}\right)n\right).$$

So we can get $t \geq \frac{n(n-6)}{6\pi}$.

466. Let $\zeta_a = e^{\frac{2\pi i}{a}}$ and then all $\zeta_a^j, 1 \leq j \leq a-1$ are different and similarly for b, c, so we can calculate

$$\frac{1}{(1-x^a)(1-x^b)(1-x^c)} = \frac{\lambda_2}{(1-x)^3} + \frac{\lambda_1}{(1-x)^2} + \frac{\lambda_0}{1-x} +$$

$$+ \sum_{j=1}^{a-1} \frac{\rho_j(a)}{1-\zeta_a^j x} + \sum_{j=1}^{b-1} \frac{\rho_j(b)}{1-\zeta_b^j x} + \sum_{j=1}^{c-1} \frac{\rho_j(c)}{1-\zeta_c^j x},$$

where λ_0, λ_1, and λ_2 are constants depending only of a, b, and c and

$$\rho_j(a) = \lim_{x \to \zeta_a^{-j}} \frac{1-\zeta_a^j x}{(1-x^a)(1-x^b)(1-x^c)} = \frac{1}{a\left(1-\zeta_a^{-jb}\right)\left(1-\zeta_a^{-jc}\right)}$$

and similarly for $\rho_j(b)$, $\rho_j(c)$. Now we calculate the coefficient of x^n in

$$\frac{\lambda_2}{(1-x)^3} + \frac{\lambda_1}{(1-x)^2} + \frac{\lambda_0}{1-x}$$

which is

$$\lambda_2 \binom{n+2}{2} + \lambda_1 \binom{n+1}{1} + \lambda_0 = \alpha n^2 + \beta n + \gamma$$

and denote the coefficient of x^n in the expansion

$$\sum_{j=1}^{a-1} \frac{\rho_j(a)}{1-\zeta_a^j x} + \sum_{j=1}^{b-1} \frac{\rho_j(b)}{1-\zeta_b^j x} + \sum_{j=1}^{c-1} \frac{\rho_j(c)}{1-\zeta_c^j x}$$

by Δ_n. We need to prove that

$$|\Delta_n| < \frac{a+b+c}{12}.$$

We have

$$\Delta_n = \frac{1}{a} \sum_{j=1}^{a-1} \frac{\zeta_a^{jn}}{\left(1-\zeta_a^{-jb}\right)\left(1-\zeta_a^{-jc}\right)} + \text{two similar terms},$$

so it suffices to show that

$$\left| \sum_{j=1}^{a-1} \frac{\zeta_a^{jn}}{\left(1-\zeta_a^{-jb}\right)\left(1-\zeta_a^{-jc}\right)} \right| \leq \frac{a^2}{12}.$$

By the Cauchy-Schwarz inequality, the above expression is bounded by

$$\left(\sum_{j=1}^{a-1}\left|1-\zeta_a^{-jb}\right|^{-2}\right)^{\frac{1}{2}}\left(\sum_{j=1}^{a-1}\left|1-\zeta_a^{-jc}\right|^{-2}\right)^{\frac{1}{2}}$$

$$=\frac{1}{4}\left(\sum_{j=1}^{a-1}\frac{1}{\left(\sin\frac{\pi jb}{a}\right)^2}\right)^{\frac{1}{2}}\left(\sum_{j=1}^{a-1}\frac{1}{\left(\sin\frac{\pi jc}{a}\right)^2}\right)^{\frac{1}{2}}.$$

When $1 \leq j \leq a-1$ varies, then $\{jb\}$, $\{jc\}$ each runs over a complete syste of residue classes mod n (except 0), so it suffices to show that

$$\sum_{j=1}^{a-1}\frac{1}{\left(\sin\frac{\pi j}{a}\right)^2} < \frac{a^2}{3}.$$

Consider the roots of the polynomial

$$P_a(\sin x) = \frac{\sin ax}{\sin x}, a \text{ odd}; \quad P_a(\sin x) = \frac{\sin ax}{\sin x \cos x}, a \text{ even},$$

and use the Viète's relations, so we can calculate

$$\sum_{j=1}^{a-1}\frac{1}{\left(\sin\frac{\pi j}{a}\right)^2} = \frac{a^2-1}{3},$$

which ends the proof.

467. The expression from the middle can be written as

$$\sum_{k=1}^{n}\sum_{dk \leq n}(\tau(k))^2 = \sum_{m=1}^{n}\sum_{k|m}(\tau(k))^2;$$

hence it suffices to prove that

$$5^{\omega(n)} \leq \sum_{d|n}(\tau(d))^2 \leq 5^{\Omega(n)}.$$

Since all the expression are multiplicative, it suffices to consider $n = p^k$, with prime.

The case $k = 0$ is obvious; for $k \geq 1$ we have to prove that

$$5 \leq \sum_{i=1}^{k+1} i^2 \leq 5^k,$$

which is easy.

468. Answer: $(x, y, z) = (45, 2, 1)$.

If $y = 1$, then $x^2 \equiv 3 \pmod 4$, impossible.

If $y = 2$, then $(x - 2)(x + 2) = 43^z \cdot 47^z$. From $\gcd(x - 2, x + 2) = 1$, it follows that $x - 2 = 43^z$, $x + 2 = 47^z$; hence $47^z - 43^z = 4$, with the unique solution $z = 1$.

If $y \geq 3$, then taking modulo 8, we deduce that z is even, $z = 2k$, $k \geq 1$, so

$$\left(x - 2021^k\right)\left(x + 2021^k\right) = 2^y.$$

We have $\gcd(x - 2021^k, x + 2021^k) = 2$, so that $2 \cdot 2021^k + 2 = 2^{y-1}$; hence $2021^k + 1 = 2^{y-2}$. It is clear that $y > 4$. Considering this equation modulo 4, we get a contradiction.

469. We have $p(q - 1)(p^q - 1) = q(p - 1)(q^p - 1)$. Assume $p > q$. Since $p | RHS$, we have $q^p - 1 \equiv q - 1 \equiv 0 \pmod p$, by the Fermat's theorem; hence $q < p \leq q - 1$, contradiction.

470. If n is even, $n = 2k$, then $a = 2k^2 - 1$, $b = 2k^2 + 1$ is a solution:

$$a^2 + n^2 = b^2 - n^2.$$

If n is odd, there are no solutions. For b odd, we have $4 | b^2 - n^2$, but $4 \nmid a^2 + n^2$, contradiction. For b even, after simplifying, we can suppose that $\gcd(a, b, n) = 1$. From $b^2 - n^2 \equiv 3 \pmod 4$, we have a prime $p \equiv 3 \pmod 4$ which divides $a^2 + n^2$. By the Fermat's sum of two squares theorem, it follows that $p | a, b$, and n, contradiction.

471. Let $\nu_p(n) = k$ for $n = p^k t$, $p \nmid t$, and let $s_p(n)$ denote the sum of digits in the base p expansion of n. Then

$$\nu_p(n!) = \frac{n - s_p(n)}{p - 1}.$$

We need to show that for every prime p,

$$\nu_p(n!) + \nu_p((n+1)!) + \nu_p((n+2)!) \leq \nu_p((3n)!)$$

or equivalently,

$$s_p(a) + s_p(b) \geq s_p(a + b)$$

for all positive integers a, b.

1. For $p > 3$,

If $p \nmid n$, then $s_p(n + 1) = s_p(n) + 1$ and $s_p(n + 2) = s_p(n) + 2$; therefore

$$s_p(n) + s_p(n + 1) + s_p(n + 2) = 3s_p(n) + 3 \geq s_p(3n) + 3.$$

If $p \nmid n$, then one of the numbers $n, n+1, n+2$, say i has the last digit greater than 2 in base p, and denote the others by j, k, i.e., $\{i,j,k\} = \{n, n+1, n+2\}$. Then

$$s_p(n) + s_p(n+1) + s_p(n+2) = s_p(i) + s_p(j) + s_p(k) =$$
$$= s_p(i-3) + s_p(j) + s_p(k) + 3 \geq s_p(3n) + 3.$$

2. For $p = 3$, let $\{i,j,k\} = \{n, n+1, n+2\}$, $i \equiv 1 \pmod 3$, $j \equiv 2 \pmod 3$, $k \equiv 0 \pmod 3$. Then

$$s_3(n) + s_3(n+1) + s_3(n+2) = s_3(i) + s_3(j) + s_3(k) =$$
$$= (s_3(i-1) + 1) + (s_3(j-2) + 2) + s_3(k) \geq s_3(3n) + 3.$$

3. For $p = 2$, if $4|n$, then $4|3n$. If there is no carry while adding a and b in base p, then $s_p(a) + s_p(b) = s_p(a+b)$. But if $4|n$, then there is at least one carry while adding n and $n+2$, so $s_2(n) + s_2(n+2) \geq s_2(2n+2) + 1$; thus

$$s_2(n) + s_2(n+2) + s_2(n+1) \geq (s_2(2n+2) + 1) + (s_2(n) + 1) \geq s_2(3n) + 3.$$

If $n \equiv 1 \pmod 4$, then

$$n = 1\ldots 01_2, n+1 = 1\ldots 10_2, n+2 = 1\ldots 11_2,$$
$$s_2(n) = s_2(n-1) + 1,$$
$$s_2(n+1) = s_2(n), s_2(n+2) = s_2(n) + 1,$$
$$s_2(n) + s_2(n+1) + s_2(n+2) = s_2(n-1) + s_2(n) + s_2(n+1) + 2 \geq s_2(3n) + 3.$$

If $n \equiv 2 \pmod 4$, then

$$n = 1\ldots 10_2, n+1 = 1\ldots 11_2, n+2 = 1\ldots 00_2,$$
$$s_2(n) = s_2(n-2) + 1, s_2(n+1) = s_2(n) + 1,$$
$$s_2(n) + s_2(n+1) + s_2(n+2) = s_2(n-2) + s_2(n) + s_2(n+2) + 2 \geq s_2(3n) + 3.$$

If $n \equiv 3 \pmod 4$, then

$$n = 1\ldots 11_2, n+1 = 1\ldots 00_2, n+2 = 1\ldots 01_2,$$
$$s_2(n) = s_2(n-2) + 1, s_2(n+2) = s_2(n+1) + 1,$$

In conclusion, $\nu_p(n!) + \nu_p((n+1)!) + \nu_p((n+2)!) \leq \nu_p((3n)!)$, for every prime $p \leq n+2$.

472. Let $2 = p_1 < p_2 < p_3 < \ldots$ be the sequence of prime numbers. For any integer $k > 2$, consoder the set $S = \{p_1, \ldots, p_k\}$. All prime factors of the numbers $\sigma(p_i)$ are smaller than p_k, since either $\sigma(p_i) < p_k$ or $\sigma(p_k) = p_k + 1$ is not a prime. There are 2^k subsets A of S, but if we consider the product of all elements of A, the sum of its divisors is divisible only by primes among $p_1, \ldots, p_{k-1}$, so if we consider the parity of the exponents of each of these primes, there are only 2^{k-1} results possible. By the Pigeonhole principle, there exist two subsets A, A' such that all exponents have the same parity. It follows that the product of all elements of the symmetric difference $A \Delta A'$ is a silly number.

Assume that there are l silly numbers $n_1, \ldots, n_l$. We will prove that there exists a silly number distinct of all these numbers. Let p_α be the largest prime factor of these l silly numbers, and let γ be the largest exponent in the prime factors decompositions of these numbers. Let p_β be a prime number larger than $\sigma(p_\alpha^{\gamma+1})$, and consider the set:

$$S = \left\{ p_1^{\gamma+1}, \ldots, p_\alpha^{\gamma+1}, p_{\alpha+1}, \ldots, p_\beta \right\}.$$

By construction, for each $a \in S$, the prime factors of $\sigma(a)$ are smaller than p_β, so we can use the same argument as before, but now the element we will construct is distinct from $n_1, \ldots, n_l$, since either it is divisible by a prime number which does not divide any of these numbers, or one of the prime numbers appear with a larger exponent than in any of the l silly numbers.

473. Consider the binary representations:

$$\sqrt{2020} = \overline{\ldots a_0.a_1 a_2 \ldots}_2, \quad \sqrt{2021} = \overline{\ldots b_0.b_1 b_2 \ldots}_2.$$

Then V_i is even if and only if $a_i = b_i$. If there are only finitely many odd V_n, then $a_i = b_i$ for every sufficiently large i. Then $\sqrt{2021} - \sqrt{2020}$ has a finite binary representation, so it is rational, contradiction.

Similarly, if there are only finitely many even V_n, then $a_i + b_i = 1$ for every sufficiently large i; hence $\sqrt{2020} + \sqrt{2021}$ has a finite binary expansion (since $1_2 = \overline{0.(1)}_2$), so it is rational, contradiction.

474. Choose distinct positive rational numbers $r_1, \ldots, r_{k+3}$ such that $r_i r_{i+1} r_{i+2} r_{i+3} = i$ for $1 \leq i \leq k$. Let $r_1 = x$, $r_2 = y$, $r_3 = z$ be some distinct primes greater than k. Then the remaining terms satisfy $r_4 = \frac{1}{r_1 r_2 r_3}$, $r_{i+4} = \frac{i+1}{i} r_i$. If r_i are represented as irreducible fractions, then the numerators are divisible by x for $i \equiv 1 \pmod 4$, by y for $i \equiv 2 \pmod 4$, by z for $i \equiv 3 \pmod 4$ and by none for $i \equiv 0 \pmod 4$. Note that $r_i < r_{i+4}$; thus the sequences $r_1 < r_5 < r_9 < \cdots$, $r_2 < r_6 < r_{10} < \cdots$, $r_3 < r_7 < r_{11} < \cdots$, $r_4 < r_8 < r_{12} < \cdots$ are strictly increasing and have no common terms; hence all terms are distinct. If $r_i = \frac{u_i}{v_i}$ is an irreducible fraction for all i, choose a prime p which divides neither v_i, $1 \leq i \leq k+1$, nor $v_i v_j (r_i - r_j) = v_j u_i - v_i u_j$ for $i < j$, and define a_i by the congruence $a_i v_i \equiv u_i \pmod p$. Then we have

$$iv_iv_{i+1}v_{i+2}v_{i+3} = r_iv_ir_{i+1}v_{i+1}r_{i+2}v_{i+2}r_{i+3}v_{i+3} =$$
$$= u_iu_{i+1}u_{i+2}u_{i+3} \equiv a_iv_ia_{i+1}v_{i+1}a_{i+2}v_{i+2}a_{i+3}v_{i+3} \pmod{p};$$

therefore $a_ia_{i+1}a_{i+2}a_{i+3} \equiv i \pmod{p}$ for $1 \leq i \leq k$.

If $a_i \equiv a_j \pmod{p}$, then $u_iv_j \equiv a_iv_iv_j \equiv u_jv_i \pmod{p}$, contradiction.

475. Case 1. z odd. If $y \geq 2$, then taking $z = 2z' + 1$, we get

$$x^2 = 5 \cdot 5^{2z'} - 4^y \implies (x - 2 \cdot 5^{2z'})(x + 2 \cdot 5^{2z'}) = 5^{2z'} - 4^y.$$

Write $x = 2 \cdot 5^{2z'} + x_1$, so we obtain

$$(5^{z'})^2 - 4x_1 \cdot 5^{z'} - x_1^2 - 4^y = 0,$$
$$\Delta' = 5x_1^2 + 4^y \equiv 5 \pmod{8}$$

for every x_1 odd, since the discriminant must be odd to get odd solutions of the equation in this case there are no solutions.

If $y = 1$, then $x = 5u + 1$, $5^{z'} = ku + 1$, with k, u integers,

$$x^2 - 1 = 5(5^{2z'} - 1) \implies u(5u + 2) = 5^{2z'} - 1 \implies (k^2 - 5)u = 2 - 2k.$$

Note that $|2 - 2k| \geq |k^2 - 5|$ only for $|k| \leq 3$. So $u = -6, -1, 0, 2$ are admissible, and after verifications we find $(x, y, z) = (1, 1, 1), (11, 1, 3)$.

Case 2. z even. Then $z = 2z'$, with z' positive integer; hence

$$4^y = (5^{z'} - x)(5^{z'} + x).$$

Note that one bracket is divisible by 2 and the other by 4; therefore $5^{z'} - x = 2$, and we obtain

$$4(5^{z'} - 1) = 4^y \implies 5^{z'} - 4^{y-1} = 1.$$

By the Mihăilescu's theorem, we find $(z', y) = (1, 2)$ and $(x, y, z) = (3, 2, 2)$.

476. Let $a \in S$ be minimum and let $b \in S$ be minimum such that $a \nmid b$. Let $x \in S$ be such that $x^2 | a(a + b)$. If $x^2 < a(a + b)$, then $x^2 \leq \frac{a(a+b)}{2} < b^2$; hence $a^2 | x^2 | a(a + b)$, so that $a | b$, contradiction. We conclude that $a(a + b) = x^2$, $x \in S$ and similarly, $b(a + b) = y^2$, $y \in S$. If $\gcd(a, b) = d$, $a = dA$, $b = dB$, then $d^2A(A + B) = x^2$, which shows that $A = s^2$, s positive integer and similarly, $B = t^2$. Also, $A + B = l^2$; hence

$$s^2 + t^2 = l^2, x = dsl, y = dtl.$$

Note that $x, y < b\sqrt{2} < 2b$. If $p \in S$ satisfies $p^2 | a(a + x)$, then either $a(a + x) = p^2, 2p^2$ or $p < b$. If $p < b$, then $a | p$, thus $a | x$, so $a | b$, contradiction.

If $q \in S$, $q^2 | b(b+y)$, then either $b(b+y) = q^2, 2q^2$ or $q < b$. If $q < b$, then

$$a | q, d^2 s^4 = a^2 | b(b+y) = d^2 t^3(t+l).$$

But $\gcd(s, t) = 1$, $s^4 | t + l$ and $(t+l)(t-l) = s^2$; hence $s = 1$, contradiction. The result follows now from the following:

Lemma. There do not exist positive integers $a, b, x,$ and y such that $a(a+b) = x^2$, $b(a+b) = y^2$ and $a(a+x), b(b+y)$ are squares or double squares.

Proof. We have shown that $a = ds^2$, $b = dt^2$, $x = dsl$, $y = dtl$, where $\gcd(s, t) = 1$ and $s^2 + t^2 = l^2$. Suppose that s is odd and t is even; then there exist u, v coprime such that $s = u^2 - v^2$, $t = 2uv$, $l = u^2 + v^2$. Note that

$$a(a+x) = d^2 s^3 (s+l) = 2d^2 s^2 u^2 s;$$

hence s is a square (s is odd), $s = m^2$. Similarly,

$$b(b+y) = d^2 t^3 (t+l) = 2d^2 t^2 (u+v)^2 t;$$

hence $t = n^2$ or $2n^2$. Then $s^2 + t^2 = l^2$ implies $m^4 + n^4 = l^2$ or $m^4 + 4n^4 = l^2$. It is well known that these diophantine equations have no nontrivial solutions.

477. Assume for contradiction that a_n has constant parity, or $\lfloor \sqrt{a_n} \rfloor$ is even for all $n > M$.

Lemma. $\lfloor \sqrt{a_{n+2}} \rfloor = 2 \lfloor \sqrt{a_n} \rfloor$.

Proof. Clearly, $a_{n+2} \leq 4a_n$, so it suffices to show that $2\lfloor \sqrt{a_n} \rfloor - 1 \leq \sqrt{a_{n+2}}$. Denote $t = \lfloor \sqrt{a_n} \rfloor, f(x) = x + (\lfloor x \rfloor)^2$; then $x > y$ iff $f(x) > f(y)$, so

$$f(f(a_n)) \geq f\left(f\left(t^2\right)\right) = f\left(2t^2\right) = 2t^2 + \left\lfloor t\sqrt{2} \right\rfloor^2 \geq 4t^2 - 2t\sqrt{2} \geq (2t-1)^2,$$

as desired. Since $\lfloor \sqrt{a_n} \rfloor$ is even for all $n > M$, it follows that $\lfloor \sqrt{a_{n+2}} \rfloor = 2 \lfloor \sqrt{a_n} \rfloor$.

Denote $b_n = \lfloor \sqrt{a_n} \rfloor$, then $a_n = b_n^2 + x_n, a_{n+1} = 2b_n^2 + x_n, a_{n+2} = (2b_n)^2 + 2(x_n - x_{n+1})$. It remains to show that $\{x_n\}$ is nonconstant: if $x_{n+1} < x_n$, then for some n large, $x_n < 0$, contradiction; if $x_n = x_{n+1}$, then $a_{n+1} = 2b_n^2 + x_n = b_{n+1}^2 + x_n$, whence $2b_n^2 = b_{n+1}^2$, contradiction.

478. Let $x = m\sqrt{2} = a + r, y = n\sqrt{7} = b + s$, where a and b are positive integers and r and s are real numbers in $[0, 1)$. Then $\lfloor x \rfloor = a$, $\lfloor y \rfloor = b$ and

$$\lfloor xy \rfloor - ab = \lfloor as + br + sr \rfloor \geq 0.$$

In fact, the inequality is strict: indeed, $r(2a + r) = 2m^2 - a^2$ is a positive integer; hence $(2a+r)s \geq \frac{s}{r}$, and similarly, $(2b+s)r \geq \frac{r}{s}$, so that $2(as + br + sr) \geq \frac{s}{r} + \frac{r}{s} \geq 2$. Finally, $as + br + sr \geq 1$.

479. If n is even, then 4 divides the denominator, but does not divide the numerator.

   If $n \equiv 1 \pmod 4$, then 5 divides the denominator, but not divides the numerator.

   If $n \equiv 3 \pmod 4$, then $\frac{3^n+47}{2} \equiv 5 \pmod 8$. Let p be a prime odd number which divides the numerator. Since the numerator is of the form $x^2 - 2$, we deduce that 2 is a quadratic residue mod p, which implies $p \equiv \pm 1 \pmod 8$; hence $\frac{3^n+47}{2} \equiv \pm 1 \pmod 8$, contradiction.

480. (a) Observe that $a_n = s_2(n)$ is the sum of digits of n in binary system. We will prove that n satisfies the required condition if and only if $n = 2^t - 1$ for some positive integer t. Let $m = \lceil \log_2 n \rceil$ be the number of binary digits of n.

   Necessity: Suppose that n has a 0 in its binary representation. If $n = \overline{a011\ldots1}$ with a arbitrary and i consecutive 1's on the right; then

   $$s_2(n) = s_2\left(n\left(2^{m-1}+1\right)\right) = s_2(a) + 1 + s_2(n) - 1 > s_2(n),$$

   contradiction.

   Sufficiency: It suffices to show that $s_2((2^t - 1)k) = t$ for all odd k, $1 \le k \le 2^t - 1$. Write $k = \overline{c1}$ and calculate

   $$s_2((2^t - 1)k) = s_2(\overline{c10\ldots0} - k) = s_2(c) + t - s_2(k) + 1 = t.$$

   (b) Take $m = 2^t$, and then $a_{km} \ge 1 = a_m$.

2.50 2022

481. If

$$g(d) = \sum_{e|d} f(e),$$

then by the Möbius inversion,

$$f(d) = \sum_{e|d} \mu\left(\frac{d}{e}\right) g(e).$$

Now, we need to prove that $k | g(k)$ for all k if and only if

$$n \Big| \sum_{d|m} \left(\frac{\frac{n}{d}}{\frac{m}{d}}\right) \sum_{e|d} \mu\left(\frac{d}{e}\right) g(e) = \sum_{e|m} g(e) h\left(\frac{m}{e}, \frac{n}{e}\right),$$

where

$$h(m,n) = \sum_{d|m} \mu(d) \binom{\frac{n}{d}}{\frac{m}{d}}.$$

If we prove that $n|h(m,n)$ for all m, n, then we see that $e|g(e)$ for all e implies that

$$n = e \cdot \frac{n}{e} \Big| g(e) h\Big(\frac{m}{e}, \frac{n}{e}\Big)$$

For all e one direction follows. Conversely, if we assume the second condition and $n|h(m,n)$ for all m, n, and then to prove $m|g(m)$ for all m by induction, we can assume that

$$n \Big| g(m) h\Big(1, \frac{n}{m}\Big) = g(m) \cdot \frac{n}{m}$$

and hence $m|g(m)$ as well.

So we have to prove that $n|h(m,n)$, that is

$$n \Big| \sum_{d|m} \mu(d) \binom{\frac{n}{d}}{\frac{m}{d}}.$$

By the inclusion-exclusion principle, the expression counts the number of choices of a set of m elements from n objects in a circle such that the n translates of this set are all distinct. But these sets fall into orbits of size n, so their number is divisible by n, as desired.

482. Let V be the set of all divisors of elements from A and W be the set of all multiples of elements from A which are in D.

Lemma. $|V| \cdot |W| \geq |A| \cdot |D|$.

Proof. Induction of the number of prime divisors of n. The case $n = 1$ is trivial. Take a prime $p|n$. For any set S, let $S_i = \{s \in S | \nu_p(s) = i\}$. Then

$$|V_0| \geq |V_1| \geq \cdots \geq |V_n|, |W_0| \leq |W_1| \leq \cdots \leq |W_n|.$$

By the Chebyshev's inequality,

$$|V| \cdot |W| = \Big(\sum_{i=0}^n |V_i|\Big) \Big(\sum_{i=0}^n |W_i|\Big) \geq (n+1) \sum_{i=0}^n |V_i| \cdot |W_i|$$

$$\geq (n+1) \sum_{i=0}^n |A_i| \cdot |D_i| = |A||D|.$$

Back to the problem, we may assume that for any $d \in D$, if d has a divisor and a multiple in A, and then $d \in A$; hence $V \cap W = A$ and $|V \cup W| = |D| - |B|$, so

$$|D| - |B| + |A| = |V| + |W| \geq 2\sqrt{|V| \cdot |W|},$$

whence the conclusion.

483. (a) We have the implications

$$n(4n+1) = m(5m+1) \Longrightarrow (n-m)(4m+4n+1) = m^2,$$

$$(n-m)(5m+5n+1) = n^2,$$

$$\gcd(m^2, n^2) = (n-m)\gcd(4m+4n+1, 5m+5n+1) = n-m \Longrightarrow n-m$$
$$= (\gcd(m,n))^2;$$

thus, $n - m$ is a square.

(b) $(n, m) = (38, 34)$.

484. Answer: $(m, n) = (-18, -2)$, (a, a) for any integer a.

We have

$$\left(2n^2 + 5m - 5n - mn\right)^2 - n^4 = m^3 n - n^4,$$

$$(n-m)(n-5)\left(3n^2 + 5m - 5n - mn\right) = n(m-n)(m^2 + mn + n^2);$$

hence $m = n$ is a solution.

If $m \neq n$, then $(n-5)(3n^2 + 5m - 5n - mn) + n(m^2 + mn + n^2) = 0$. Solve it for m, and then the discriminant $\Delta = (2n-5)^2(25 - 4n^2)$ is nonnegative only for $-2 \leq n \leq 2$. But $25 - 4n^2$ is a square; hence $n = 0, \pm 2$. If $n = 0$, then $m = 0$; if $n = -2$, then $m = -18$; if $n = 2$, then $m = 2$.

485. Answer: 31.

Note that $402 = 2 \cdot 3 \cdot 67$; hence $n \equiv 1 \pmod 2$. Assume that $402 | n^2 - 1$. Then $n \equiv \pm 1 \pmod 3$ and $n \equiv \pm 1 \pmod{67}$. There are four such residue classes mod 402. If $402 | n^4 - 1$, then $67 \nmid n^2 + 1$, since $67 \equiv 3 \pmod 4$, so we have the same residue classes as above. If $402 | n^3 - 1 = (n-1)(n^2 + n + 1)$, then $n \equiv 1 \pmod 3$. If $n \not\equiv 1 \pmod{67}$, then $67 | (2n+1)^2 + 3$; thus $2n + 1 \equiv \pm 8 \pmod{67}$; whence $n \in \{29, 37\} \pmod{67}$. Then we have $4 + 2 = 6$ residue classes.

Since $\lfloor \frac{2022}{402} \rfloor = 5$, we have $5 \cdot 6 = 30$ such numbers among $\{1, 2, \ldots, 2010\}$. In the set $\{2011, \ldots, 2022\}$, only 2011 satisfies the $3 \cdot 67$ condition.

486. Let $x_1 < x_2 < \cdots < x_{4k+2}$ be the given integers. We will prove that there is an integer $k \leq 4n$ such that $x_{4k+2} + 2x_k > 3x_{k+1}$. Suppose not, and denote $y_k = x_{4n+2} - x_k$ for all k. Then

$$3y_{k+1} = 3x_{4n+2} - 3x_{k+1} \leq 2x_{4n+2} - 2x_k = 2y_k \Rightarrow y_k \geq \frac{3}{2} y_{k+1}$$

for all $k \leq 4n$; hence

$$5^n \geq y_1 \geq \left(\frac{3}{2}\right)^{4n} y_{4n+1} \geq \left(\frac{3}{2}\right)^{4n} = \left(5 + \frac{1}{16}\right)^n,$$

contradiction.

487. Answer: $n = 1, 3$.

Let n be a positive integer having the required property and $d_1, d_2, \ldots, d_k$ a permutation of the divisors of n as in the enounce. Denote $s_i = \sqrt{d_1 + d_2 + \cdots + d_i}$ for all $i \leq k$. Call an integer l good if $s_i = i$ and $d_i = 2i - 1$ for all $i \leq l$. We will prove that every integer $l \leq k$ is good.

Note that

$$d_i = s_i^2 - s_{i-1}^2 \geq s_i + s_{i-1} \geq 2$$

or all $i \geq 2$; hence $d_1 = s_1 = 1$. Let $l \leq k - 1$ be a good integer. We will show that $l + 1$ is also good. Indeed, as $s_{l+1} + s_l$ divides $s_{l+1}^2 - s_l^2 = d_{l+1}$, it divides also n. Then there is a positive integer m such that

$$d_m = s_{l+1} + s_l \geq 2s_l + 1 \geq 2l + 1.$$

As $d_m > d_i$ for all $i \leq l$, it follows that $m \geq l + 1$. But then

$$s_{l+1} + s_l = d_m = (s_m - s_{m-1})(s_m + s_{m-1})$$
$$\geq (s_m - s_{m-1})(s_{l+1} + s_l) \geq s_{l+1} + s_l;$$

hence the inequalities are in fact equalities, so that $m = l + 1$ and $s_m - s_{m-1} = 1$; whence $s_{l+1} = s_l + 1 = l + 1$. In conclusion,

$$d_{l+1} = (s_{l+1} - s_l)(s_{l+1} + s_l) = 2l + 1,$$

so that $l + 1$ is good.

We deduce that n is divisible by $1, 3, \ldots, 2k - 1$. Conversely, if n is divisible by $1, 3, \ldots, 2k - 1$, then it has the required property. If $k = 1$, then $n = 1$; if $k = 2$, then $n = 3$; if $k > 2$, then d_{k-1} is an odd divisor of $n - d_{k-1} = 2$, hence $d_{k-1} = 1$, contradiction.

488. Suppose that there is an integer $k \geq 9$ such that $a_i < a_k$ for all $i \leq k-1$. Then for $i < \frac{k}{2}$ we know that $a_{k-2i} + a_{k-i}$ is a multiple of a_k and $a_{k-2i} + a_{k-i} < 2a_k$; hence $a_{k-2i} + a_{k-i} = a_k$. Taking $i = 1, 2, 4$, we deduce

$$a_{k-8} + a_{k-4} = a_{k-4} + a_{k-2} = a_k = a_{k-2} + a_{k-1},$$

so that $a_{k-8} = a_{k-2}$ and $a_{k-4} = a_{k-1}$. Induction on l shows that a_{k-1} divides a_{k-1-3l} for all $l \leq \frac{k-2}{3}$, if $a_{k-1} \leq \max\{a_1, a_2, a_3\}$ and a_{k-2} divides a_{k-2-6l} for all $l \leq \frac{k-3}{6}$, if $a_{k-2} \leq \max\{a_1, \ldots, a_6\}$. We deduce $a_k = a_{k-1} + a_{k-2} \leq 2\max\{a_1, \ldots, a_6\}$, hence the sequence $(a_n)_{n \geq 1}$ is bounded.

Let λ be the maximum value which the sequence takes infinitely many times, and let $\kappa \geq 0$ an integer such that $a_n \leq \lambda$ for each $n \leq \kappa + 1$. Omitting, if necessary, the first κ terms of the sequence, that is, replacing a_n by $\widehat{a}_n = a_{n+\kappa}$, which does not change the enounce, we can suppose that $\kappa = 0$, that is, $a_n \leq \lambda$ for all $n \geq 1$.

Consider the set $\mathcal{E} = \{k \geq 1 \mid a_k = \lambda\}$ and $k < l < m$ three consecutive elements from $\mathcal{E}$. If $k \equiv l \pmod{2}$, then $\lambda = a_l$ divides $a_k + a_{\frac{k+l}{2}} = \lambda + a_{\frac{k+l}{2}}$; hence $a_{\frac{k+l}{2}} = \lambda$, contradiction. It follows that $k \not\equiv l \pmod 2$ and similarly, $l \not\equiv m \pmod 2$. But then $\lambda = a_m$ divides $a_k + a_{\frac{k+m}{2}} = \lambda + a_{\frac{k+m}{2}}$; hence $a_{\frac{k+m}{2}} = \lambda$, $\frac{k+m}{2} \in \mathcal{E}$, so that $\frac{k+m}{2} = l$; therefore $l - k = m - l$ and more generally, $\mathcal{E}$ is an arithmetic progression. Let d be the common difference. If m, n are positive integers and $m + n \in \mathcal{E}$, then $\lambda = a_{m+n}$ divides $a_n + a_m$ and $m + n + d \in \mathcal{E}$; hence $\lambda = a_{m+n+d}$ divides $a_{n+d} + a_m$. It follows that $a_n \equiv -a_m \equiv a_{n+d} \pmod{\lambda}$. As a_n, a_{n+d} are at most λ, we deduce $a_n = a_{n+d}$.

In conclusion, the sequence $(a_n)_{n \geq 1}$ is d-periodic.

489. There exists $a \in S$ such that $\{\gcd(a, s) | s \in S - \{a\}\}$ contains at least two elements. Since a has only finitely many divisors, there is $d | a$ for which the set

$$B = \{b \in S \mid \gcd(a, b) = d\}$$

is infinite. Choose $c \in S$ such that $\gcd(a, c) \neq d$. Choose $b_1, b_2 \in B$ such that

$$\gcd(b_1, c) = \gcd(b_2, c) = d'.$$

If $d = d'$, then $\gcd(a, b_1) = \gcd(c, b_1) \neq \gcd(a, c)$. If $d \neq d'$, then either

$$\gcd(a, b_1) = \gcd(a, b_2) = d, \gcd(b_1, b_2) \neq d,$$

or

$$\gcd(c, b_1) = \gcd(c, b_2) = d', \gcd(b_1, b_2) \neq d'.$$

490. We will prove that for any prime $p|a$, there exists a prime $q|b$ such that $\nu_q(b) = \nu_p(a)$. Let m be the number of distinct prime factors of b and consider a prime factor p of a. By the Zsigmondy's theorem, we can choose distinct primes r, s such that the order of p mod r is s and r, s do not divide $\nu_p(a)$. Then choose

$$d \equiv 1 \pmod{\nu_p(a)}, d \equiv 0 \pmod{r^{mx}}, d \equiv 0 \pmod{s},$$

and write $d = \nu_p(a) \cdot n + 1$. By the Lifting the exponent lemma, we have

$$\nu_r(\sigma(a^n)) \geq 1 + \nu_r(d) - \nu_r(s) \geq mx + 1.$$

From the Pigeonhole principle, we have a prime $q|b$ with $\nu_r(q^{\nu_q(b) \cdot n+1} - 1) \geq x$. If l is the order of q mod r, then by the Lifting the exponent lemma, we have

$$\nu_r(n \cdot \nu_q(b) + 1) \geq x - \nu_r(q^l - 1) + \nu_r(l).$$

Choosing x large enough, we get

$$\nu_p(a) \cdot n \equiv \nu_q(b) \cdot n \pmod{r^x},$$

which gives $\nu_p(a) = \nu_q(b)$.

491. Let p be a prime divisor of n, and choose S to be the set of all numbers with all prime factors $>p$. Then no prime at most p can divide $\frac{a^c - b^c}{a - b}$. Indeed, if q divides it, then $a^c \equiv b^c \pmod{q}$, $a \equiv g^m \pmod{q}$, $b \equiv g^n \pmod{q}$ for some primitive root g, therefore $cm \equiv cn \pmod{q-1}$. But since $\gcd(c, p-1) = 1$, we must have $m \equiv n \pmod{p-1}$ and so $q|a-b$; hence

$$\nu_q(a^c - b^c) = \nu_q(a-b) + \nu_q(c) = \nu_q(a-b),$$

so $q \nmid \frac{a^c - b^c}{a - b}$.

Finally, since $p|n$, then $n \notin S$, as desired.

492. Answer: $(a, b, c) = (1, 1, n! - 1)$, $n \geq 2$ and all permutations.

If $a \leq b \leq c$ and $ab > 5$, then

$$(abc)^2 = ab \cdot bc \cdot ca \equiv (-1)^3 \equiv 2 \pmod{3},$$

contradiction. We deduce $ab \in \{1, 5\}$.

If $ab = 1$, then $a = b = 1$ and $c = n! - 1$, for some positive integer $n \geq 2$.

If $ab = 5$, then $a = 1$, $b = 5$ and $5c + 1 \geq 26$ cannot be a factorial, since $5 \nmid 5c + 1$.

493. If $P(x) = x^{5n-1} + x^{5n-2} + x^{5n-3} + x + 1$ and ω is a primitive root of unity of order 5, we see that $\{\omega^{5n-1}, \omega^{5n-2}, \omega^{5n-3}, \omega, 1\} = \{1, \omega, \omega^2, \omega^3, \omega^4\}$; therefore $P(\omega) = 0$, which implies $n^4 + n^3 + n^2 + n + 1 | P(n), \forall n \geq 2$. If $n \geq 2$, then $n^{5n-1} > n^4$, hence $a = P(n)$ is a composite number.

494. Answer: $p = 3$.

Note that $f(n) \equiv n(n-6) \pmod 9$.

If $3 | p^2 + 32$, then the sum of digits of $f(p^2 + 32)$ is ≥ 9.

If $3 \nmid p^2 + 32$, then $p = 3$ and $f(3^2 + 32) = 1102$, whose sum of digits is 4.

495. Note that $x_2 = 7 \equiv 1 \pmod{2 \cdot 3}$ and $x_3 = 7 + 3 \cdot 1 = 10 \equiv 1 \pmod{3 \cdot 3}$.

Consider $p > 3$ and observe that $x_n \equiv 1 \pmod 3$ for all n, so it suffices to show that $p | x_p - 1$ for $p > 3$. The general term of the sequence is

$$x_n = \left(\frac{1+\sqrt{13}}{2}\right)^n + \left(\frac{1-\sqrt{13}}{2}\right)^n, \forall n \in \mathbb{N}^*.$$

By the binomial expansion,

$$\left(\frac{1+\sqrt{13}}{2} + \frac{1-\sqrt{13}}{2}\right)^p - \left(\frac{1+\sqrt{13}}{2}\right)^p - \left(\frac{1-\sqrt{13}}{2}\right)^p = 1 - x_p \equiv 0 \pmod p.$$

496. Answer: yes.

Set $a = x^{2023}$, $b = x^{2021}$, $c = y^{2024}$, $d = y^{2022}$ for some positive integers x and y, and let $q = \frac{m}{n}$ in lowest terms. Then

$$\frac{a^{2021} + b^{2023}}{c^{2022} + d^{2024}} = \frac{2x^{2021 \cdot 2023}}{2y^{2022 \cdot 2024}} = \frac{x^{2021 \cdot 2023}}{y^{2022 \cdot 2024}} = \frac{m}{n}.$$

Consider $x = m^{x_1} n^{x_2}, y = m^{y_1} n^{y_2}$. Then it suffices to solve the equations in integers:

$$2021 \cdot 2023 x_1 - 2022 \cdot 2024 y_1 = 1, \; 2021 \cdot 2023 x_2 - 2022 \cdot 2024 y_2 = -1.$$

From $\gcd(2021 \cdot 2023, 2022 \cdot 2024) = 1$, we conclude that these equations have solutions in positive integers.

497. Answer: $n+1$ prime.

From

$$\sum_{i=1}^{n} \sum_{j=1}^{n} \frac{ij}{n+1} = \frac{(1+2+\cdots+n)^2}{n+1} = \frac{n^2(n+1)}{4}$$

it follows that the given equation is equivalent to

$$\sum_{i=1}^{n}\sum_{j=1}^{n}\left\{\frac{ij}{n+1}\right\} = \frac{n^2}{2}, \quad (*)$$

where $\{x\}$ is the fractional part of x. Since

$$\left\{\frac{ij}{n+1}\right\} + \left\{\frac{i(n+1-j)}{n+1}\right\} = \begin{cases} 1, n+1 \nmid ij \\ 0, n+1 \mid ij \end{cases},$$

By pairing up terms, we find that the equality $(*)$ only if the 0 case never occurs. Thus n is a solution exactly if $n+1$ is not the product of two smaller positive integers, i.e., $n+1$ is a prime.

498. We have $(19b+1)^2 = (a^2 - a + 1)(a^6 - a^4 - a^3 + a + 1)$. Also,

$$\gcd(a^2 - a + 1, a^6 - a^4 - a^3 + a + 1) \in \{1, 19\}.$$

But $19 \nmid (19b+1)^2$; therefore the gcd is 1, which implies that both numbers are squares. Then $(2a-1)^2 + 3$ is a square; hence $a = 0, 1$ and $b = 0$.

499. Answer: $(p, q) = (3, 3), (7, 5)$.

Assume that $q \geq 7$ and $p \geq 13$. Working modulo 7 gives us $p \equiv q - 2 \pmod{3}$. Since $p, q \neq 3$, we have $p \equiv 2 \pmod{3}$ and $q \equiv 1 \pmod{3}$. Comparing the exponents of 2 in both sides, we have $q - 2 = \nu_2(q!) = q - s_2(q)$; hence $s_2(q) = 2$, that is, $q = 2^k + 1$. But $q \equiv 1 \pmod{3}$ is a contradiction.

500. The polynomial $P = (9x^2 + 3x + 1)(27x^3 + 2)$ satisfies the condition: primes $p \equiv 1 \pmod{3}$ divide the first factor (for some values of x), and primes $p \equiv 2 \pmod{3}$ divide the second factor. From $P \equiv 2 \pmod{3}$, it follows that P excludes 3.

2.51 2023

501. Answer: $n = 4$; n square-free.

If $n = 4$, then only $8 = 4 \cdot 2$ has four divisors.

If n is square-free and has k distinct prime divisors, then any number with n divisors can have at most k distinct prime divisors, so these primes must be the primes dividing n, which are in finite number.

Conversely, let $n = p_1^{e_1} \ldots p_k^{e_k}$ be the canonical representation of n. If $e_i \geq 2$ for some odd p_i, then $p_i^{e_i - 1} \geq e_i + 1$ and $p_j^{e_j} \geq e_j + 1$ for all others, so

$$p_i^{p_i^{e_i-1}-1} q^{p_i - 1} \prod_{j \neq i} p_j^{p_j^{e_j}-1}$$

has n divisors for an arbitrary $q \nmid n$ and is a multiple of n; hence there are infinitely many such multiples. If $p_1 = 2$, $e_1 \geq 3$, then $2^{e_1 - 1} \geq e_1 + 1$ and a similar argument applies. Finally, if $n = 4p_1\ldots p_k$, $k \geq 1$, then n has infinitely many multiples

$$2^{p_1 - 1} p_1 p_2^{p_1 - 1} \ldots p_k^{p_k - 1} q,$$

with q prime, $q \nmid n$.

502. Answer: 2^{2021}.

Observe that $b = 1 + 2^1 e_1 + 2^2 e_2 + \cdots + 2^{2022} e_{2022}$, with $e_i \in \{-1, 1\}$, $1 \leq i \leq 2022$ and $e_{2022} = 1$, since $b > 0$. Note that this representation is unique, because

$$b - 2^{2022} - 1 = \sum_{1 \leq k \leq 2021} 2^k e_k$$

implies

$$b - 2^{2022} - 1 + \sum_{1 \leq k \leq 2021} 2^k = 2 \sum_{1 \leq k \leq 2021} 2^k f_k, f_k \in \{0, 1\}, 1 \leq k \leq 2021.$$

The binary representation is unique, whence the conclusion.

503. 1. Let $f(a, b) = \lambda$. Induction on $a + b$, we will show that

$$f(a, b) = \frac{1}{4} + \frac{1}{12ab}.$$

Define $g : [0, 1] \to \mathbb{R}$,

$$g(x) = \begin{cases} 1, \{ax\} > \{bx\}, \\ 0, \{ax\} \leq \{bx\} \end{cases}.$$

We have

$$f(1, 1) = f(a, a) = \frac{1}{3}, f(a, a) + f(a, b - a) - f(a, b) =$$

$$= \int_0^1 \{ax\}(\{ax\} + \{(b-a)x\} - \{bx\})dx = \int_0^1 \{ax\}g(x)dx =$$

$$= \frac{1}{2a}\left(a + \sum_{j=1}^{b-a-1} \left(\frac{j}{b-a}\right)^2 - \sum_{j=1}^{b-1}\left(\frac{j}{b}\right)^2\right) = \frac{1}{3} + \frac{1}{12b(b-a)};$$

hence

$$f(a,b) = f(a, b-a) - \frac{1}{12b(b-a)}$$
$$= \frac{1}{4} + \frac{1}{12a(b-a)} - \frac{1}{12b(b-a)} = \frac{1}{4} + \frac{1}{12ab}.$$

2. Let ω be chosen in $\{1, 2, \ldots, p-1\}$ such that each number has a $\frac{1}{p-1}$ chance of being chosen. Let

$$X(\omega) = \left\{\frac{a\omega}{p}\right\} + \left\{\frac{b\omega}{p}\right\} + \left\{\frac{c\omega}{p}\right\} + \left\{\frac{-(a+b+c)\omega}{p}\right\}$$

be a random variable. Then X can take only values 1, 2, and 3, and it takes 1 and 3 equally many times, because $X(\omega) + X(p-\omega) = 4$. We need to show that

$$\mathcal{P}(X \neq 2) = \text{Var}(X) = \text{Cov}(X, X) \geq \frac{1}{6},$$

where $\text{Cov}(X, Y) = \mathbb{E}(XY) - \mathbb{E}(X)\mathbb{E}(Y)$. By the bilinearity of covariance, we can see that if $j_1 = a, j_2 = b, j_3 = c, j_4 = -(a+b+c)$, and then

$$\text{Cov}(X, X) = \sum_{k=1}^{4} \sum_{l=1}^{4} \text{Cov}\left(\left\{\frac{j_k x}{p}\right\}, \left\{\frac{j_l x}{p}\right\}\right).$$

Lemma. Let $1 \leq u, v \leq 2p^{\frac{2}{3}}$. Then

$$\mathbb{E}\left(\left\{\frac{ux}{p}\right\}\left\{\frac{vx}{p}\right\}\right) - \int_0^1 \{ux\}\{vx\} dx < 32 p^{\frac{-1}{3}}.$$

Proof. Define $f(x) = \{ax\}\{bx\}$. Observe that

$$\frac{1}{p-1} \sum_{j=1}^{p-1} f\left(\frac{j}{p}\right)$$

is a Riemann sum for

$$\int_0^1 f(x) dx.$$

It suffices to show that in an interval I of length $\frac{1}{p}$,

$$\sup_I f(x) - \int f(x)dx \le 4p^{-\frac{1}{2}}.$$

This is at most

$$\sup_I\{ax\}\sup_I\{bx\} - \inf_I\{ax\}\inf_I\{bx\}.$$

If $\{ax\}$ or $\{bx\}$ doesn't pass through an integer, then this is at most

$$\frac{(a+b)p + ab}{p^2};$$

otherwise this is at most 1, and there are at most $8p^{\frac{2}{3}}$ such intervals where $\{ax\}$ or $\{bx\}$ passes through an integer.

Back to the problem, by Dirichlet's approximation, there exists k such that

$$\left\|\frac{ka}{p}\right\|, \left\|\frac{kb}{p}\right\|, \left\|\frac{kc}{p}\right\| < 2p^{-\frac{1}{3}}.$$

We need to check that

$$\text{Cov}(X,X) = \sum_{k=1}^{4}\sum_{l=1}^{4} \text{Cov}\left(\left\{\frac{j_k x}{p}\right\}, \left\{\frac{j_l x}{p}\right\}\right) = \sum_{k,l} \text{sgn}(j_k j_l)\frac{(\gcd(j_k, j_l))^2}{12 j_k j_l} \ge \frac{1}{6}.$$

The only equality cases are $(1, -2, -3, 4)$ and $(1, -3, -4, 6)$, and we can check these cases individually.

504. Choose $m = r + d, n = 1 + d\prod_{p<m} p^2$. Then $|S(m, 1)| = r + d$ and

$$S(m, n) = \sum_{\substack{k \le m \\ k \text{ squarefree}}} (-1)^{\omega(k)} \left\lfloor\frac{m}{k}\right\rfloor \left\lfloor\frac{n}{k}\right\rfloor;$$

hence

$$S(m, n) - S(m, 1) = \sum_{\substack{k \le m \\ k \text{ squarefree}}} (-1)^{\omega(k)} d \prod_{p \le m} \frac{p^2}{k}$$

is divisible by d.

505. Answer: no.

Let $\{lx\} = \min\limits_{1 \leq k \leq n-1} \{kx\}$ and suppose that $\{nx\} = \min\limits_{1 \leq k \leq n} \{kx\}$. Assume that for all $n > N$, we have

$$\min_{1 \leq k \leq n} \{kx\} < \frac{1}{n+1}.$$

Then

$$\{nx\}, \{lx\} < \frac{1}{n}, \{nlx\} = l\{nx\} = n\{lx\}.$$

From $l < n$, it follows that $\{nx\} < \{lx\}$, contradiction.

506. Define

$$A = \{(x,y) \in S \times T \mid kx \equiv y \pmod{p}, 0 \leq x, y \leq p-1\}.$$

We will prove that all the points in A are collinear. Indeed, suppose that A contains three noncollinear points, and let B be the convex hull of A. Then B contains three points (x_i, y_i), $i = 1, 2, 3$, with $y_i = kx_i + pv_i$, $v_i \in \mathbb{Z}$, so

$$(x_i, y_i) = (x_i, kx_i + pv_i) = (1, k)x_i + (0, p)v_i.$$

The points (x_i, v_i) can be obtained from (x_i, y_i) through a linear transformation φ^{-1}. The Jacobian of φ is

$$\begin{pmatrix} 1 & k \\ 0 & p \end{pmatrix},$$

with determinant p. Then

$$\text{area}(\text{Int } \varphi^{-1}(B)) = \frac{1}{p} \text{area}(\text{Int } B).$$

The Pick's theorem implies

$$\text{area}(\text{Int } \varphi^{-1}(B)) = |\text{interior points}| + \frac{|\text{border points}|}{2} - 1.$$

From $|A| \geq 1 + \lambda s$, we deduce $\text{area}(\text{Int } \varphi^{-1}(B)) \geq \frac{\lambda s - 1}{2}$, and so $\text{area}(\text{Int } B) \geq \frac{p(\lambda s - 1)}{2}$. But $\text{Int } B \subset S \times T$ and $\text{area}(S \times T) = st$, so

$$st < \frac{s\lambda p}{6} < \frac{(\lambda s - 1)p}{2},$$

contradiction.

We have proved that all points from A are collinear; whence it follows that these points form an arithmetic progression. Denote by (x_i, y_i) these points and observe that $\max x_i - \min x_i \leq s - 1$; hence

$$x_i - x_{i-1} \leq \frac{s-1}{\lambda s} \leq \frac{1}{\lambda}$$

and analogously,

$$y_i - y_{i-1} \leq \frac{t}{\lambda s}.$$

We can take $a = x_i - x_{i-1}$, $b = y_i - y_{i-1}$.

507. By the Fermat's theorem, we have $x^p - y^p \equiv x - y \pmod{p}$. Let S be the given sum. If $x \equiv y \pmod{p}$, then $S \equiv p x^{p-1} \equiv 0 \pmod{p}$.

If $x \not\equiv y \pmod{p}$, then $S \equiv \frac{x^p - y^p}{x - y} \equiv 1 \pmod{p}$, because $\gcd(x - y, p) = 1$, so $x - y$ has an inverse modulo p.

508. Let $c(k)$ be the number of apparitions of the digit k among $a_0, a_1, \ldots, a_{n-1}$. If $a_i \geq 2$, then $i \geq 1$ and $a_{i-1} = \lfloor \frac{a_i}{2} \rfloor$. For each $k \leq 5$, we have $a_i = k$ if and only if $a_{i+1} = 2k$ or $a_{i+1} = 2k+1$; hence $c(k) \geq c(2k) + c(2k+1)$. Let $m = \min_{1 \leq k \leq 9} \{c(k)\}$. Then

$$n \geq c(1) + c(2) + c(3) + c(4) + c(5) + c(6) + c(7) + c(8) + c(9) \geq$$
$$\geq 2c(2) + 2c(3) + c(4) + c(5) + c(6) + c(7) + c(8) + c(9) \geq$$
$$\geq 3c(4) + 3c(5) + 3c(6) + 3c(7) + c(8) + c(9) \geq$$
$$\geq 3c(5) + 3c(6) + 3c(7) + 4c(8) + 4c(9) \geq 17m.$$

Note. The best possible estimation is $m \leq \frac{5n}{94}$, attained from $n = 94$, when

$$c(1) \geq c(2) \geq \cdots \geq c(9) = 5.$$

More generally,

$$\lim_{n \to \infty} \frac{m}{n} = 1 - \log_{10} 9.$$

509. Answer: 1344.

The number $2 \cdot 2019$ is nippon, since it has divisors 1, 2, and 2019. Let n be the smallest nippon number. If $a_k = \frac{n}{d_k}$, then $1 \leq a_3 < a_2 < a_1$, $\gcd(a_1, a_2, a_3) = 1$ and

$$n\left(\frac{1}{a_1} + \frac{1}{a_2} + \frac{1}{a_3}\right) = 2022 = 2 \cdot 3 \cdot 337.$$

In order to minimize n, we will choose a_i as small as possible. For $a_3 = 1$, $a_2 = 2$, the above equality becomes $(3a_1 + 2)d_1 = 2^2 \cdot 3 \cdot 337$. The divisors of RHS congruent to 2 modulo 3 are 2 and 674. The first case is impossible; hence

$$a_1 = \frac{674 - 2}{3} = 224, d_1 = 2 \cdot 3 = 6, n = 6 \cdot 224 = 1344.$$

Suppose that there is a nippon number $n < 1344$ and write

$$\frac{1}{a_1} + \frac{1}{a_2} + \frac{1}{a_3} = \frac{p}{q}, \gcd(p, q) = 1.$$

Then $np = 2 \cdot 3 \cdot 337q$. We deduce $p | 2 \cdot 3 \cdot 337$ and, excluding the preceding case, $a_2 \geq 3$; therefore

$$1 + \frac{1}{3} + \frac{1}{a_1} \geq \frac{p}{q} = \frac{2022}{n} > \frac{2022}{1344} > 1 + \frac{1}{3} + \frac{1}{6};$$

whence $a_1 \leq 5$. Now we have to investigate the triples

$$(a_1, a_2, a_3) = (4, 3, 1), (5, 3, 1), (5, 4, 1), (4, 3, 2), (5, 3, 2), (5, 4, 2), (5, 4, 3),$$

and we see that none of the numerators of the corresponding fractions

$$\frac{p}{q} = \frac{19}{12}, \frac{23}{15}, \frac{29}{20}, \frac{13}{12}, \frac{31}{30}, \frac{19}{20}, \frac{47}{60}$$

divides $2 \cdot 3 \cdot 337$.

510. Answer: $(1, 1), (2, 2), (5, 17)$.

Denote $A_k = 1! + 2! + \cdots + k!$, $B_n = 1 + 2 + \cdots + n$. We see that $B_n = \frac{n(n+1)}{2}$; hence

$$n^2 \leq 2B_n = 2A_k < (n+1)^2,$$

and so $n = \lfloor \sqrt{2A_k} \rfloor$. Now,

$$A_1 = 1, A_2 = 3, A_3 = 9, A_4 = 33, A_5 = 153,$$

$$n = \lfloor \sqrt{2A_k} \rfloor = 1, 2, 4, 8, 17,$$

$$B_1 = 1, B_2 = 3, B_4 = 10, B_8 = 36, B_{17} = 153,$$

and we find the above solutions.

It is easy to see that $A_k \equiv A_6 \equiv 5 \pmod{7}$ for $k \geq 6$, whereas $B_n \equiv 0, 1, 3, 6 \pmod{7}$; whence we see that there are no solutions (k, n) with $k \geq 6$.

511. Answer: 21, 46.

If $4^n = 10^{\alpha_n} a_n, 5^n = 10^{\beta_n} b_n, 1 \leq a_n, b_n < 10$, then $a_n b_n^2 = 10^t, 0 \leq t < 3$. From $s = \lfloor 10 a_n \rfloor = \lfloor 10 b_n \rfloor$, it follows that

$$10^3 \leq s^3 \leq 10^{t+3} < (s+1)^3 \leq 10^6;$$

hence $s = \left\lfloor \sqrt[3]{10^{t+3}} \right\rfloor, 0 \leq t < 3$. If $t = 0$, then $10 = s = 10 a_n = 10 b_n$, so that 4^n and 5^n are powers of 10, contradiction.

The only possible values of s are $s = \left\lfloor \sqrt[3]{10^4} \right\rfloor = 21$ and $s = \left\lfloor \sqrt[3]{10^5} \right\rfloor = 46$.

512. Answer: 24.

We have

$$\frac{39}{1428} = \frac{13}{2^2 \cdot 7 \cdot 17}.$$

Write this fraction in binary expansion as a sum of the non-repeating part and the repeating part. Then $119 = 7 \cdot 17$ divides the denominator, and it suffices to find the order of 2 modulo 119. From $2^3 - 1 = 7, 2^4 + 1 = 17$, it follows that this order is

$$3 \cdot 8 = 24.$$

513. Let $i^4 + (\tau(i))^4 = p_i, 1 \leq i \leq n$, with p_i primes. Then

$$p_1 + p_2 + \cdots + p_n = 2(1^4 + 2^4 + \cdots + n^4).$$

so $p_1 + p_2 + \cdots + p_n$ is even. For $i \geq 2$, we have $p_i > i^4 > 2$, so p_i is odd.

Case 1. n even. Then $p_1 + n - 1$ is even, so p_1 is odd; hence $\tau(1) \neq 1$. Consider a graph with vertices $\{1, 2, \ldots, n\}$ and connect $i \neq j$ if and only if $i^4 + j^4$ is a prime number. If $i - j$ is even, then $i^4 + j^4$ is an even number greater than 2. It follows that the graph is bipartite, with partition $A = \{1, 3, \ldots, n-1\}, B = \{2, 4, \ldots, n\}$. We want to find $a_1, a_3, \ldots, a_{n-1}$ a permutation of B such that $i^4 + a_i^4$ is a prime for all odd i. Similarly, we want to find $a_2, a_4, \ldots, a_n$ a permutation of A such that $i^4 + a_i^4$ is a prime for all even i. Let N be the number of applications $\tau : A \to B$, which is the same as the number of applications $\tau : B \to A$ that satisfy the above conditions. Then the number of permutations τ of $\{1, 2, \ldots, n\}$ that satisfy the required condition is N^2.

Case 2. n odd. As above, p_i is odd for all $i \geq 2$, so $p_1 + (n-1)$ is even; hence $p_1 = 2$, which implies $\tau(1) = 1$. Excluding 1, we can proceed similarly, taking

$$A = \{3, 5, \ldots, n\}, B = \{2, 4, \ldots, n-1\}.$$

We see that $S(n)$ is a square in all cases.

514. Let $p_1 < p_2 < \cdots < p_m$ be the primes less than 1402, and consider $n = AB$, where $A = (p_1 p_2 \ldots p_m)^{1402}$, $B = p_{m+1} p_{m+2} \ldots p_k$. We will prove that it suffices to choose k sufficiently large. Take $i \leq 1402$ and any sign: we need to prove the inequality

$$\tau(n) > 1401 \tau(n \pm i).$$

Observe that $\tau(n) \geq 2^k$. Since $\tau(N) \leq 2^{\Omega(N)}$ for all positive integers N, where $\Omega(N)$ is the number of prime factors of N with repetition, it suffices to prove that

$$\Omega(n \pm i) \leq k - 12.$$

By construction, each of $p_1, \ldots, p_m$ divides $n \pm i$ at most once, and none of $p_{m+1}, \ldots, p_k$ divides $n \pm i$. If $n \pm i = ab$, $\gcd(b, p_1 \ldots p_k) = 1$, then $b \leq n + i \leq 2n \leq A p_1 \ldots p_k$, and it suffices to show that $\Omega(b) \leq k - 2023$. If this inequality is not true, then

$$b \geq p_k^{k-2023}, \quad p_k^{k-2023} \leq A p_1 \ldots p_k \leq A p_1 \ldots p_{2024} p_k^{k-2024};$$

hence $p_k \leq A p_1 \ldots p_{2024}$, which is false for large k.

515. It suffices to show that each positive integer k is a product of factors of the form

$$f(n) = \frac{\left\lfloor \frac{n}{2} \right\rfloor + 1}{\left\lfloor \frac{n}{3} \right\rfloor + 1}.$$

We will prove this assertion by induction:

$$f(0) = 1, f(2) = 2, f(4) = \frac{3}{2}, f(6t) = \frac{3t+1}{2t+1}, f(6t+2) = \frac{3t+2}{2t+1}.$$

Suppose that the assertion holds for all $n < k$. Then

$$k = 3t = f(2) f(4) t$$
$$k = 3t + 1 = (2t+1) f(6t)$$
$$k = 3t + 2 = (2t+1) f(6t+2)$$

which ends the proof by induction.

516. Answer: $c = \frac{\sqrt{2}}{2}$.

The condition $c \leq \frac{\sqrt{2}}{2}$ is necessary, as shown in the pair $(d,n) = (1,2)$. Now we prove that $c = \frac{\sqrt{2}}{2}$. If $n = p_1 \ldots p_r$ is squarefree and r is even, and then let d be a prime divisor of n, $d < \sqrt{n}$. We have

$$\tau(d) \geq 2^{\frac{r}{2}} > 2^{\frac{r-1}{2}} = \sqrt{\frac{\tau(n)}{2}}.$$

If r is odd, then let d as above, so

$$\tau(d) \geq 2^{\frac{r-1}{2}} = \sqrt{\frac{\tau(n)}{2}};$$

hence in both cases there is such d. For $n = PQ$,

$$P = \prod_{\nu_p(n)=1} p, \quad Q = \prod_{\nu_p(n) \geq 2} p^{\nu_p(n)}.$$

Since P is squarefree, there exists $P_d | P$ for which $P_d \leq \sqrt{P}$ and $\tau(P_d) \geq \sqrt{\frac{\tau(P)}{2}}$. Define

$$Q_d = \prod_{\nu_p(n) \geq 2} p^{\left\lfloor \frac{\nu_p(n)}{2} \right\rfloor}.$$

Then $Q_d | Q$ and $Q_d \leq \sqrt{Q}$, since

$$Q_d \leq \prod_{\nu_p(n) \geq 2} p^{\frac{\nu_p(n)}{2}} = \sqrt{Q}.$$

From the inequality $\lfloor \frac{x}{2} \rfloor + 1 \geq \sqrt{x+1}, x \geq 2$, it follows that $\tau(Q_d) \geq \sqrt{\tau(Q)}$. Now, $P_d Q_d | PQ = n$, $P_d Q_d \leq \sqrt{PQ} = \sqrt{n}$. Since

$$\gcd(P,Q) = \gcd(P_d, Q_d) = 1, \tau(P_d) \geq \sqrt{\frac{\tau(P)}{2}}, \tau(Q_d) \geq \sqrt{\tau(Q)},$$

it follows that

$$\tau(P_d Q_d) = \tau(P_d)\tau(Q_d) \geq \sqrt{\frac{\tau(P)\tau(Q)}{2}} = \sqrt{\frac{\tau(PQ)}{2}} = \sqrt{\frac{\tau(n)}{2}};$$

hence this divisor satisfies the required condition.

517. Suppose that $a > b > c$. If $\frac{st}{r}$ is an integer, then $\left(\frac{st}{r}\right)^2$ is also an integer. From

$$\left(\frac{st}{r}\right)^2 = \frac{(ac+1)(bc+1)}{ab+1} = c^2 + 1 + \frac{(a-c)(b-c)}{ab+1},$$

it follows that $ab+1 | (a-c)(b-c)$, which is a positive integer, so

$$(a-c)(b-c) \geq ab+1 \implies c^2 \geq ac+bc+1 > 2c^2+1,$$

contradiction.

518. Answer: no such pairs exist.

If n is odd, then

$$m = 1^{2k+1} + \left(2^{2k+1} + n^{2k+1}\right) + \left(3^{2k+1} + (n-1)^{2k+1}\right) + \cdots \equiv 1 \pmod{(n+2)},$$

so $n+2 \nmid m$.

If n is even, then $n = 2t$, with t positive integer, and

$$m = 1^{2k+1} + \left(2^{2k+1} + n^{2k+1}\right) + \left(3^{2k+1} + (n-1)^{2k+1}\right) + \cdots \equiv 1 + \left(\frac{n+2}{2}\right)^{2k+1} \equiv$$
$$\equiv 1 + (t+1)^{2k+1} \pmod{(n+2)},$$

so $n+2 \nmid m$.

519. Since S is nonempty, it contains some integer a. Since $1 | 2^a + 1$, it follows that $1 \in S$, and thus $2^1 + 1 = 3 \in S$. We call a pair (m, n) of positive integers k-valid if $m, n \in S$, $n | m | 2^n + 1$ and $\frac{m}{n}$ is divisible by k distinct primes. We will show by induction that there is a k-valid pair for any positive integer k.

Lemma. If (m, n) is k-valid and p is a prime divisor of $\frac{m}{n}$, then $(p^\alpha m, p^\alpha n)$ is k-valid for any nonnegative integers α.

Proof. By the lifting the exponent lemma, we have $\nu_p(2^{pm}+1) = \nu_p(2^m+1)+1$, hence $pn | pm | 2^{pn} + 1$. Also, $pn \in S$ implies $pm \in S$. From $\frac{pm}{pn} = \frac{m}{n}$, we deduce that (pm, pn) is k-valid. Repeating this argument α times, we get the desired result.

Back to the problem, for $k = 1$ take $(m, n) = (3, 1)$. Assume that (m, n) is k-valid for some positive integer k, and let p be a prime divisor of $\frac{m}{n}$. Then for any positive integer t,

$$0 < \nu_p(2^{tm}+1) = \nu_p(2^m+1) + \nu_p(t) < n + \log_p t.$$

Thus, the k primes we know to divide $\frac{m}{n}$ make up at most a fraction

$$\prod_p p^{n+\log_p t} < \left(\frac{m}{n}\right)^n t^k$$

of $2^{tn} + 1$. Using the lemma, we can pick t arbitrarily large such that (tm, tn) is k-valid. It follows that $2^{tn} + 1$ is divisible by a prime q that does not divide $\frac{m}{n}$; hence (tmq, tn) is $k + 1$-valid.

Choose a N-valid pair (m, n) with N large enough to ensure that $\frac{m}{n}$ has at least 2023 prime divisors greater than 2023. By the lemma, we can pick N such that the powers of all these primes are at least 2023 in the prime factorization of m.

520. Answer: $f(x) = cx$, c positive integer.

We see that f is injective, since 0 is not a divisor of any positive integer.

Take $m = 1$, and then $f(n + 1) - f(1) | f(n)$; take $n = 1$, and then $f(m + 1) - f(m) | f(1)$.

Take $n > N$ such that $f(n + 1) > f(1)$, and then $f(n) = k(f(n+1) - f(1))$ for some positive integer k,

$$f(1) \geq |f(n+1) - f(n)| = |k(f(n+1) - f(1)) - f(n+1)| \geq$$
$$\geq (k-1)f(n+1) - kf(1),$$

so $k = 1$ for n large enough. Hence $f(n+1) = f(n) + f(1)$ for large n, so

$$f(n) = f(1)n + C.$$

Then $nf(1) | nf(1) + C$, so $C = 0$. For $m > N$ we have the implication

$$(m+n)f(1) - f(m) | nf(1) \implies (m+n)f(1) - f(m) | f(1)m - f(m),$$

so $f(n) = f(1)n$ for all n.

521. By the AM–GM inequality,

$$a + b + c \geq 2\sqrt{ab} + c \geq 3\sqrt[3]{abc},$$

where LHS − RHS = 1, so that $a = b$ or $c^2 = ab$.

Case 1. $a = b$. Then $\gcd(a, c) = 1$ and a^2c is a perfect cube; hence $a = x^3$, $b = y^3$ and

$$1 = 2x^3 + y^3 - 3x^2y = (x - y)^2(2x + y),$$

which is impossible.

Case 2. $c^2 = ab$. Then $a + b = 2c + 1$; hence $(\sqrt{a} - \sqrt{b})^2 = 1$, so that $\sqrt{a} - \sqrt{b} = \pm 1$. We deduce that a and b are both perfect squares.

There are two families of solutions:

$$\left\{k^2, (k+1)^2, k(k+1)\right\} \text{ and } \left\{(k+1)^2, k^2, k(k+1)\right\}, k \in \mathbb{N}^*.$$

522. Answer: $(2, 5)$.

We have $p(p^4 + p^2 + 10q) \equiv 0, 2, 3 \pmod 5$, $q(q^2 + 3) \equiv 0, 1, 4 \pmod 5$; whence it follows $q = 5$, since q is a prime. The equation

$$p(p^4 + p^2 + 50) = 5(5^2 + 3) = 140$$

has only the solution $p = 2$ in prime numbers.

2.52 2024

523. Let S be the set of cool polynomials with degree ≤ 2023 and coefficients in $[0, n]$. We need to show that $|S|$ is even. For each $P \in S$, write $P = x^t Q$, with $Q(0) \neq 0$. This is possible, since $0 \notin S$. Define $\tau : S \to S$ by $\tau(P) = x^{\deg P} Q(\frac{1}{x})$. The application τ is well defined and is an involution on S.

It suffices to show that $\tau(P) \in S$ for each $P \in S$. The coefficients of $\tau(P)$ are in $[0, n]$, and the degree of $\tau(P)$ is at most 2023, so it remains to show that $\tau(P)$ is cool. Observe that

$$\prod_{2 \leq x \leq p-2} x^{\deg P} P\left(\frac{1}{x}\right) \equiv (2 \cdot 3 \cdot \ldots \cdot (p-2))^{\deg P} \prod_{2 \leq x \leq p-2} P\left(\frac{1}{x}\right) \equiv$$

$$\equiv \prod_{2 \leq x \leq p-2} P\left(\frac{1}{x}\right) \equiv \prod_{2 \leq x \leq p-2} P(x) \equiv 1 \pmod p,$$

where the last congruence is true, since P is cool.

Now, we will show that the number of fixed points of τ is even. Let $\mathcal{F}$ be the set of the fixed ponts of τ. Define an equivalence relation on $\mathcal{F}$ by $P \sim Q$ if there exists an integer m such that $P(x) = x^m Q(x)$. It is easy to see that this is indeed an equivalence relation. We prove that each equivalence class has an even number of elements. Indeed, each equivalence class is of the form $\{Q, xQ, \ldots, x^m Q\}$, with $Q \in \mathcal{F}, Q(0) \neq 0$ and $m = 2023 - \deg Q$. It follows that this equivalence class has $2024 - m$ elements, so we need to show that $\deg Q$ is even for each $Q \in \mathcal{F}$ with $Q(0) \neq 0$. Let Q be such a polynomial and suppose that $\deg Q$ is odd. The polynomial Q is reciprocal; hence

$$Q(x) = (x - \varepsilon)^{\deg R} R\left(x + \frac{1}{x}\right),$$

with $\varepsilon \in \{-1, 1\}$, $R \in \mathbb{Z}[x]$. Since Q is cool,

$$1 \equiv \prod_{2 \leq k \leq p-2} Q(k)$$
$$\equiv \prod_{2 \leq k \leq p-2} (k-\varepsilon) \prod_{2 \leq k \leq p-2} R\left(k + \frac{1}{k}\right) \cdot (2 \cdot 3 \cdot \ldots \cdot (p-2))^{\deg R} \pmod{p}.$$

From $\varepsilon \in \{-1, 1\}$, it follows that the first product is $-\frac{1}{2}$; hence

$$\prod_{2 \leq k \leq p-2} R\left(k + \frac{1}{k}\right) \equiv -2 \pmod{p}.$$

The numbers $k + \frac{1}{k}$ occur in pairs, so the above product is a quadratic residue modulo p. But $p \equiv 5 \pmod 8$, so

$$\left(\frac{-2}{p}\right) = -1,$$

contradiction. We deduce that each such Q has even degree.

Since each equivalence class of $\mathcal{F}$ has an even number of elements, we deduce that $|\mathcal{F}|$ is even; therefore $|S|$ is even.

524. If $x = 1$, then $y^2 + 1 = 2^z$; whence working modulo 4 we obtain the solution $(x, y, z) = (1, 1, 1)$.

Now, $x, y \geq 2$. If x and y are odd, then modulo 4 we get a contradiction. Assume that x and y are even, $x = 2^k x_1$, $y = 2^k y_1$, with x_1 odd. If y_1 is odd, then from

$$\nu_2\left(x^{y^2+1}\right) = \nu_2\left(y^{x^2+1}\right),$$

we deduce $x = y = 2^t$, so $(x, y, z) = (2^t, 2^t, (2^{2t} + 1)t + 1)$, $t \in \mathbb{Z}$.

If y_1 is even, then $y_1 = 2^{2^{2k}t} m_1$. From

$$\nu_2\left(x^{y^2+1}\right) = \nu_2\left(y^{x^2+1}\right), 2^{2k} k \left(y_1^2 - x_1^2\right)$$
$$= \nu_2(y_1)\left(2^{2k} x_1^2 + 1\right) \text{ and } \gcd\left(2^{2k}, 2^{2k} x_1^2 + 1\right),$$

it follows that $2^{2k} | \nu_2(y_1)$, contradiction.

Appendix

This section contains some introductory information on the subject. More material can be found in the recommended books.

Chinese Remainder Theorem If $n_1, \ldots, n_k$ are pairwise coprime positive integers and $a_1, \ldots, a_k$ are integers, then the system of congruences $x \equiv a_1 \pmod{n_1}, \ldots, x \equiv a_k \pmod{n_k}$ has a solution. If x_1 and x_2 are solutions, then $x_1 \equiv x_2 \pmod{N}$ where N is the product of the n_i.

Example. The system of congruences

$$x \equiv 1 \pmod{2}, x \equiv 2 \pmod{3}, x \equiv 3 \pmod{5}, x \equiv 4 \pmod{7}$$

has the solution $x \equiv 53 \pmod{210}$.

From $2|x+1, 3|x+1$, it follows $6|x+1$; hence $x = 6k - 1$. Now, $x + 2 = 6k + 1 \equiv k + 1 \equiv 0 \pmod{5}$, so that $k = 5l + 4$ and so $x = 30l + 23 \equiv 2l + 2 \equiv 4 \pmod{7}$, hence $l \equiv 1 \pmod{7}$, so that $l = 7n + 1$ and $x = 30(7n + 1) + 23 = 210n + 53$.

Dirichlet's Theorem on Arithmetic Progressions If a and d are positive coprime integers, then there are infinitely many primes that are congruent to a modulo d.

An arithmetic progression of length 5 consisting entirely of primes is 5, 11, 17, 23, and 29. An arithmetic progression of length 6 with prime terms is 7, 37, 67, 97, 127, and 157. More generally, the Green-Tao theorem asserts that the set of prime numbers contains finite arithmetic progressions of any length.

Arithmetic progressions with odd d are often ignored, because half of the numbers are even and the other half is the same numbers as a progression with common difference $2d$. The largest d for which Dirichlet's theorem can be proved by Euclidean methods for all a with $\gcd(a, d) = 1$ is $d = 24$ (Murty). The proof of the general case is not elementary.

Euler's Theorem If n and a are coprime positive integers and $\varphi(n)$ is Euler's totient function, then $a^{\varphi(n)} \equiv 1 \pmod{n}$.

Examples

1. Find the last two digits of 2^{2013}.
   From $\varphi(100) = 40$, it follows $2^{2013} = (2^{40})^{50} \cdot 2^{13} \equiv 8192 \equiv 92 \pmod{100}$.
2. Define $f(n)$ for n a positive integer by $f(1) = 3$ and $f(n+1) = 3^{f(n)}$. What are the last two digits of $f(2012)$? (*Virginia Tech Regional Mathematics Contest* 2012).

We first show that $f(n) \equiv 3 \pmod 4$ for all $n \geq 1$. We have $f(1) \equiv 3 \pmod 4$. Since $f(n)$ is odd for all n, we see that $f(n+1) \equiv 3^{f(n)} \equiv 3^{f(n-1)} \equiv f(n) \pmod 4$, and we deduce that $f(n) \equiv 3 \pmod 4$ for all $n \geq 1$.

Now we show that $f(n) \equiv f(3) \pmod{25}$ for all $n \geq 3$. First observe that

$$f(n+1) \equiv 3^{f(n)} \equiv 3^{f(n-1)} \equiv f(n) \pmod 5$$

For $n \geq 2$, provided $f(n) \equiv f(n-1) \pmod 4$. It follows that $f(n+1) \equiv f(n) \pmod{20}$ for all $n \geq 2$. Therefore, $f(n+1) \equiv 3^{f(n)} \equiv 3^{f(n-1)} \equiv f(n) \pmod{25}$ for all $n \geq 3$. Since the last two digits of $f(3)$ are 87, the last two digits of $f(2012)$ are also 87.

Fermat's Theorem If p is a prime number, then $a^p \equiv a \pmod p$ for any integer a.

The converse is not generally true, however a slightly stronger form is true:

Lehmer's theorem. If there exists an integer a such that $a^{p-1} \equiv 1 \pmod p$ and for all primes $q | p-1$ one has $a^{\frac{p-1}{q}} \not\equiv 1 \pmod p$, then p is prime.

Kobayashi's Theorem Let $P(n)$ be the set of all prime divisors of the positive integer n. Define

$$P(K) = \bigcup_{k \in K} P(k)$$

for any nonempty set K of positive integers. If $P(K)$ is a finite set for any infinite set K, then for any positive integer m, the set $P(K+m)$ is infinite.

Example. Let $n > 1$ be an integer. Prove that there are infinitely many integers $k \geq 1$ such that $\left\lfloor \frac{n^k}{k} \right\rfloor$ is odd. (*IMOSL 2014*).

If n is odd, then we can pick any prime p dividing n and select $k = p^m$ for sufficiently large integers m. If n is even, then by Kobayashi's theorem, there exist infinitely many primes p dividing some number of the form $n^{n^r - 1} - 1$ for some integer r. Let $p > n$ be such a prime with corresponding integer r. Then

$$n^{n^r p} \equiv n^r \pmod{n^r p}.$$

For $k = n^r p$, it follows

$$\left\lfloor \frac{n^k}{k} \right\rfloor = \frac{n^k - n^r}{n^r p},$$

which is odd.

Appendix

Legendre's Formula For any prime number p and any positive integer n, let $\nu_p(n)$ be the exponent of the largest power of p that divides n, and let $s_p(n)$ be the sum of the digits in the base p expansion of n. Then

$$\nu_p(n!) = \sum_{k \geq 1} \left\lfloor \frac{n}{p^k} \right\rfloor = \frac{n - s_p(n)}{p - 1}.$$

Legendre's formula can be used to prove Kummer's theorem: For given integers $n \geq m \geq 0$ and a prime number p, the value $\nu_p\left(\binom{n}{m}\right)$ is equal to the number of "carries" when m is added to $n - m$ in base p (equivalently, the number of "borrows" required when subtracting the base p representations of m from n).

Applications

1. The number $\binom{1000}{500}$ is not divisible by 7. (*USSR Math Olympiad*).
2. 2^n does not divide $n!$. If $2^{n-1} | n!$; then n is a power of 2.
3. There is no integer m such that $2^{n-m} | n!$ for every positive integer $n \geq m$.
4. Let m be a positive integer and $f(m)$ be the greatest k such that $2^k | m$. Then there exist infinitely many m such that $m - f(m) = 1989$. (*IMOSL Columbia 1989*).
5. $\binom{2n}{n}$ is a multiple of $n + 1$.
6. If $n > 1$, then $2 | \binom{2n}{n}$ and $4 \nmid \binom{2n}{n}$ if and only if n is a power of 2.
7. If n and q are positive integers such that all prime divisors of q are greater than n, then

$$(q-1)(q^2-1)\cdots(q^{n-1}-1) \equiv 0 \pmod{n!}.$$

(*AMM*)

8. Let p be a prime number. The number of binomial coefficients $\binom{n}{0}, \binom{n}{1}, \ldots, \binom{n}{n}$ that are multiples of p is $n + 1 - (n_0 + 1)(n_1 + 1) \cdots (n_r + 1)$, where $n_0, \ldots, n_r$ are the digits in the expansion of n in base p.

(*I. Tomescu*).

9. The number of odd entries in any row of Pascal's Triangle is a power of 2.

(*Glaisher*)

10. All the binomial coefficients $\binom{n}{0}, \ldots, \binom{n}{n}$ are odd if and only if $n = 2^s - 1$.
11. A prime p divides the binomial coefficients $\binom{n}{k}$, $1 \leq k \leq n-1$, if and only if $n = p^r$.
12. Let p be a prime and n be a positive integer. Then $p \nmid \prod_{k=0}^{n} \binom{n}{k}$ if and only if $n = p^s m - 1$ with $1 \leq m < p$. (*Luxembourg Math Olympiad 1980*).

13. If $n = n_0 + n_1 p + n_2 p^2 + \ldots + n_r p^r$ and $m = m_0 + m_1 p + m_2 p^2 + \ldots + m_r p^r$ are the expansions of n and m in base p, then

$$\binom{n}{m} \equiv \binom{n_0}{m_0}\binom{n_1}{m_1}\cdots\binom{n_r}{m_r} \pmod{p}.$$

(*Lucas*)

14. The gcd of $\binom{n}{1}, \binom{n}{2}, \ldots, \binom{n}{n-1}$ is either 1 (if n is not a power of a prime) or a prime.

Legendre's Symbol Let p be an odd prime number. An integer a is a quadratic residue modulo p if it is congruent to a square modulo p and is a quadratic nonresidue modulo p otherwise. The Legendre symbol $\left(\frac{a}{p}\right)$ is a multiplicative function with values 1 for a nonzero quadratic residue a modulo p, -1 for a quadratic nonresidue modulo p and 0 for $a \equiv 0 \pmod{p}$. By the Euler's criterion,

$$\left(\frac{a}{p}\right) \equiv a^{\frac{p-1}{2}} \pmod{p}.$$

Examples

1. The positive integers a and b are such that the numbers $15a + 16b$ and $16a - 15b$ are both squares of positive integers. Find the least possible value that can be taken on by the smaller of these two squares.

   Let $15a + 16b = k^2$, $16a - 15b = l^2$; hence

   $$a = \frac{15k^2 + 16l^2}{481}, b = \frac{16k^2 - 15l^2}{481}.$$

   Then $481 = 13 \cdot 37 \Rightarrow 15k^2 + 16l^2 \equiv 0 \pmod{13}$, $2k^2 \equiv -3l^2 \pmod{13}$, and $k^2 \equiv 5l^2 \pmod{13}$. From $\left(\frac{5}{13}\right) = -1$, it follows $13 \mid k, l$. Now $15k^2 + 16l^2 \equiv 0 \pmod{37} \Rightarrow l^2 \equiv 6k^2 \pmod{37}$. From $\left(\frac{6}{37}\right) = -1$, it follows $37 \mid k, l$, whence $481 \mid k, l$. For $k = l = 481$, it follows $a = 31 \cdot 481$, $b = 481$.

2. Let p be an odd prime, $a_i = \left(\frac{i}{p}\right)$, $1 \leq i \leq p - 1$. Consider the polynomial

   $$f = a_1 + a_2 x + \ldots + a_{p-1} x^{p-2}.$$

   (a) Prove that 1 is a simple root of f if and only if $p \equiv 3 \pmod{4}$.
   (b) Prove that if $p \equiv 5 \pmod 8$, then 1 is a root of order exactly two of f.

   (*Romania IMO test* 2004)

Appendix

(a) Let $R = \{1, 2, \ldots, p-1\}$ and denote by Q and N the sets of quadratic residues, respectively, quadratic nonresidues from R. Since from $x, y \in R$ the congruence $x^2 \equiv y^2 \pmod{p}$ is equivalent to $x = y$ or $x + y = p$, it follows that $Q = \{x^2 | x \in R\}$ has $\frac{p-1}{2}$ elements, that is, $|Q| = |N|$; hence $f(1) = 0$ and f is divisible by $x - 1$. We have

$$f'(1) = \sum_{i=1}^{p-1}(i-1)a_i = \sum_{i=1}^{p} i a_i - \sum_{i=1}^{p} a_i = \sum_{i=1}^{p} i a_i = \sum_{i \in Q} i - \sum_{i \in N} i.$$

It follows that $f'(1) = 0$ if and only if the sum of quadratic residues equals the sum of quadratic nonresidues. Since

$$\sum_{i \in Q} i + \sum_{i \in N} i = \frac{p(p-1)}{2},$$

the latter is possible only if $p \equiv 1 \pmod{4}$. In this case $p - 1 \in Q$ and $x \in Q$ implies $p - x \in Q$; therefore, the $\frac{p-1}{2}$ elements of Q can be grouped in $\frac{p-1}{4}$ pairs, each pair having sum p; hence

$$\sum_{i \in Q} i = \frac{p(p-1)}{4} = \sum_{i \in N} i.$$

(b) In this case, $(x-1)^2 | f$. Notice that

$$f''(1) = \sum_{i=1}^{p-1}(i-1)(i-2)a_i = \sum_{i=1}^{p-1} i^2 a_i - 3\sum_{i=1}^{p-1} i a_i + 2\sum_{i=1}^{p-1} a_i = \sum i^2 a_i.$$

We will use the fact that for a prime $p \equiv 5 \pmod{8}$, we have $2 \in N$. Consider the sets

$$Q_< = \left\{r \in Q \middle| r < \tfrac{p}{2}\right\}, Q_> = \left\{r \in Q \middle| r > \tfrac{p}{2}\right\}, N_< = \left\{r \in N \middle| r < \tfrac{p}{2}\right\},$$
$$N_> = \left\{r \in N \middle| r > \tfrac{p}{2}\right\}.$$

Since $-1 \in Q$, sending $r \in R$ to $p - r \in R$ defines a one-to-one function from Q to Q, from N to N, from $Q_<$ to $Q_>$, and from $N_<$ to $N_>$. Therefore both $Q_<$ and $Q_>$ have $\frac{p-1}{4}$ elements and

$$\sum_{r \in Q_>} r^2 = \sum_{r \in Q_<} (p-r)^2 = \frac{p^2(p-1)}{4} - 2p \sum_{r \in Q_<} r + \sum_{r \in Q_<} r^2.$$

so,

$$\sum_{r \in Q} r^2 = \frac{p^2(p-1)}{4} - 2p \sum_{r \in Q_<} r + 2 \sum_{r \in Q_<} r^2.$$

Analogously,

$$\sum_{r \in N} r^2 = \frac{p^2(p-1)}{4} - 2p \sum_{r \in N_<} r + 2 \sum_{r \in N_<} r^2.$$

By the same arguments, sending r to $2r$ (respectively, to $p - 2r$) define one-to-one maps from $Q_<$ to the sets N_{even} (respectively, N_{odd}) of even (respectively, odd) quadratic nonresidues. Therefore

$$\sum_{r \in N} r^2 = \sum_{r \in Q_<} (p-2r)^2 + 4 \sum_{r \in Q_<} r^2 = \frac{p^2(p-1)}{4} - 4p \sum_{r \in Q_<} r + 8 \sum_{r \in Q_<} r^2.$$

A similar argument yields

$$\sum_{r \in Q} r^2 = \frac{p^2(p-1)}{4} - 4p \sum_{r \in N_<} r + 8 \sum_{r \in N_<} r^2.$$

Combining these equalities in pairs, we get

$$\sum_{r \in Q} r^2 - \sum_{r \in N} r^2 = 2 \left(p \sum_{r \in Q_<} r - 3 \sum_{r \in Q_<} r^2 \right),$$

$$\sum_{r \in Q} r^2 - \sum_{r \in N} r^2 = 2 \left(p \sum_{r \in N_<} r - 3 \sum_{r \in N_<} r^2 \right).$$

If $f''(1) = 0$, then

$$\sum_{r \in Q} r^2 = \sum_{r \in N} r^2. \quad (*)$$

Taking account of the preceding equalities, we have

Appendix

$$\sum_{r \in Q} r^2 = \frac{p^2(p-1)}{4} - \frac{4p}{3}\sum_{r \in Q_<} r = \frac{p^2(p-1)}{4} = \frac{4p}{3}\sum_{r \in N_<} r,$$

which implies

$$\sum_{r \in Q_<} r = \sum_{r \in N_<} r.$$

On the other hand,

$$\sum_{r \in Q_<} r + \sum_{r \in N_<} r = \sum_{r < \frac{p}{2}} r = \frac{p^2-1}{8}$$

is odd, contradiction.

Lifting-the-Exponent Lemma For any integers x and y, a positive integer n and a prime number p such that $p \nmid xy$, the following statements hold:

If p is odd, $p | x - y$, then $\nu_p(x^n - y^n) = \nu_p(x - y) + \nu_p(n)$; if n is odd and $p | x + y$, then $\nu_p(x^n + y^n) = \nu_p(x + y) + \nu_p(n)$.
If $p = 2$ and $2 | x - y$, then $\nu_2(x^n - y^n) = \nu_2(x - y) + \nu_2(x + y) + \nu_2(n) - 1$ for n even; if $2 | x - y$, then $\nu_2(x^n - y^n) = \nu_2(x - y)$ for n odd.

Corollary. If $4 | x - y$, then $\nu_2(x + y) = 1$; thus, $\nu_2(x^n - y^n) = \nu_2(x - y) + \nu_2(n)$.

Examples

1. Show that 2 is a primitive root mod 3^k for all positive integers k.

    We seek the smallest positive n such that $2^n \equiv 1 \pmod{3^k}$. If $3^k | 2^n - 1$, note that $2^n \equiv 1 \pmod 3$, so n is even. Write $n = 2m$ and apply LTE:

    $$k \le \nu_3(4^m - 1^m) = \nu_3(4-1) + \nu_3(m) = 1 + \nu_3(m),$$

    so $\nu_3(m) \ge k - 1$. The smallest possible such m is 3^{k-1}, so the smallest possible n is $2 \cdot 3^{k-1} = \varphi(3^k)$.

2. Let n be the least positive integer for which $149^n - 2^n$ is divisible by $3^3 \cdot 5^5 \cdot 7^7$. Find the number of positive divisors of n. (*AIME 2020*).

    Note that $149 - 2 = 147 = 3 \cdot 7^2$. Using the LTE lemma, since $3 \nmid 149$ and 2 but $3 | 147$, $\nu_3(149^n - 2^n) = \nu_3(149 - 2) + \nu_3(n) = 1 + \nu_3(n)$. Thus $3^3 | 149^n - 2^n \iff 3^2 | n$. Similarly, $7 \nmid 149$, 2 but $7 | 147$, so $\nu_7(149^n - 2^n) = \nu_7(147) + \nu_7(n) = 2 + \nu_7(n)$ and $7^7 | 149^n - 2^n \iff 7^5 | n$.

    Since $5 \nmid 147$, the factors of 5 are addressed by noticing that since the residues of 149^n modulo 5 are 4, 1, 4, 1, ... and those of 2^n are 2, 4, 3, 1, ..., the residues of $149^n - 2^n$ modulo 5 cycle through the sequence 2, 2, 1, and 0. It follows that $5 | 149^n - 2^n$ if and only if $n = 4k$ for some positive integer k. By the LTE lemma,

$$\nu_5\left(149^{4k} - 2^{4k}\right) = \nu_5\left(\left(149^4\right)^k - \left(2^4\right)^k\right) = \nu_5\left(149^4 - 2^4\right) + \nu_5(k).$$

Since $149^4 - 2^4 \equiv (-1)^4 - 2^4 \equiv -15 \pmod{25}$, $\nu_5(149^4 - 2^4) = 1$, hence

$$5^5 | 149^n - 2^n \Longleftrightarrow 5^4 | k \Longleftrightarrow 4 \cdot 5^4 | n.$$

From the above three results, it is found $n = 2^2 \cdot 3^2 \cdot 5^4 \cdot 7^5$. Then $\tau(n) = 270$.

Quadratic Reciprocity Law Let p and q be distinct prime numbers. Then

$$\left(\frac{p}{q}\right)\left(\frac{q}{p}\right) = (-1)^{\frac{p-1}{2} \cdot \frac{q-1}{2}}.$$

First supplement:

$$\left(\frac{-1}{p}\right) = (-1)^{\frac{p-1}{2}}.$$

Second supplement:

$$\left(\frac{2}{p}\right) = (-1)^{\frac{p^2-1}{8}}.$$

Applications

1. If the positive integer a is not a perfect square, then $\left(\frac{a}{p}\right) = -1$ for infinitely many primes p.

   Suppose by contradiction that the assertion is false, which means that there exists a number r such that for every prime $q > r$, $\left(\frac{a}{q}\right) = 1$. Because a is not a perfect square, we can write $a = b^2 p_1 p_2 \ldots p_k$, where $p_1 < p_2 < \ldots < p_k$ are prime numbers. Take a prime $p > r$, $p \equiv 5 \pmod 8$. We have $\left(\frac{a}{p}\right) = \left(\frac{p_1}{p}\right)\left(\frac{p_2}{p}\right) \ldots \left(\frac{p_k}{p}\right)$. If p_i is odd, then $\left(\frac{p_i}{p}\right) = \left(\frac{p}{p_i}\right)$, from quadratic reciprocity law. If $p_1 = 2$, then $\left(\frac{2}{p}\right) = -1$ and

$$\left(\frac{a}{p}\right) = \left(\frac{p}{p_1}\right) \ldots \left(\frac{p}{p_k}\right) \text{ or } \left(\frac{a}{p}\right) = -\left(\frac{p}{p_2}\right) \ldots \left(\frac{p}{p_k}\right).$$

We can take $r_2, r_3, \ldots, r_k$ residues (mod $p_2 p_3 \ldots p_k$) such that $\left(\frac{r_2}{p_2}\right) \ldots \left(\frac{r_k}{p_k}\right)$ is 1 or -1 as we wish. By Chinese remainder theorem, there are infinitely many numbers t with $t \equiv 5 \pmod 8$, $t \equiv r_i \pmod{p_i}$, $2 \leq i \leq k$. By Dirichlet's theorem,

Appendix 287

there are infinitely many prime numbers $q \equiv t \pmod{8p_2 \ldots p_k}$, and we take $q > r$. Then $\left(\frac{a}{q}\right) = 1$, but we have seen that we can select $r_2, r_3, \ldots, r_k$ such that $\left(\frac{a}{q}\right) = -1$, contradiction.

2. The equation $y^2 = x^3 - 5$ has no integral solutions.

Reducing mod 4, we get that y is even $x \equiv 1 \pmod{4}$. Rewrite the equation as $y^2 + 4 = (x - 1)(x^2 + x + 1)$. Then $x^2 + x + 1 \equiv 3 \pmod{4}$, so there exists a prime p such that $p | x^2 + x + 1$; hence $p | y^2 + 4$. It follows that -4 is a quadratic residue modulo p, contradiction, since $\left(\frac{-4}{p}\right) = \left(\frac{-1}{p}\right)\left(\frac{4}{p}\right) = -1$.

Vieta Jumping This is a nickname for a particular kind of descent method that has become quite popular in higher level math Olympiad number theory problems. The method is already known in more advanced number theory as *reduction theory of quadratic forms*. Standard Vieta jumping is a proof by contradiction and consists in three steps:

1. Assume toward a contradiction that some solution exists that violates the given requirements.
2. Take the minimal such solution according to some definition of minimality.
3. Show that this implies the existence of a smaller solution, hence a contradiction.

Example. Let a and b be positive integers such that $ab + 1$ divides $a^2 + b^2$. Prove that $\frac{a^2+b^2}{ab+1}$ is a perfect square. (*IMO* 1988)

1. Fix some value k that is a non-square positive integer. Assume there exist positive integers (a, b) for which $k = \frac{a^2+b^2}{ab+1}$.
2. Let (A, B) be positive integers for which $k = \frac{A^2+B^2}{AB+1}$ and such that $A + B$ is minimized and without loss of generality assume $A \geq B$.
3. Fixing B, replace A with the variable x to yield $x^2 - kBx + B^2 - k = 0$. We know that one solution of this equation is $x_1 = A$. Since the equation is quadratic, the other solution is

$$x_2 = kB - A = \frac{B^2 - k}{A}.$$

4. The first expression shows that x_2 is an integer, while the second expression implies that $x_2 \neq 0$, since k is not a perfect square. From $\frac{x_2^2+B^2}{x_2B+1} = k > 0$, it follows that $x_2 B > -1$ and hence x_2 is a positive integer. Finally, $A \geq B$ implies that

$$x_2 = \frac{B^2 - k}{A} < \frac{B^2}{A} \leq A;$$

hence $x_2 < A$ and thus $x_2 + B < A + B$, which contradicts the minimality of $A + B$.

Wilson's Theorem A positive integer $n > 1$ is a prime number if and only if

$$(n-1)! \equiv -1 \pmod{n}.$$

Example. For each positive integer n, let $f(n)$ denote the greatest common divisor of $n! + 1$ and $(n+1)!$. Find a formula for $f(n)$. (*Ireland 1996*)

If $n + 1$ is prime, then it divides $n! + 1$, so $f(n) = n + 1$.

If $n + 1$ is not prime, then all prime factors of $n + 1$ are less than $n + 1$ and thus divide $n!$, so $f(n) = 1$.

Zsigmondy's Theorem If $a > b > 0$ are coprime integers, then for any integer $n \geq 1$, there is a prime number p (called a *primitive* prime divisor) that divides $a^n - b^n$ and does not divide $a^k - b^k$ for any positive integer $k < n$, with the exceptions:

$n = 1$, $a - b = 1$; then $a^n - b^n = 1$ has no prime divisors.
$n = 2$, $a + b$ is a power of two; then any odd prime factor of $a^2 - b^2$ must be contained in $a - b$, which is also even.
$n = 6$, $a = 2$, $b = 1$; then $a^6 - b^6 = 63 = 3^2 \cdot 7 = (a^2 - b^2)^2(a^3 - b^3)$.

Similarly, $a^n + b^n$ has at least one primitive prime divisor, with the exception $2^3 + 1^3 = 9$.

Examples

1. Find all positive integers a, $n > 1$ and k for which $3^k - 1 = a^n$.

    Because -1 is a quadratic nonresidue mod 3, we have that n is odd. From $a + 1$ divides $a^n + 1$, we have that $3 | a + 1$. If $a \neq 2$ or $n \neq 3$, $a^n + 1$ has a prime divisor different from 3, hence $a^n + 1$ cannot be a power of 3.

    It remains the only solution $(a, n, k) = (2, 3, 2)$.

2. Let $p_1, p_2, \ldots, p_n$ be distinct primes greater than 3. Show that $2^{p_1 p_2 \cdots p_n} + 1$ has at least 4^n divisors. (*IMOSL 2002*).

    Let $a = p_1 p_2 \ldots p_n$ and $b = 2^a + 1$. It suffices to prove that b has at least 2^n prime divisors. This is true, because Zsigmondy's theorem for sums implies that $2^d + 1$ introduces a new prime, for every divisor $d | a$, as $3 \nmid a$. Because a has 2^n divisors, b has at least 2^n prime divisors.

Bibliography

1. Becheanu, M.: Problems from Mathematical Olympiads 1987–1994 (in Romanian), GIL, Zalau (1995)
2. Camp Selection Problems 2010–2018. New Zealand Mathematical Olympiad Committee. https://www.mathsolympiad.org.nz
3. Chandrasekharan, K.: Introduction to Analytic Number Theory. Springer, Berlin (1968) (reprint 2012)
4. Djukić, D., Janković, V., Matić, I., Petrović, N.: The IMO Compendium, 2nd edn. Springer, New York (2011)
5. Estonian Math Competitions 1998–2021. University of Tartu Youth Academy (2021)
6. India IMO Training Camp. https://olympiads.hbcse.in
7. Li, K.Y.: Math Problem Book I. Hong Kong Mathematical Society, Hong Kong (2001)
8. Lősungen zur IMO Selektion. https://mathematical.olympiad.ch
9. Mathematical Olympiad in China (2007–2014). World Scientific, Singapore
10. Niven, I., Zuckerman, H.S., Montgomery, H.L.: An Introduction to the Theory of Numbers, 5th edn. Wiley, New York (1991)
11. Olimpiadas Matemáticas Españolas 1963–2004. Real Sociedad Matemática Española (2004)
12. Romanian Mathematical Competitions 1996–2008. Societatea de Științe Matematice din România, București (1996, 1997, . . . , 2008)
13. Sierpiński, W.: A Selection of Problems in the Theory of Numbers, Pergamon, Oxford (1964) (reprint 2014)
14. Sierpiński, W.: 250 Problems in Elementary Number Theory. Elsevier/Polish Scientific Publishers, New York/Warszawa (1970)
15. Sierpiński, W.: Elementary Theory of Numbers. North-Holland Mathematical Library, PWN-Polish Scientific Publishers, Warszawa (1988)
16. Team Selection Tests. https://artofproblemsolving.com
17. Uță, M., Catzaiti, A., Maftei, I. V. (eds.): Team Selection Tests for the IMO 1975–1983 (in Romanian). Piatra Neamț (1984)
18. Vietnamese Mathematical Competitions 2017–2019. Hanoi (2019)
19. Vinogradov, I. M.: Elements of Number Theory. Dover Publications, Inc., New York (1954) (reprint 2014)

Index

C
Chinese remainder theorem, 69, 106, 115, 126, 161, 195, 211, 212, 219, 221, 233, 279, 286

D
Diophantine equation, 80, 92, 93, 100, 257
Dirichlet's theorem, 37, 86, 156, 221, 225, 234, 279, 286
Divisibility, 69, 212, 227
Divisors, 4, 8–16, 18–28, 32–34, 36, 37, 39–53, 61, 64, 66, 67, 81, 85, 92, 98, 105, 106, 109, 113, 118, 119, 123, 126, 129–133, 136–138, 142, 145–150, 161, 163, 170, 171, 174, 176, 181–184, 196, 197, 201, 204, 206–208, 216–220, 223–226, 229, 230, 232, 242, 245, 250, 255, 259–263, 265, 266, 270, 271, 274–276, 280, 281, 285, 288

E
Euler's theorem, 98, 105, 165, 224, 280
Euler's totient function, 224, 280

F
Fermat's theorem, 64, 67, 72, 81, 90, 101, 106, 112, 119, 126, 129, 136, 137, 141, 166, 182, 194, 206, 225, 245, 253, 270, 280

L
Last digit, 3, 63, 136, 223, 254
Legendre's formula, 220, 281
Legendre's symbol, 243, 282
Lifting the exponent lemma, 93, 169, 177, 190, 196, 225, 230, 239, 263, 275, 285

M
Mersenne number, 26, 155
Mihăilescu's theorem, 149, 256

N
Number of divisors, 42, 52, 81, 201, 245

P
Pell equation, 107, 110, 149, 172, 173, 182, 222
Perfect cube, 18, 39, 52, 276
Perfect square, 7–12, 14, 17–21, 23, 25, 27–32, 34, 35, 38–40, 42, 43, 48, 49, 52, 80, 102, 115, 133, 163, 202, 217, 276, 286, 287
Prime counting function, 44
Pythagorean triple, 12, 78, 96, 99, 188, 207

Q
Quadratic reciprocity law, 97, 286

© The Editor(s) (if applicable) and The Author(s), under exclusive license to Springer Nature Switzerland AG 2024
C. Mănescu-Avram, *Selection Tests in Number Theory for Mathematical Olympiads*, Problem Books in Mathematics, https://doi.org/10.1007/978-3-031-59742-8

S
Sequence of integers, 28, 38, 95
Sum of digits, 21, 57, 65, 72, 82, 93, 181, 253, 258, 264
Sum of divisors, 30

V
Vieta jumping, 287

W
Wilson's theorem, 55, 114, 158, 216, 245, 288

Z
Zsigmondy's theorem, 88, 96, 110, 114, 154, 171, 174, 183, 207, 229, 230, 263, 288

SPRINGER NATURE

GPSR Compliance

The European Union's (EU) General Product Safety Regulation (GPSR) is a set of rules that requires consumer products to be safe and our obligations to ensure this.

If you have any concerns about our products, you can contact us on ProductSafety@springernature.com

In case Publisher is established outside the EU, the EU authorized representative is:

Springer Nature Customer Service Center GmbH
Europaplatz 3
69115 Heidelberg, Germany

The manufacturer's authorised representative in the EU is Springer Nature Customer Service Centre GmbH, Europaplatz 3, 69115 Heidelberg, Germany. If you have any concerns regarding our products, please contact ProductSafety@springernature.com

Printed and bound by CPI Group (UK) Ltd, Croydon, CR0 4YY

26/03/2026

02078916-0008